The Vertebrate Story

A revised and enlarged edition of "Man and the Vertebrates"

The Vertebrate Story

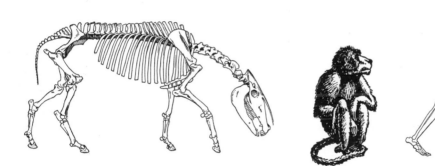

By *Alfred S. Romer*

Alexander Agassiz Professor of Zoölogy and Director, Museum of Comparative Zoölogy, Harvard University

THE UNIVERSITY OF CHICAGO PRESS

Library of Congress Catalog Number: 58-11957

THE UNIVERSITY OF CHICAGO PRESS, CHICAGO 37
Cambridge University Press, London, N.W. 1, England
The University of Toronto Press, Toronto 5, Canada

© *1933, 1939, 1941, and 1959 by The University of Chicago
Fourth Edition 1959. Composed and printed by* THE
UNIVERSITY OF CHICAGO PRESS, *Chicago, Illinois, U.S.A.*

Preface

This book contains an account of the backboned animals; of the nature of the highly varied types from fish to men included within this successful animal group; of their structure, function, and ways of life. Throughout, the evolutionary story is a leading theme, and an outline is given of their fossil history, extending back over a period of several hundred millions of years. There is, I trust, enough substance to the story as told here to make it of value to the student. But I further hope that it will be of interest to still other readers who may desire some acquaintance with their vertebrate relatives and with the human pedigree, which is treated in much greater detail than that of other animal types.

The present work represents a radical metamorphosis of an earlier book of mine, *Man and the Vertebrates*, first published in 1933 and last revised in 1941. This included a much briefer account of the vertebrates and their evolution and, in addition, an account of human anatomy and embryology on a comparative basis. With the passing of the years it became obvious to me that revision was called for. On consideration, I determined to make the revision a very radical one. The entire section on the human body—a third of the original work—has been discarded, and the story of the vertebrates as a whole is expanded to include a variety of topics of interest which had previously been omitted because of lack of space. In this process much of the account was rewritten, so that more than half of the text is new. The book is thus in great measure a new work. Further, while there are substantial chapters on human origins and races, the fraction of the work devoted to man himself is much reduced in its proportions. The former title, with top "billing" given to man, is hence inappropriate and misleading, and a new name, indicating that our story is that of the vertebrates-as-a-whole seems called for.

In the various editions of *Man and the Vertebrates*, I gratefully acknowledged my indebtedness to numerous individuals and institutions for aid with both text and illustrations; despite the many changes made in the new work, this indebtedness continues. To Drs. G. W. Bartelmez, Merle C. Coulter, Ralph W. Gerard, Carl R. Moore, H. H. Newman, H. H. Strandskov, and Griffith Taylor, I am deeply grateful.

The writer has been strongly influenced in his treatment of human races by the stimulating ideas of Dr. Griffith Taylor; his views as to racial evolution are expressed, with slight modification, in the diagram showing the phylogeny of human races.

The New York Zoölogical Society and the American Museum of Natural History, New York, have most courteously allowed me the use of numerous illustrations of animal types. A majority of the restorations of fossil animals are from a series of murals in the Field Museum of Natural History, Chicago, painted by Charles R. Knight. I am much indebted to Appleton-Century-Crofts, Inc., for the use of a large number of illustrations from Snider's *Earth History*, published by them, and to the American Medical Association for similar borrowings from Harvey's *Simple Lessons in Human Anatomy*.

I am indebted to Associate Curators Harry Raven and Harry L. Shapiro, of the American Museum of Natural History, Professor Sherman Bishop, of the University of Rochester, and Dr. George L. Streeter, of the Department of Embryology, Carnegie Institution of Washington, for the loan of photographs. Mr. Llewellyn I. Price .contributed a number of illustrations, and Mr. Brandon Grove aided in photographic work. For illustrations in the present volume I owe further thanks to Professor Camille Arambourg, Dr. Charles M. Bogert, Dr. Edwin H. Colbert, Mrs. Isabelle Hunt Conant, Dr. Roger Conant, Dr. D. Dwight Davis, Mr. Harold V. Green, Dr. William J. Hamilton, Jr., Dr. Robert F. Inger, Mr. G. E. Kirkpatrick, Dr. Frederick Medem, Professor J. Millot, Mrs. Rachel N. Nichols, Mr. Clifford H. Pope, Mr. Michael Ramus, Dr. John T. Robinson, Dr. Charles E. Snow, Mr. R. Van Nostrand, Mr. Walker Van Riper, and Miss Patricia Washer, and to the New York Zoölogical Society, the Zoölogical Society of Philadelphia, the American Museum of Natural History, the Chicago Museum of Natural History, the San Diego Zoölogical Society, the McGraw-Hill Company, and the W. B. Saunders Company. Miss Nelda Wright has aided very greatly in the preparation of the manuscript.

I am greatly indebted to my wife, Ruth Hibbard Romer, for her aid throughout the preparation of this volume.

ALFRED SHERWOOD ROMER

CAMBRIDGE, MASSACHUSETTS
    March 10, 1958

Contents

CHAPTER 1

Introduction
to the Vertebrates

It has been said that "the proper study of mankind is man." This is, in a sense, true. But, taken literally, the statement leads to an extremely narrow intellectual outlook. To understand the meaning of any other natural phenomenon we try to place it in its proper setting, its relation to other phenomena in time and space. There is no reason to treat ourselves otherwise. Man does not live in a vacuum, has not come out of a void. A fact of prime importance for an understanding of man is his position in the world of life, his relationships to other living things.

Man is a member of that series of living creatures known as the vertebrates, or animals with a backbone, a group including not only all the other warm-blooded hairy creatures, to which man is closely allied, but such varied forms as birds, reptiles, frogs and salamanders, and fishes. Their history is our history; and we cannot properly understand man, his body, his mind, or his activities, unless we understand his vertebrate ancestry. In the pages that follow we shall give a brief account of the backboned animals. Throughout, we shall pay attention to the evolutionary lines leading in the direction of mammals and man, and at the end will discuss human evolution in some detail. But we will, nevertheless, note the characters of the other types of vertebrates which we encounter in our progress up the vertebrate family tree. Man may pride himself on his current dominance (if limited dominance) over the rest of the animal kingdom. However, it is well for him to be aware that there are many groups of his vertebrate relatives which may lack his intellect but are even more highly endowed in other regards and live equally successful lives.

In this first, introductory, chapter we shall discuss the areas of scientific work and thought contributing to the story which we wish to tell; we shall note some of the general features of vertebrate structure and give a brief account of what is known—or deduced—as to the origin of backboned animals.

SYSTEMATICS—CLASSIFICATION

In olden times the natives of a region knew the limited number of larger and more striking local animals by names familiar to all the neighborhood. But with increasing communications between regions and the international development of scientific work, this use of local "popular" names was not enough. Many animals lack vernacular names, and such names, when present, vary in meaning. Even within the English-speaking world, for example, we find that the European red deer is called the elk in America, and the European elk becomes the moose; to most of the world a salamander is an amphibian, but in Florida it is the burrowing rodent which the rest of the world calls a gopher. Further, with the wide exploration of the world from the fifteenth century onward, it not only became necessary to have some logical system of nomenclature for the varied animals constantly being discovered, but also, because of the immense numbers of them, to have some systematic way of classifying them, of arranging them in categories so that the human mind could gain some broad picture of animal life as a whole.

Our modern methods of scientific naming and classifying animals are in the main due to a great Swedish naturalist of the eighteenth century, Linnaeus.

The basic unit in nomenclature is the species. How a species is to be defined has been, and will long continue to be, debated. Essentially, however, it may be considered as an interbreeding population of animals whose members may differ in various individual regards but possess characteristics in common, setting them off from other populations of somewhat similar character. Thus, for example, all domestic dogs have many common characters, although man, through breeding and artificial selection, has created races of dogs as varied as the great Dane, bulldog, and Mexican hairless. Despite their individual race differences, all dogs can breed with one another (sometimes to the dismay of owners of pedigreed stock); they form a population which can properly be considered a species.

To those familiar with mammals it is obvious that the wolves form a comparable population, similar but distinct and seldom breeding with dogs. Still other populations which can be considered as closely related species are those of the coyotes of North America and the jackals of the Old World. This series of closely related species is considered as forming a unit in classification—a genus.

As to nomenclature, each species of animal is given two names, cast in Latin form—that of the genus followed by that of the species. Thus the genus including the dogs and their close relatives is given the name *Canis;* the domesticated dog specifically is *Canis familiaris*, the coyote is *Canis latrans*, and so on. One can use a generic term such as *Canis* without necessarily specifying any one of the included species, but the specific name is never used alone.

Higher categories in classification are needed above the level of the genus in order to show the similarities between animal types and, as we now realize, to indicate the degree of relationship between various groups. Thus (to continue with our example) there are various sorts of foxes and tropical doglike or wolflike forms which are obviously related to the species included in *Canis* but not closely enough to be included in that genus. These can, however, all be included with the more typical dog relatives in a common *family*, termed the Canidae. (Family names are compounded of the Latin or Greek root of the name of a typical genus plus the ending *idae*.) Similar in various regards to the members of the dog family are other animals, generally carnivorous in habits—raccoons, bears, cats, and so on. All these are considered as forming the *order* Carnivora. The carnivores, together with all other warm-blooded, hair- or fur-bearing animals, form a still higher category, the *class* Mammalia. Again, the mammals, together with other backboned animals and lowly relatives, form a *phylum* termed the Chordata, a major subdivision of the *kingdom* Animalia. We thus have, in classifying an animal or plant a minimum sequence, in descending order, of several terms—kingdom, phylum, class, order, family, genus, and species. In addition, further steps may be introduced into the classification by using the prefixes super-, sub-, or infra-, as in superfamily, infraorder, subclass, etc.; the vertebrates proper, for example, are the subphylum Vertebrata of the phylum Chordata.

When systems of classification were first established, they were regarded in general as merely a convenient sort of filing system, animals with common features being "filed" in the same compartment. But with the general acceptance of the evolutionary theory in the last century, it became evident that a classification, if based on fundamental structural features rather than on superficial resemblances, had real meaning in expressing the blood relationships of the forms classified. The species included in a genus have descended from a common ancestor; the genera within a family are more closely related to one another than to other animals. Evolution thus proved a stimulus to workers in systematics, and study of the problems of relationships encountered by the systematist have contributed to the development of evolutionary theory.

THE LIFE OF ANIMALS

It is not enough to name an animal; we want to know everything about him: what sort of a life he leads, his habits and instincts, how he gains his food and escapes enemies, his relations to other animals and his physical environment, his courtship and reproduction, care of his young, home life (if any). Some aspects of these inquiries are dignified by such names as *ecology* and *ethology;* for the most part they come broadly under the term *natural history.* Many workers who may study deeply—but narrowly— the physiological processes or anatomical structure of animals are liable to phrase this, somewhat scornfully, as "*mere* natural history." But, on re- flection, this attitude is the exact opposite of the proper one. No anatomi- cal structure, however beautifully designed, no physiological or biochemi- cal process, however interesting to the technical worker, is of importance except insofar as it contributes to the survival and welfare of the animal. The study of the functioning of an animal in nature—to put it crudely, how he goes about his business of being an animal—is in many regards the highest possible level of biological investigation. Mere laboratory study of the anatomy and physiology of an animal tells us as little of his true na- ture as would be the case if man were studied only in the medical school laboratories of anatomy and physiology, neglecting those aspects of his nature and life that make up the larger part of a university curriculum.

Much of the study of animal life has been in the nature of field observa- tion. But if, with regard to the creature's activities, we ask why as well as what, answers can often be found by experiment in field or laboratory. We may cite some homely examples: Is it the *red* rag that incites the bull, as popular belief would have it? (No; tests show that bulls are practically insensitive to color.) Birds fly south when the days grow short. Does length of daylight influence a bird's activities? (Yes; that is why one sees the lights burning in chicken houses of an autumn evening.) Tadpoles turn into frogs at the time when the thyroid gland in their throats develops and becomes active. Is this a coincidence, or does the thyroid hormone ac- tually stimulate the change? (The latter, definitely; if tiny tadpoles are fed thyroid extract, they turn into tiny frogs long before their proper time.)

EVOLUTION

The story we are to tell is essentially an evolutionary one. We will be concerned mainly with the facts of the case, but one's thoughts naturally turn to the problem of how these many marvelous changes and adapta- tions seen in vertebrate history have been brought about. A full discussion of problems of evolutionary theory would take a volume in itself, but we may here give the bare outlines of the situation.

Philosophers and other non-scientists have often suggested that evolution may have been due to some supernatural agency or some mysterious "drive" within the animal itself. No one can prove, of course, that this is not the case. But as scientists we attempt to explain the phenomena of nature in terms of natural laws before resorting to supernatural interpretations.

The first attempt at an explanation of the causes of evolutionary changes was made by Lamarck at the beginning of the last century. He suggested that the changes brought about in an individual's body during his lifetime might be transmitted to his offspring. A blacksmith's sons, says this theory, are liable to have stronger arms because of their father's use of his muscles; the giraffe's short-necked ancestors, by continually reaching upward into the trees for higher leaves, produced descendants with longer necks. This is a plausible theory, but it failed of acceptance; to this day, despite much research, there has never been found any evidence that what an individual does during his lifetime has the slightest effect in producing related changes in his offspring.

Half a century later Darwin gave an answer to the problem which has proved much more acceptable and which, with various additions and modifications, is the basic element in modern evolutionary theory. He pointed out that individual variations constantly occur, apparently more or less at random, within a race or species, and that many of these variations are inheritable. Man has been able, by selection among variants, to produce marked changes in domesticated animals and plants. Selection also occurs in nature, if more slowly. Only a fraction of all the animals that are born or hatched survive to maturity and reproduce. Given variation among these individuals, it is clear that those with advantageous variations are the ones most likely to survive. By a process of natural selection, those best fitted generally triumph in this constant struggle for existence. Each advantageous variation may be small, but over the many millions of years of geologic time, major evolutionary changes are brought about.

In Darwin's day nothing was known of the nature of hereditary mechanisms or of the causes of variation among individuals. Today, the science of genetics has given us a clear understanding of inheritance and considerable knowledge of the origin of variations. It was once thought that inheritance was a sort of general blending process, the offspring splitting the difference between the sum of the characters of the two parents. Now this is known not to be the case. Every individual—in fact, every cell of every individual—has some thousands of discrete units, termed "genes." The form of each new individual is determined—aside from environmental shaping—by the genes inherited from his parents. Each gene may affect the development of a number of structures or parts of a body; conversely, every organ is affected in its development by a considerable number of genes. Each individual contains genes-in-pairs, one derived from either parent, and he in turn passes on to his offspring one gene from each pair.

A given gene, however, does not always work in exactly the same fashion; it may vary. Let us assume, as an example, that in a certain mammal a gene associated with eye color customarily works to produce brown eyes. There may, however, crop up blue-eyed individuals; a change, a *mutation*, has occurred. A gene may be conceived of as a large and complex molecule; presumably there has been some slight change in its chemical composition. Such mutations are known in many cases to be brought about by radiations of various sorts or by action on germ cells by chemical agents; the effects produced on the offspring appear as random variations, for better or for worse, quite indiscriminately and (contrary to the Lamarckian belief) without any relation to the life led by the parents. Mutant genes often produce detrimental characters and may be rapidly eliminated from the race. Many, however, linger on in individuals, giving to the species added variability upon which selection may act.

We have there, then, mechanisms of heredity and variation suitable for the working of natural selection as Darwin conceived of it. But modern studies, with genetics as a base, show that the evolutionary process is much more complex than he imagined. We may note, as an example, that with genes present in pairs in an individual, if one member of a pair is a mutant, tending to produce a result different from that produced by its more normal team-mate, there will be a conflict of interests, and one will tend to dominate over the other in the structures or functions which it controls. It is obvious that selection can act readily on a dominant gene, since the characters which it produces are present whenever one member of a gene pair is of this nature, but can have no influence over the "weaker" of such a pair of genes—technically termed "recessive"—whether its effect be good or bad, unless by chance both members of the pair of genes concerned are of the same recessive nature.

THE FOSSIL RECORD

An excellent line of attack on the history of living things lies in the domain of paleontology, the study of fossil animals. Comparative anat-

omy, studies of function, and embryology may enable us to deduce probable relationships and hypothetical ancestors. But it is infinitely more satisfactory to find, if we can, the concrete remains of the ancestors themselves in the record preserved by rocks laid down in past ages of the earth.
While the greater emphasis in this book will be placed on living animals,
fossil forms will receive a fair share of attention. After all, living animals
are but a small fraction of all that have ever lived, and the hosts of animals which preceded and gave rise to those living at the present moment
cannot, in all fairness, be excluded from our survey.

For many forms of life the fossil record is none too satisfactory. Plant
tissues are in great measure soft structures which are comparatively seldom fossilized, and, although we have considerable knowledge of the
evolution of various types of trees, much of the botanical story cannot be
confirmed by direct fossil evidence. Among the animals lacking a backbone there are a number of groups which have shells or other hard parts.
Of these we have abundant records. But most of the important connecting
links were apparently soft-bodied creatures which are rarely preserved as
fossils. Further, the history of many invertebrate groups has covered a
vast stretch of time, and the older rocks which should contain the early
chapters of the story have been lost from the earth's crust or so much
modified by crushing or heating that the fossil story they once carried has
been obliterated.

With the backboned animals paleontological inquiry is much more
fruitful, for several reasons. Almost all vertebrates have hard skeletal
parts—bones, armor plates, or teeth—capable of fossilization. Further,
vertebrates are a comparatively modern group. Our known history
stretches back less than half-a-billion years. This (while a rather good
stretch of time) is but a fraction of the known history of the earth, and for
all this time we have available for study numerous series of fossil-bearing
rocks.

The study of fossils is a comparatively new branch of investigation. It is
only within the last century and a half that fossils have been generally
recognized to be the actual remains of once living creatures rather than
freaks of nature of some kind. It was not until the general acceptance of
the evolutionary concept in the middle of the last century that their importance as ancestors of living animals was realized and a serious study of
extinct forms begun. Today we have a fairly good outline of the lines of
descent of many groups of backboned animals, but none is absolutely
complete. There are many gaps, many perplexities. It is obvious that we
shall never fill in every little chink in this historical structure, for it is quite

improbable that we shall ever find the remains of all the countless animals that have inhabited the earth. But year by year new fossils come to light which help fill the blanks in our puzzle. We have already progressed far in our study of the evolution of vertebrate life in past ages, and it is not too much to hope that, in the not too distant future, we may have a well-rounded picture of all the main points in the past history of our vertebrate ancestors.

A brief mention of the type of work done by the student of fossils may not be amiss. The average museum visitor who sees a mounted skeleton of an extinct animal has little idea of the great amount of work which lies back of this exhibit. Fossil-bearing rocks of any particular age are exposed only in scattered districts of the earth, and long trips are often necessary to reach them. Then, too, rocks must be not only present in a region but must be exposed at the surface. This means that most collecting must be done in treeless, grassless, sparsely settled, arid regions such as the deserts of our own West, those of Mongolia, or the Karroo Desert of South Africa. "Bone-hunting" is not the most comfortable of occupations; even in good deposits, fossils may be very hard to find, and there may be many weary days of fruitless tramping over barren rocks before the collector finds a specimen worth excavating.

Then the work begins. If the find is of value, every piece must be removed with care and its position noted. Broken bones are often bandaged and plastered as carefully as a broken human leg, so that the fossil may not become irreparably shattered. The specimen is carefully packed to prevent damage on its long trip back to the laboratory, where work continues. The rock surrounding the bones must be removed (often it is exceedingly hard), all broken pieces cemented together and their positions determined, and, finally, the whole skeleton mounted. The mounting of a dinosaur may take years, and even some specimens no larger than a man's fist may be so delicate that months may be spent in preparing them.

Parallel with the laboratory work on the skeleton goes, in the case of new or rare forms, the study of the bones as they are prepared and the determination of the anatomy of the creature and its proper position in the family tree—sometimes a task of difficulty in the case of strange new forms.

Parenthetically, we may mention the popular belief that the paleontologist often restores a fossil animal from one toe bone. A century ago it was thought that if (for example) one found the body of an animal built like a horse, but with the legs missing, one could assume that the creature had

Bone-hunters in the field. A group of paleontologists from Harvard and the Carnegie Museum of Pittsburgh examining a newly discovered fossil quarry in the Eocene deposits of eastern Utah. From this quarry were obtained numerous skulls of small deerlike mammals. In the background are seen some 300 feet of strata formed in a relatively short period at the end of the Eocene.

Mass-burial (*right*). In most cases only one or a few individuals are found together in fossil deposits. This slab contains skulls and other remains of 12 large flatheaded Triassic amphibians as found in a layer where perhaps 100 animals were buried together. (Specimen in Museum of Comparative Zoölogy, Harvard.)

Discovery (*below*). Looking into an excavation in which Paul C. Miller, of the University of Chicago, uncovers the skeleton of a primitive reptile in the Karroo Desert of South Africa.

Resurrecting a fossil. These are figures of *Dimetrodon*, a long-spined reptile from the Texas redbeds. *Above*, a remarkably perfect specimen of the body shown as found in the field (a specimen at the University of Chicago). *At the left*, Bob Witter, of Harvard, is shown "bandaging" such a specimen in the field with burlap and plaster for safer transportation to the laboratory. *Below*, the same specimen, mounted by George Nelson. On the next plate is a scene showing characteristic animals found in these beds; *Dimetrodon* plays a prominent role.

Texas in late Paleozoic times. A scene based on fossils of the early part of the Permian period, found most plentifully in the redbeds of western Texas. The four-footed animals of the time were primitive reptiles and amphibians. Most of the forms illustrated are archaic relatives of mammalian ancestors, the pelycosaurs. Some of the members of this group (as *Casea*, left) had a rather normal reptilian appearance. Others, however, specialized in the development of a sail-like structure on the back. Five individuals are shown of the genus *Dimetrodon*, the most aggressive carnivore of its day and, except for its spines, close to the ancestry of the mammals. Into this group (*center*) has blundered an *Edaphosaurus*, a form which likewise bore a sail (with short crossbars on the spines) but was an inoffensive herbivorous animal.

In the right foreground are specimens of *Diplocaulus*, a water-dwelling amphibian with a grotesque flat, horned skull. (From a mural by Charles R. Knight; photograph courtesy Chicago Natural History Museum.)

An enemy of the ancient vertebrates. In Silurian deposits are often found the skeletons of Eurypterids, predaceous water-dwelling arachnids related to the scorpions and the horseshoe crabs. *Pterygotus*, shown here, reached a length of 6½ feet, whereas most of the fishes then were but a few inches in length. (From Clarke and Ruedemann.)

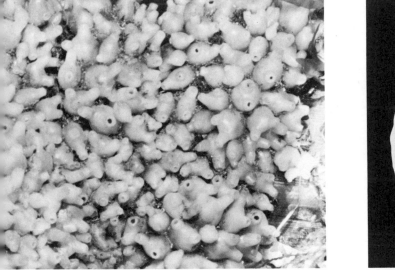

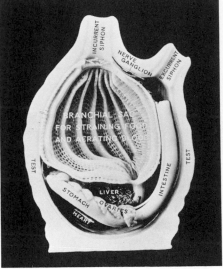

Sea squirts. Members of this primitive chordate group, known as ascidians or tunicates, are found attached to rocks or other objects in marine waters. *Upper left*, part of a colony of tiny tunicates (*Molgula manhattensis*) growing on a wharf pile. *Right*, a model to show the internal anatomy. Water is brought in through the opening at the top, strained through an apparatus corresponding to the gills of vertebrates, and expelled through a lateral orifice. Food particles pass down into stomach and intestine. The nervous system consists only of a small ganglion from which nerve fibers radiate. (Photographs courtesy American Museum of Natural History, New York.)

The larva of a sea squirt (*Botryllus*) (*below*). About 90 times natural size. This tiny animal contains in the large "head" most of the structures found in the adult. The tail contains a nerve cord and notochord, visible in the photograph. After swimming about for some time, the "head" attaches, the tail is resorbed, and the adult tunicate develops. (Photograph copyright General Biological Supply House, Chicago.)

hoofs. But—to follow this particular case—it was found that some such creatures (chalicotheres, chap. 13) in reality had feet armed with huge claws with which to dig up roots. Paleontologists have become chastened and more conservative. Parts of skeletons are restored in plaster in almost every fossil specimen, for few finds are absolutely complete; but, except for minor features, all work of restoration is based on our knowledge of the bones concerned in other specimens of the same or closely related types.

The scientist as well as the lay reader is curious as to how these old animals would have appeared and acted when alive and in the flesh, and numerous life-restorations of fossil animals have been made; a considerable number of them are shown in the present work.

Some of the restorations which we see in the Sunday supplements are caricatures which never existed except in the artist's imagination; but many of the more carefully prepared restorations probably give us a fair approach to the actual appearance of the forms they portray. From the skeletons we get many clues as to the posture of the body, the style of walking, the bulk and position of the muscles which give the major contours to the body. Of the skin we know comparatively little, and in colored restorations the pigment given the skin is a guess, although based whenever possible on the colors assumed by living relatives of the animal or forms which lead similar lives.

In order to use our knowledge of fossil forms effectively we must have some idea of the comparative times of their appearance and disappearance as measured by the scale set up by the geologists. A very considerable portion of the scientific work done in the field of geology has been devoted to the study of the comparative age of the various layers, or strata, laid down in the seas or lowland regions of the world in ancient days. On the basis of these studies has been erected a series of subdivisions of world-history, an outline of which is given in the accompanying table. We cannot, of course, tell the exact length in years of any of the geologic units, but the figures indicated are fairly good "guesses," based on such data as the thickness of the rocks formed during any particular period of the earth's history and the present degree of decomposition of radioactive minerals laid down in them. The major subdivisions of the earth's history are known as "eras." These are divided into "periods," which may be further subdivided into "epochs." With this last type of division we shall be concerned only in our later study of mammalian evolution (chap. 11).

We shall have nothing to do with the first two of the five great eras in the known history of the earth, the Archeozoic and Proterozoic, for although there must have been much life in the form of plants and lower animals, our records of them are very poor, and vertebrates are quite unknown.

Our interest in this story begins in the third era, the Paleozoic, or time of ancient life. In the Cambrian, first period of this era, almost all invertebrate groups abounded, but vertebrates were still unknown. In the following Ordovician period are found rare fragments of vertebrates, and toward the end of the Silurian period we meet for the first time with good skeletal remains of early water-living, backboned animals.

TABLE OF GEOLOGIC PERIODS*

Era (and Duration)	Period	Estimated Time since Beginning of Each Period (in Millions of Years)	Life
Cenozoic (age of mammals; about 70 million years)	Quaternary	1	Modern species of mammals, extinction of large forms; dominance of man
	Tertiary	70	Rise of placental mammals and birds
Mesozoic (age of reptiles; lasted about 120 million years)	Cretaceous	120	Dominance of angiosperms commences; extinction of large reptiles and ammonites by end of period
	Jurassic	155	Reptiles dominant on land, sea, and in air; first birds; archaic mammals
	Triassic	190	First dinosaurs, turtles, ichthyosaurs, plesiosaurs; cycads and conifers dominant
Paleozoic (lasted about 360 million years)	Permian	215	Radiation of reptiles, which displace amphibians as dominant group; widespread glaciation
	Carboniferous	300	Fern and seed fern coal forests; sharks and crinoids abundant; radiation of amphibia; first reptiles
	Devonian	350	Age of fishes (mostly fresh-water); first trees and first amphibians
	Silurian	390	Invasion of the land by plants and arthropods; brachiopods; primitive jawless vertebrates
	Ordovician	480	Appearance of vertebrates (ostracoderms); brachiopods and cephalopods dominant
	Cambrian	550	Appearance of all invertebrate phyla and many classes; dominance of trilobites and brachiopods; diversified algae

* The older eras of earth history, before fossils became abundant, are omitted. The Carboniferous is generally subdivided into two periods, Mississippian (earlier) and Pennsylvanian (later). The time estimates are based on the rate of disintegration of radioactive materials found in a number of deposits.

The Devonian period, next in order, is often called the Age of Fishes, for these primitive vertebrates had by then become diversified and abundant. Here too we meet with the remains of the oldest-known land dwellers among vertebrates, the first amphibians. In the Carboniferous period (often divided into two periods, Mississippian and Pennsylvanian) amphibians had reached the peak of their development, and the first reptiles are encountered. The Permian, last of the Paleozoic periods, witnessed a decline in the amphibians and a beginning of the story of reptilian diversification.

The Mesozoic era is often called the Age of Reptiles, for it was during this time that there flourished many reptilian types now extinct. Most of the interesting dinosaurian and other groups had their origin in the Triassic. At the end of that period appeared the first of the mammals, destined to remain inconspicuous, however, during this era. In the Jurassic the reptiles appear to have been at the peak of their development, while from them had evolved the oldest-known birds. The end of the Cretaceous period marked the extinction of the great reptilian dynasties.

The Cenozoic, terminating with present times, is a comparatively short era, but one of great importance, for it was during this time that there took place the interesting evolutionary history of the mammals, and man himself appears in the final stage of Cenozoic history.

GEOGRAPHICAL DISTRIBUTION

No vertebrate is ubiquitous. All species are restricted in their area of distribution. Many factors are involved—the presence or absence of food or enemies, for example, proper conditions of temperature, and other environmental factors. Still further, the general topographic relations of the land masses and oceans of the globe often set sharp bounds to the potential range of a species.

On the whole, the distribution of marine vertebrates has less restrictions placed upon it than is the case with land or fresh-water forms. The temperatures of ocean waters, even at the surface where directly exposed to the sun's rays, vary much less, north and south, than do those of the land areas, and many fishes of the Atlantic and Pacific range widely over much of the extent of those oceans. South America and Africa do not extend nearly so close to Antarctica as do the northern continents toward the North Pole; hence in the south there is relatively free communication between Atlantic, Indian, and Pacific Ocean areas, whereas fishes of the northern temperate Atlantic and Pacific must pass through the cold

Arctic to get from one ocean to the other. There is, however, much variation in fish habitats in relation to water depths and distance from land.
The small plants and animals which form the basic food supply of fishes
and other marine vertebrates are more abundant near the shores where
useful materials—particularly nitrates and phosphates—are poured out
from the land by rivers, and the invertebrates, found abundantly on the
bottom of the shallower ocean areas, are a favorite fish food. Hence there
is a great concentration of fish life in coastal waters. Farther out, in the
high and deep seas, food supplies—and consequently fishes—are much less
abundant, even at the surface. And the depths of the seas, although inhabited, are but sparsely populated. Life is very difficult. Pressures in the
deeps are immense but do not greatly interfere with life, as long as the
animal does not venture far upward or downward. However, food is to be
found only as it descends from above; it is always bitterly cold; and there
is absolute, utter darkness.

Turning to land areas, here, too, climatic and ecological factors are of
great importance. Temperature not only affects vertebrate life directly
but is even more important as affecting the nature of the vegetation that
is the basic food supply. Elevation above sea level is, of course, in itself
important, causing in southern mountains "islands" of life resembling
those of more northern lowlands. And rainfall is, on the whole, even more
important than temperature in determining the amount of vegetation
growing in a given region and hence its potentiality for supporting animal
life.

We live today at a time in the earth's history when there is a very sharp
temperature gradient between very cold polar regions and hot tropics. In
northern continental areas one can distinguish a series of life zones or
bands running from north to south, each with its characteristic fauna—the
nearly barren Arctic tundra; the cold evergreen forests; temperate zones
of deciduous forests, grasslands and wastes; hot subtropical and tropical
regions, grading from dense rain forests through savannas and grasslands
to dry deserts. In the Southern Hemisphere much of the zonation is repeated in reverse order, except that (apart from ice-covered Antarctica)
land stops too far from the pole to show any great development of the
colder life zones.

Topography is more important in defining vertebrate distribution on
land than in the sea. Continental relations are of major importance, but
before discussing them, we may note two specific problems of smaller
scale—the distribution of fresh-water fishes and the nature of island
faunas.

A typical flying bird has few topographic difficulties in attaining wide distribution; in complete contrast, a fresh-water fish has the hardest problem to solve in becoming widely distributed, for it is not at home either on land or in the sea. If a species of fish evolved in the Mississippi River system, for example, how can it enter other systems of the Atlantic coast or the Pacific—or even another river emptying into the Gulf? Some stream inhabitants can enter salt water briefly and thus make short coastal migrations, but many fresh-water forms lack this potentiality. There are occasional possibilities. Two rivers may (but rarely do) have a connection near their sources. In the course of geological history one stream may "capture" the headwaters of another—and capture its fishes with it. By an outside chance a fish egg might stick to the muddy feet of a bird in cross-country flight. And there are verified accounts of "rains of fishes," where small fish have been swept up by a tornado and carried for many miles. These are all most unlikely events. And yet, such fishes have often become distributed over wide areas. The pike, that abundant fish of northern lakes, does not appear to be a fish type of unusual antiquity, but it has successfully spread over the entire Northern Hemisphere.

Islands separated from mainlands by stretches of sea water usually have some sort of fauna, large or small, of terrestrial vertebrates. How have these been acquired? If the island is near shore or separated only by shallows, the problem is not too difficult. Most land animals can swim for at least short distances. Fluctuations in sea levels are known to have occurred in recent geological times, so that in many cases it would have been possible to reach areas that are now islands without suffering even wet feet.

But with islands distant in the oceans the problem is more puzzling. It seems certain that most of them have never had any connection with a continental area. Yet in the case of the Galapagos Islands off the western coast of South America, for example, where we have islands lying the better part of a thousand miles from the coast of Ecuador, we find, peacefully at home, large tortoises and iguanid lizards that are purely terrestrial in nature. Their presence there is little short of a miracle.

One of the few possibilities for such an "outside chance" of migration having occurred lies in their having been landed there aboard a raft—a natural one, we must hasten to say. In river systems flood waters on occasion bring together a tangled and compact mass of trees and bushes, along with a freight of small animals. Such rafts, with living cargo, have been seen as much as a thousand miles out at sea and, barring storms, may float on indefinitely. The chances that a particular animal should board such a raft and that this Noah's Ark should make a successful landing are extremely small. But given enough time, the almost impossible becomes the almost certain.

Turning now to large-scale distribution, the arrangement of the great continental areas, together with associated temperature differences, led a century ago to recognition of the fact that the whole world can be reasonably divided into some six major geographical regions. There is no absolutely sharp division between the faunas of these areas, but in a broad way each region has its characteristic fauna, differing markedly from that of other areas. In the Northern Hemisphere lies the great Eurasian land mass and, separated from it by a narrow and shallow strait, the North American continent. Pendant from these northern areas are the three southern con-

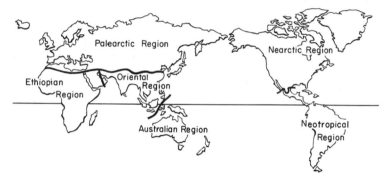

A sketch map of the world to show the main zoögeographic regions

tinents, of which two—Africa and South America—are connected with the north; the third—Australia—is separated, but somewhat linked, by the East Indies, a series of difficult steppingstones. With a major exception in the Orient, these continental areas are basically the major zoögeographic regions.

Over Europe and the Mediterranean region, we find a relatively uniform fauna of vertebrates (with some changes, of course, from north to south) and much the same animal types range on eastward through northern and central Asia to Manchuria, Japan, and Siberia. This region is termed the Palearctic. In most of North America the fauna is also very similar so that an American naturalist or hunter would feel fairly at home in the woods or plains of Europe. The two regions combined are sometimes considered as the Holarctic region; but there are some differences, and hence North America (down to Mexico) is generally treated as a separate if related region, the Nearctic.

Temperatures as well as continental boundaries, however, enter the regional picture, and in the Old World there are two distinctive tropical zoögeographic regions. South of the great mountain chain of the Himalayas, in India and extending eastward into China and southeast to the Malay region and the nearer East Indies, lies the Oriental region. Here is a fauna markedly different from that of the Palearctic; any reader of the literature on this region will readily recognize the distinction in the presence of such animals as the tiger, elephant, numerous monkeys, the orangutan, and so on. Africa likewise has its distinctive tropical fauna. This resembles to some degree that of the Oriental region with elephant, monkeys, and great apes, although of different types. The African fauna, however, is on the whole a more rich and varied one, having especially a great variety and abundance of antelopes. This is the Ethiopian region. Its boundaries do not exactly coincide with those of the African continent, for the Mediterranean shores have a fauna closer to that of Europe than to the tropics south of the Sahara; and Arabia, although essentially a transitional region, is on the whole closer faunally to Africa than to the other Old World regions.

Even more distinctive are the other two southern continents. South America (with Central America thrown in for good measure) is the Neotropical region. It is connected by an isthmus with North America; but the passage is a narrow one, and we know that it has not existed long geologically. In consequence we would expect—and do find—that its fauna differs greatly from that of North America. A small proportion of its animals, from deer to mice, are like those of the north. But it has a whole series of rodents of unique types and a large population of monkeys, very different from those of other continents. Most distinctive of all are the so-called edentates—tree sloths, armadillos, and anteaters. These constitute an entire order of mammals not present in other regions.

Most isolated of all regions is the Australian. Apart from bats in the air and seals on its shores, its mammalian population (extending to the New Guinea region) is unique. The only "higher" land mammals in its population are rats and mice of a rather special sort. Most remarkable of all inhabitants are the only two egg-laying mammals in the world, the duckbill and the spiny anteater. Nearly as distinctive, and forming the bulk of the mammal population, is a host of pouched mammals—marsupials—which are unknown elsewhere except for the American opossums and one or two obscure South American forms.

There are many curious examples of discontinuous distribution of animal types in these different regions. Why, for example, are tapirs found in the Malay region and South America but in no intervening region? Why are the only relatives of the camels of the Old World deserts to be found in the llamas of the South American Andes? The answers, surely, are to be found in the past history of the continents and their connections. Some of the attempts to solve distributional problems in this fashion, however, have been rather on the naïve side. For example, to explain similarities between the fresh-water fishes of South America and Africa, a land bridge has been assumed to have once been present between the two continents. But in many other regards African and South American faunas are very different. What sort of a land bridge could there have been over which most animals could not travel but with streams in which fresh-water fishes could swim happily? Were there "no trespassing" signs? If all the hypothetical land bridges conceived of by various authors were plotted on one map, they would probably form a spider web criss-crossing the oceans. Actually, for at least the extent of the Cenozoic era—the Age of Mammals, some 70 million years or so back from the present—most problems of distribution can be much more simply solved by assuming that land masses have long been in much their present shape, with an occasional make-and-break of continental connections, but that, however, there have been climatic changes during this time.

Throughout this period the northern continents appear to have formed essentially a single broad east-west land mass. At times (as at present) there has been a break between Siberia and Alaska, so that for a while evolution went its own way in the two northern regions—this accounting for some of the differences—but for much of the time the entire Holarctic area was a unit, thus accounting for the many similarities between the faunas of Eurasia and North America.

During early Cenozoic times climates appear to have differed greatly from present ones. Then, it would seem, the Arctic was less cold than now, the tropics less hot, and much of Eurasia and North America was a possible, pleasantly warm, habitat for animals which cannot withstand the present northern winters. In consequence the Holarctic region was once the home of many animals—from various fishes to tapirs—now absent from these areas and present only in the sheltered warmth of discrete tropical regions.

From the great northern land masses the southern continents hang like pendants; the connecting links have had a variable history. Africa may have been separate for a bit in early Cenozoic days, so that certain mammal groups—elephants and conies—may have developed in isolation there. But for most of the era it has had good land connections with the north and is now the home of many animals whose ancestors roamed Eurasia in climatically happier times but were driven south by the colder weather of the late Cenozoic.

The history of South America has been more distinctive. The North American connection existed, it appears, at the very beginning of the Age of Mammals but was shortly afterward broken, so that only a few forms had entered. In isolation there developed a very distinctive fauna of mammals. Finally, at about the beginning of the Ice Age, a million or so years ago, the bridge was re-established. Various northern animals (such as llamas) entered, and much of the native fauna perished. Others, however, survived, such as the native rodent and monkey types and, especially, the sloths and other edentates.

Still more isolated has been Australia. The evidence suggests that this region has been separated from Asia for the entire course of the Cenozoic. Before that, egg-laying mammals and pouched relatives of the opossums were present but none of the higher mammal groups. Some rat and mouse types finally reached Australia by gradually migrating down the East Indies, but otherwise the native fauna developed in complete isolation until the coming of man.

Back of the Cenozoic our knowledge of continental areas and faunas is much less complete. Since these more ancient times are of relatively little importance in regard to the modern picture, in which we are primarily interested here, we will mention them but briefly. Very likely the basic geographical and climatic picture for much of the Age of Reptiles, the Mesozoic, was not too dissimilar to that of the early Cenozoic. But, still further back, toward and to very ancient Paleozoic days, the world may have been very different. The more conservative geologists believe stoutly that the continents and poles have been as they are today since the earth began. But such facts as the great Paleozoic glaciers in Africa and Australia suggest that the earth may have once spun on a different axis than the present one, and there is, further, some evidence that the great continents are not as stable as they seem but may have, over long eons of time, shifted their positions. We shall later mention one of these heretical ideas but will for the most part carefully shun these areas of controversy.

DEVELOPMENT

The development of the individual from a seemingly simple egg to the highly complex adult organism is one of the most important of biological phenomena. We shall not be concerned in this volume with the details of embryological processes, but some of the major features of development are significant in tracing the relationships and evolutionary history of vertebrate groups. For example, reptiles, as we shall see, evolved a shelled egg with a large amount of yolk, and the embryo developing in it grows a complex series of membranes around itself. The egg of a typical mammal is, in contrast, a tiny structure, but its ancestors once had an egg of reptilian type, as is shown by the fact that the embryo develops a similar series of membranes, even including, rather absurdly, a yolk sac—although an empty one.

In the early days of the study of animals from an evolutionary point of view, an enthusiastic German, Haeckel, claimed that higher vertebrates in early developmental stages resembled members of lower classes; during the early development of man and other mammals, for example, there is a time when the embryo passes through a fish stage, with gills and so on. The embryo, so to speak, climbs its own family tree. This, however, was an overstatement of the case. Mammal embryos do have fishlike features at one stage; but they resemble, not adult fishes, but fish embryos. Despite this necessary qualification, the study of embryos does afford interesting indications of relationships; similarities in early development between various organs and structures which differ greatly in the adult forms furnish strong evidence of evolutionary changes. For example, every mammal has inside its eardrum three little bones which aid in sound conduction. Study of embryos, however, shows that these tiny auditory ossicles are identical in nature with much of the jaw apparatus of such a lower vertebrate as a shark; and this conclusion, first reached on the basis of the study of embryos a century and more ago, has over the years found complete confirmation through the study of fossils.

Among vertebrates which produce large-yolked eggs or bear their young alive—sharks and skates, reptiles, birds, and mammals—the offspring at hatching or birth are generally small replicas of the adult, and are soon able to live, in modest fashion, somewhat the sort of life enjoyed by their parents. Not so among many fishes and amphibians. The eggs are of small to minute size and contain little yolk on which the embryo can grow. As a result the young when hatched are in general too small to live as the adults do, to eat the same sort of food, or to escape enemies in the same fashion. In consequence the young of many of these forms are adapted to live for a time in a manner quite different from the mature form, and they are frequently very different in appearance and structure. Such a form is

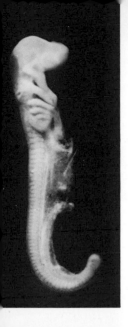
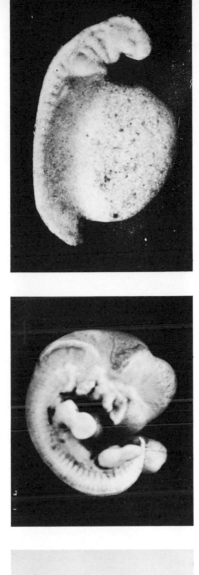
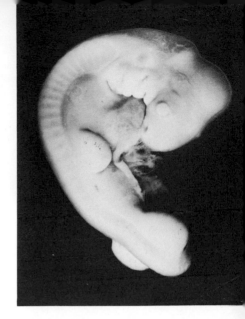
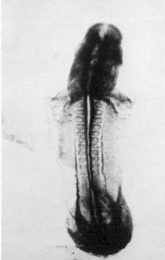

A comparison of vertebrate embryos. *Above, left,* five embryos of approximately the same age: a shark (*Mustelus*), a salamander (*Cryptobranchus*), a chick, a rat, and a man. All have the same structural features, but there are variations in pattern. In most the yolk is contained in a sac which has been removed; in the salamander figured, the yolk expansion of the gut is, however, preserved. Rat and human embryos are very similar, although the specimen of the latter is somewhat more advanced. (Photographs from Department of Anatomy, University of Chicago, courtesy G. W. Bartelmez.) *Lower left* (seen from above), earlier embryos of chick and rabbit. The dark mass at the front (upper) end is the brain, which grows precociously. Farther back, the spinal cord lies in the midline; on either side, blocks of tissue (mesodermal somites) from which muscle and skeleton arise. (Photographs copyright General Biological Supply House, Chicago.)

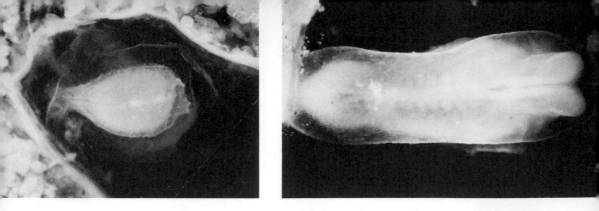

Human embryos in their membranes. *Above*, two early embryos seen from the upper surface. *Left*, an embryo of about 3 weeks, showing little apparent differentiation of structure. At the back end (*left*) is the body stalk through which the embryo is attached to the surrounding chorion. *Above right*, a similar view of a 4-week-old embryo. The body has grown out greatly anteriorly, i.e., to the right, where folds of nervous tissue are forming the brain and the spinal cord; on either side are somites of mesoderm.

Below, side views of two early embryos. *Left*, an embryo of $3\frac{1}{2}$ weeks; *right*, the same embryo of 4 weeks shown above. At the left in both cases is the body stalk. The embryos are surrounded above and at the sides by the translucent amnion. Beneath is the large (but nearly empty) yolk sac. (Embryos from the collections of the Department of Anatomy, University of Chicago; photographs courtesy Dr. G. W. Bartelmez.)

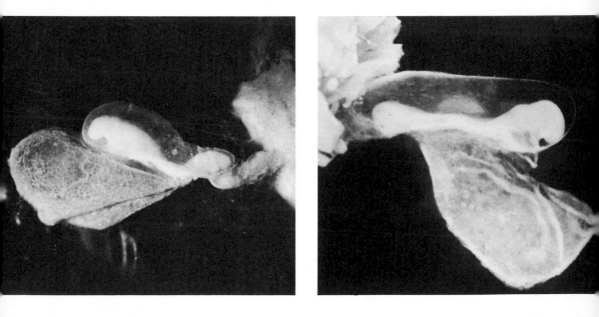

An album of human embryos (*opposite*). A series from about $3\frac{1}{2}$ weeks after fertilization to about 3 months. The youngest stages are very similar to those of other mammals; only gradually do distinctive human features appear. Noticeable during the second month is the relatively enormous growth of the brain. The limbs appear relatively late. The earliest stages show a distinct tail. At first the active heart projects prominently from the chest. A body stalk at the right gradually narrows to form the umbilical cord. The embryos are shown of the same absolute size to facilitate comparison; there is, however, great growth during this period. The smallest shown is actually about $\frac{1}{10}$ of an inch, the last about $2\frac{1}{4}$ inches long. (Embryos in the collections of the Department of Anatomy, University of Chicago; photographs courtesy Dr. G. W. Bartelmez.)

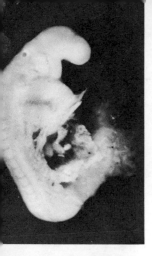
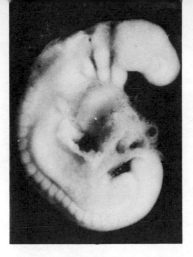
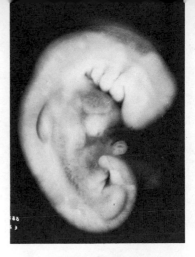

ree and one-half weeks   Four weeks   Five weeks

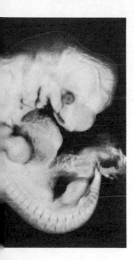
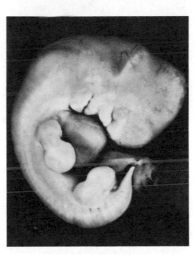

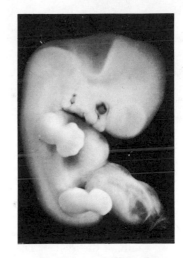

e and one-half weeks   Six weeks   Six and one-half weeks

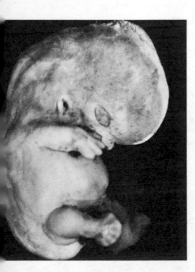
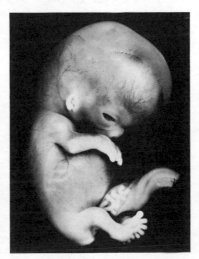
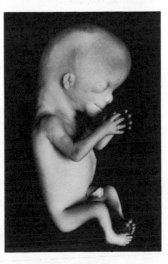

Seven weeks   Eight weeks   Twelve weeks

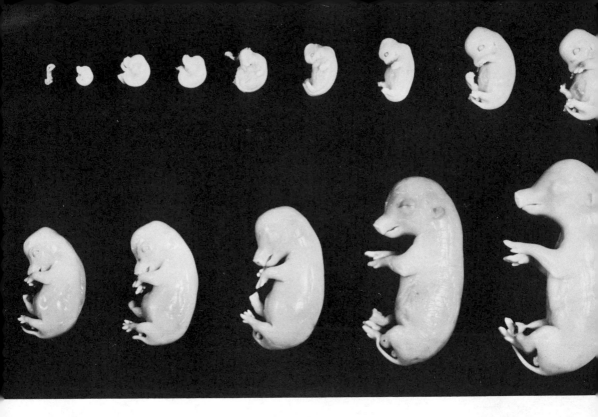

Pig embryos. A series of 14 specimens ranging from about 2½ to 6 weeks after fertilization, about twice natural size. The first half-dozen specimens are closely comparable to the earlier members of the human series shown on the pages following (much more enlarged); in the older specimens, however, pig features gradually emerge. (Photographs copyright General Biological Supply House, Chicago.)

Human embryos in the uterus. *Left,* an early embryo surrounded by membranes and lying in the uterus, or womb, of the mother. The spongy tissues at the bottom are the maternal tissues of the uterus; closer in toward the embryo are rather loose tissues which later fuse with those of the mother to form the placenta, through which nourishment reaches the growing embryo. The tiny embryo is inclosed in a water-filled sac—the amnion. *Right,* a later stage (about 7 weeks) with the amnion cut open to show the embryo. The embryo is connected with the outer tissues by a body stalk, which later narrows to become the umbilical cord. (From *Carnegie Institution Publications in Embryology.*)

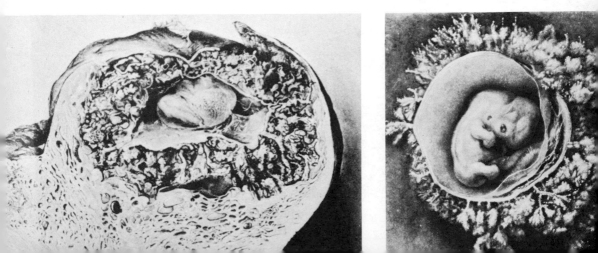

termed a "larva." After the larva has lived for a period (weeks, months, or even years) and attained a fair amount of growth, it changes markedly in structure and habits—the process of "metamorphosis"—to become an adult. The most familiar larval form is, of course, the frog tadpole, which differs notably from the adult in such obvious features as its swimming tail and lack of limbs and differs equally greatly in internal structure. Salamanders have a comparable aquatic larval stage, although the transformation is less radical and less rapid. Lampreys have a larval form very different from the adult and, as we shall see later in this chapter, so do some of the lower chordate relatives of the vertebrates. The teleost fishes which constitute the vast majority of modern fish types, lay very tiny eggs and hence produce very tiny young. In many teleosts there are larval forms very different from the adults. In a number of cases the young are little translucent leaf-shaped animals that one cannot correlate with the adults unless the intermediate stages are known. Again, as shown in an accompanying figure, the marine sunfish are, as adults,

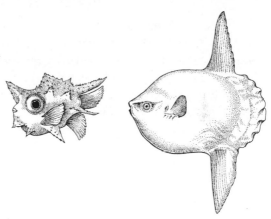

At the left, the tiny larva of the oceanic sunfish (*Mola*); at the right, the adult, of very different proportions and many thousands of times the larval size. (After Norman.)

enormous slab-sided fishes; the larva, however, is a little spherical ball of a fish, covered with spines which help to protect him from becoming a morsel of food for some fish cousin.

Normally, only a small fraction of an animal's life is spent as a larva. But in some cases (some fresh-water lampreys, for example) most of its life is spent in a long drawn-out period as a larva, and although it changes eventually into adult form, it does little afterward except reproduce and shortly die. In other cases (some salamanders are good examples) an animal may, à la Peter Pan, refuse to grow up and, except that its reproductive organs mature, remains a larva all its life—a situation termed "paedogenesis."

If this condition long persists in the race, the potentiality of ever assuming adult form may disappear from its germ plasm—the animal has, so to speak, forgotten how to grow up. And—a disturbing but very interesting thought—if evolutionary changes were to take place in the descendants of an animal of this sort, the direction of evolution would be a new one, starting from the structure of the one-time larva, with the forgotten ancestral adult completely wiped out of the picture. It is possible that evolution through the development of paedogenetic forms has played an important part in animal history. And it is not unlikely, as we shall see, that the whole vertebrate story may have had its start in this fashion.

VERTEBRATE STRUCTURE

While we shall not enter into the complicated details of vertebrate anatomy, the story of vertebrate evolution and classification is based mainly on morphological features, and we shall merely note the broad outlines of basic vertebrate structure, to which details may be added here and there later as necessary.

An obvious basic fact is that vertebrates are bilaterally symmetrical animals, with one side the mirror image of the other. This is important, for it implies (as in the case of some highly developed invertebrates) that vertebrates are active animals, whereas forms which have radial symmetry or little symmetry at all—such as corals, starfishes, and mollusks—are essentially sedentary types. In higher vertebrates the proportions of the body may vary considerably, but the primitive shape is that of typical fishes—a rather elongated, spindle-shaped body (fusiform is the more technical word) capable of rapid locomotion through the water, with a muscular trunk and, beyond it, a highly muscular tail. This is a form which man has imitated in creating fast-moving bodies, from submarine torpedos to airplane fuselages.

A word here about vertebrate locomotion. In most land vertebrates this is accomplished chiefly by the limbs. But in fishes the fins which represent the limbs are tiny and generally used only for balancing and steering; trunk and tail are alone responsible for progression. Waves of contraction of the muscles of one side and then the other pass backward along the body and, with increased amplitude, the tail; each wave as it passes backward along the body pushes against the water and thus moves the animal forward. Most land animals have, as we said, shifted to propulsion by the limbs; but apart from the fact that some land forms, such as the snakes, have reverted to a sinuous body motion with loss of limbs, it would seem that much of the original wave motion was carried over into all land animals. If, for example, such an amphibian as a salamander be observed walking, it can be seen that (shown ón our plates following p. 98) its limbs do little but keep its belly off the ground; a fishlike wriggling of the body is the main propulsive force.

A diagnostic feature of primitive vertebrates is their method of water-breathing. Many invertebrates have gills of some sort—structures filled with blood and exposed to water so that an exchange of oxygen and carbon

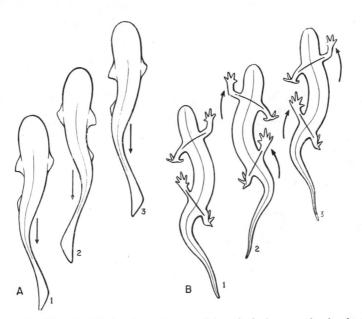

A, dorsal views of a fish swimming, to show the essential method of progression by the backward thrust of the body on the water, resulting from successive waves of curvature traveling backward along trunk and tail. The curve giving the thrust (indicated by arrow) in *1* has passed down the tail in *2* and is replaced by a succeeding wave of curvature; the thrust of this wave carries the fish forward to the position seen in *3*, and so on. *B*, dorsal views of locomotion in a salamander. Although limbs are present, much of the forward progress is still accomplished by throwing the body into successive waves of curvature. In position *1*, the right front and left hind feet are kept on the ground, the opposite feet raised; a swing of the body in *2* carries the free feet forward, as indicated by arrows. If these feet are now planted and the other two raised, a following reversed swing of the body will carry them forward another step, as seen in *3*. (From Romer, *The Vertebrate Body*, W. B. Saunders Company.)

dioxide waste from the body can take place. But the usual invertebrate type of gill projects outside the body; only vertebrates and their close kin have a unique system of internal gills. A stream of water enters, usually through the mouth, continues into the gill region of the gut—the pharynx —and then passes out through a series of slits or pouches on either side to reach the surface again. On its outward passage, the water flows over a series of blood-filled membranous structures, the gills, where the exchange of gases occurs.

A further unique feature of vertebrates and their relatives is the nature of the nervous system. Many invertebrates have nerve cords and brain-like structures of some sort or other, but here alone the nerve cord (the spinal cord) runs along the upper side of the body and, in contrast to invertebrates generally, the cord—and the expanded brain at its front end—is a hollow structure.

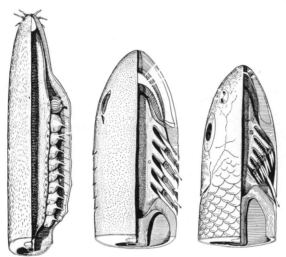

The head region of a hagfish, a shark, and a higher bony fish. In each, the right half has been sliced through horizontally to show the gill system. In the hagfish the gills are pouches opening to the surface by rounded tubelike openings. In the others the gill membranes lie along slits running from the throat (pharynx) to the surface. In the shark each gill opens separately; in the bony fish the gills are protected by a covering of skin reinforced by bone (the operculum). (After Dean.)

All familiar vertebrates have the same types of special sense organs that we possess: nose, eyes, and ears. But even at that, we may point out here, there are variations among primitive forms. Nostrils were, at first, simple pockets lying near the front end of the head for detecting odors in the water and had originally no connection with the interior of the mouth, as they do in man and all air-breathers. Well-developed paired eyes are found in all vertebrate groups, but in addition there was present in primitive forms a third eye, the "pineal," opening on the top of the head. This has been reduced to a small organ concealed within the skull in most living forms. Ears we think of as organs of hearing; but our ears are also organs of equilibrium, and this was quite probably their original function. Fish can hear but only through vibrations set up within their bodies, for they lack external ears, eardrums, and all the mechanisms by which land animals receive and amplify sounds. The only portion of the ear which they have is the essential inner structure, a double series of liquid-filled sacs and canals lying within the braincase, which mainly register the pose of the body and its movements.

We need, for our purposes, say little here about many other features of
the internal anatomy of vertebrates—digestive, excretory, and reproductive systems. We may, however, emphasize the point mentioned earlier,
that there is in typical vertebrates a distinct tail region into which none of
the internal organs pass—this in contrast with such invertebrates as annelid worms and the great arthropod groups, such as crustaceans and insects. Among the vertebrates the major muscles of lower forms, particularly fishes, are arranged from early embryonic stages in segmental blocks
down either flank in trunk and tail to form the main "motive power" of
the animal in its undulatory swimming movements. In correlation with
this arrangement of the muscles, the nerves and trunk skeleton take on a
segmental pattern. A vertebrate is thus partially segmented; but there is

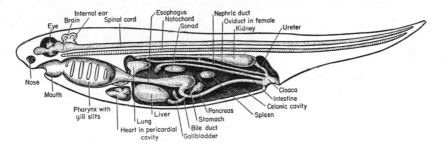

Diagrammatic longitudinal section through an "idealized" vertebrate, to show the relative position
of the major organs. (From Romer, *The Vertebrate Body*, W. B. Saunders Company.)

not segmental arrangement of skin, on the one hand, nor of the internal
body organs on the other, whereas in an annelid or arthropod the segmen-
tation affects every part and structure of the body.

The skeletal system of vertebrates is distinctive. Its materials consist of
cartilage ("gristle") and bone. Cartilage is present in some invertebrates,
but bone—a tissue made solid by deposits of calcium phosphate—is
unique. Many early vertebrates, as we shall see, were completely covered
by bony armor laid down in the skin and hence termed "dermal" (skin)
bone; much of this is reduced or lost in most living vertebrates, but many
of the skull bones originated as part of this armor pattern. Quite in contrast with invertebrates, much of the skeleton of vertebrates is developed
deep within the body. In most vertebrates these internal elements are first
laid down in the embryo as tiny cartilages which are later replaced by
bone. In some fishes—cyclostomes and sharks and their relatives—the
skeleton consists of cartilage alone. It was formerly thought that this was

a primitive condition and that, as in the development of the individual, cartilage appeared first in the evolutionary series, bone later. Current evidence, however, indicates that bone appeared at a very early stage in vertebrate history and that its absence in lampreys and sharks is a degenerate feature—the retention of an embryonic condition in the adult.

The primary element in a vertebrate skeleton is the backbone, a segmental series of elements beneath and around the spinal cord, running the length of the body and tail, with ribs usually extending out from the sides of each segment and with supports for the median fins. In most fishes, with paired fins, and their descendants with limbs developed from such fins, girdle and fin or limb skeletons are, of course, present.

At the front end of the trunk the backbone ends in contact with a braincase surrounding the brain, nose, and inner ear. A special set of cartilages or bones form the bars found in the mouth and throat region and associ-

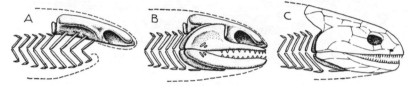

Diagrams to show successive stages in skull formation. *A*, a primitive vertebrate with braincase and gill bars but lacking special jaw elements and dermal bones. *B*, the shark stage. Jaws have formed from gill bars. *C*, the bony fish stage; dermal bones have been added.

ated with the gills. Most of these are, of course, reduced in gill-less land animals, but persist as bones of the base of the tongue and the "Adam's apple." Elements at the front, however, develop into bars forming part or all of the jaw structures. Lampreys and sharks have no skull in a proper sense of the term; other fishes and all higher vertebrates, however, have a true skull, formed from the braincase and upper jaw bars together with a series of dermal bones which form a roofing shield over the head and weld these internal structures together.

Older, apparently, as a skeletal element than either bone or cartilage is the structure termed the "notochord." It is a long rod-shaped affair running the length of the body in the position of the backbone of higher vertebrates. It consists internally of soft tissue but is covered with a stout sheath which gives it firmness as well as flexibility. In most vertebrates it is reduced or absent in the adult, having been replaced by the vertebrae; but even in such advanced forms as mammals the notochord is prominent in the early embryo, dwindling and vanishing later. In a lamprey the vertebrae are little developed, and the notochord is prominent throughout life; and in some lowly vertebrate relatives, discussed in the next section, there is no true backbone, but a prominent notochord is present.

LOWER CHORDATES

Amphioxus.—The characters mentioned in the preceding section are those of a typical member of the vertebrate stock. If, however, we desire to look into the origin of vertebrates, we naturally wish first to descend the evolutionary ladder as far as we can, to find as simple a form as possible to compare with lower animals. Some fishes are more primitive than others; the lampreys and hagfishes (discussed somewhat later) lack such typical vertebrate structures as jaws and paired limbs. But a far more simple form is the little animal, the lancelet, usually known by the scientific name of *Amphioxus*.

This is a small translucent creature found in shallow tropical marine waters, sometimes swimming freely about, but spending much of its life partly buried in the sand of the bottom with only the front end of the body projecting. Generally rare, it is so common in the Amoy region of the Chinese coast that it is sold in bulk as food in the markets.

Its general appearance is rather fishlike, but in structure it is much more primitive than any true fish. There are no paired fins or limbs of any sort, no jaws, and no bones or cartilages, not even a backbone. There is nothing that can be called a brain in any ordinary sense of the term, no ears, no eyes (although pigment spots in the brain tube appear to be sensitive to light), and only a tiny pit which may be the rudiment of a nostril. This form is obviously a much more primitive creature than any typical vertebrate.

There are, however, several important characters which show that *Amphioxus* is really a primitive relative of the vertebrates. Although this little animal has no backbone, it does have a highly developed notochord; more highly developed, in fact, than in any vertebrate, for it extends forward to the tip end of the "nose," where it provides an effective stiffening device when the animal burrows. Another vertebrate feature of *Amphioxus* is the well-developed hollow nerve cord running the length of the body above the notochord. And a third good feature of resemblance is the presence of gills of the vertebrate type. Indeed, it would seem as if this simple little creature were trying to emphasize this point of agreement, for whereas a normal vertebrate usually has but five or six gill openings, *Amphioxus* may have several dozen. The animal has an important functional reason for this emphasis on gills, for they are used not merely for breathing but for food-collecting as well. The animal feeds on tiny food particles present in the currents of sea water drawn constantly into its mouth and passed out through the gill slits. These particles are strained out in their passage through the "throat," imbedded in bands of a sticky material secreted there, and carried by ciliary action back into the intestines. It is highly probable that this function of the gills is the primary one with breathing probably secondary, for many small soft-skinned animals can breathe successfully through their skins without the need for special organs.

The lancelet is clearly a relative of the vertebrates, a point strengthened, incidentally, by the fact that the larva of the lamprey is very similar to it in structure. But it is a much more primitive type than any vertebrate and, lacking a backbone, cannot very well be included in that group. However, every vertebrate has a notochord at some stage of its development, and to express this relationship, naturalists have erected a major group, or phylum, of the animal kingdom, known as the Chordata, or "animals possessing a notochord." Of this phylum the vertebrates are the major subdivision, while *Amphioxus* and a few other "poor relations" of the backboned animals make up several lower chordate groups.

It is not impossible that the lancelet is fairly close to the primitive types from which the vertebrates have arisen, although this living creature has some specialized features.

Tunicates.—The tunicates, or sea squirts, comprise a considerable number of small inhabitants of modern seas. Some are free-floating little creatures, some form branching colonies. Perhaps the most generalized of the group are the forms found as adults attached to rocks or other objects in shallow water. They are quite motionless and, in this respect, quite unlike the typically active vertebrates; nor is there any bodily resemblance to vertebrates. The sea squirt appears to be a nearly shapeless lump of matter enveloped in a tough leathery skin—the tunic which gives its name to the group. At the top of the animal is an opening into which water is drawn. Inside is a large barrel-shaped strainer through which the water passes to flow out a second opening at the side. Food particles in the water are collected at the bottom of the straining device and pass into the stomach and intestine. In the adult tunicate there is very little nervous system of any sort, not to speak of a brain or spinal cord. Nor is there any trace of a notochord or any skeletal system. Nothing more unlike a typical chordate could, it would seem, be imagined.

But examination of the straining barrel opens up a different conception of the tunicate's position in the animal world. The animal, like *Amphioxus*, not only strains its food with this structure but breathes by means of it; it is, in reality, a very complicated chordate gill system, the true throat of the animal. We have here one good chordate character. And while that is the only resemblance to *Amphioxus* in the adult, the young tunicate

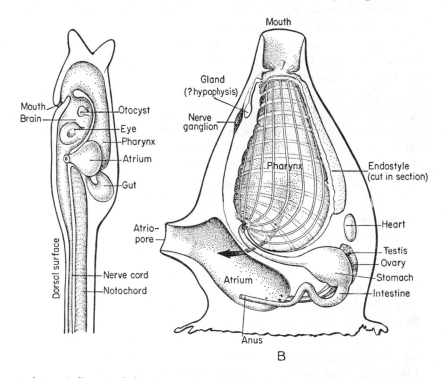

B

A solitary tunicate. *A*, diagram of the structures seen in the free-swimming larva (head end above, and only a short section of tail figured). The otocyst is a simple ear structure. *B*, the sessile adult, formed by elaboration of the structures at the anterior end of the larval body. Water passing through the latticework gills of the "throat," or pharynx, enters the atrium and, as indicated by the arrow, streams out through the atriopore. (From Romer, *The Vertebrate Body*, W. B. Saunders Company.)

rounds out the picture. In many tunicates there is a larval form rather like a tadpole in shape, with a large head region and a slim tail. In the tail is a well-developed notochord and a typical dorsal nerve cord as well. The tunicate starts life as a small free-swimming larva, fairly similar to the lancelet in construction. Presently it becomes attached, in the head region, to the bottom; active life is abandoned, the gill barrel expands, and the tail—and with it nerve cord and notochord—disappears.

Which is the more primitive type, the sessile adult tunicate or the active larva? Many students, impressed with the active nature of vertebrates, have tended to the opinion that the larva represents the more primitive type and that the sessile nature of the adult is a degenerate development. But careful consideration of the whole chordate picture suggests the reverse. It is more probable that the chordate ancestor of the vertebrates was, rather like the adult tunicate, a sessile feeder on tiny food particles (as we shall see is the case in even more simple chordates) which had, at this stage, developed a motile larva with a swimming tail furnished with muscle, notochord, and nerve cord. The primary function of this larva was to move about and "set up shop" for the adult in a favorable spot. Living tunicates have gone no farther with this development. Very likely the evolution of *Amphioxus* and the vertebrates came about by a process of paedogenesis (discussed earlier in this chapter). Under some conditions it was presumably advantageous if the larva never "grew up" and became sessile but, rather, retained its tail and its motile habits throughout life. The ancestral larva, not the ancestral adult, opened the door to advance and freedom.

Acorn worms.—Quite different, again, are the acorn worms. These seashore burrowers are somewhat like the ordinary annelid worms in general appearance but have as characteristic structures a "collar," in front of which is a tough burrowing snout or proboscis, the two sometimes resembling an acorn in its cup (hence the name). Despite the wormlike appearance of these forms, their internal structure is quite different from that found in the annelids. Strong proof that they are, on the other hand, related to the chordates is shown by the presence of numerous and typical gill slits. The nervous system is not highly developed, but there is a hollow dorsal nerve cord in the collar region, and a small structure in the proboscis is thought to be a rudimentary notochord. The acorn worms are assuredly far below the lancelet in the level of their organization but are highly specialized and hence not on the main line of vertebrate ascent.

Pterobranchs.—In studying the acorn worms we have descended the tree of life to a point where there is little left of vertebrate characters or even the basic chordate characters of *Amphioxus* apart from the presence of gill slits as a feeding (and breathing) device. One further, final, step brings us to forms which lack, or nearly lack, even this last remaining vertebrate resemblance and would not be suspected of relationship were it not for an arrangement of body parts comparable to that of acorn worms. These are the pterobranchs. There is no popular name for these tiny, rare, marine animals. Apart from a few doubtful types, there are but two known genera. They form little plantlike colonies, whose individuals project like small flowers at the ends of a branching series of tubes. The short body is doubled back on itself, so that the anus opens anteriorly back of the head. Proof of relationship to the acorn worms lies in the fact that there is a snoutlike anterior projection beyond the mouth, corresponding to the acorn worm proboscis, and back of this, a short collar region. But other

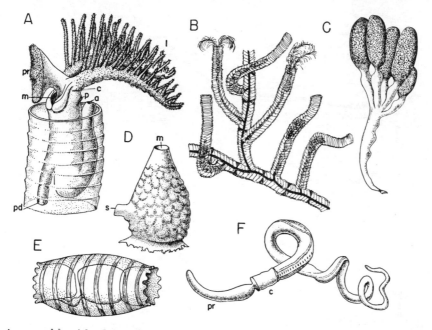

Tunicates and hemichordates. *A*, an individual of the pterobranch genus *Rhabdopleura* projecting from its inclosing tube. *B*, a part of a colony of the same. *C*, a colonial sessile tunicate; each polygonal area is a separate individual of the colony. *D*, external view of a solitary tunicate. *E*, a free-floating tunicate, or salp. *F*, an acorn worm (*Balanoglossus*). *a*, Anus; *c*, collar region; *l*, lopophore—ciliated collecting arms; *m*, mouth; *p*, pore opening from body cavity; *pd*, stalk (peduncle) by which individual is attached to remainder of colony; *pr*, proboscis or anterior projection of body; *s*, siphon which carries off water and body products. (From Romer, *The Vertebrate Body*, W. B. Saunders Company.)

resemblances to acorn worms—to say nothing of more highly developed chordates—are almost lacking. There is little development of a nervous system, no trace of a hollow nerve cord, and not the slightest suggestion of a notochord. And the feeding mechanisms are of a very different type. True, these plantlike animals feed, as do other lower chordates, on food particles drawn in by ciliary action. But the gill mechanism, which is so important in the filter-feeding of acorn worms, tunicates, and *Amphioxus*, is almost absent. One of the two pterobranchs has a single pair of small gill openings, the other none at all. Instead, there projects from the collar region large branching tentacle-like structures, termed "lophophores." These are richly supplied with cilia, which set up water currents, bringing food particles to the mouth.

So unchordate-like are these small creatures that one is tempted to suggest that they are degenerate forms, perhaps relatively modern in develop-

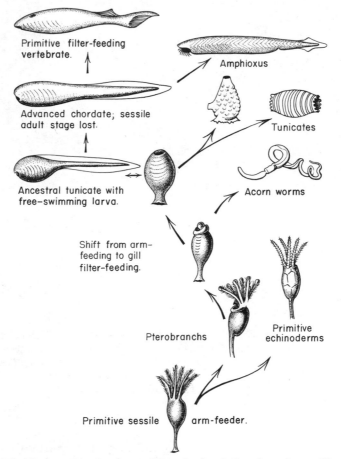

Primitive filter-feeding vertebrate.

Amphioxus

Advanced chordate; sessile adult stage lost.

Tunicates

Ancestral tunicate with free-swimming larva.

Acorn worms

Shift from arm-feeding to gill filter-feeding.

Pterobranchs

Primitive echinoderms

Primitive sessile arm-feeder.

A diagrammatic family tree suggesting the possible mode of evolution of vertebrates. The echinoderms may have arisen from forms not too dissimilar to the little pterobranchs; the acorn worms, from pterobranch descendants which had evolved a gill-feeding system but were little more advanced in other regards. Tunicates represent a stage in which, in the adult, the gill apparatus has become highly evolved, but the important point is the development in some tunicates of a free-swimming larva with advanced features of notochord and nerve cord and free-swimming habits. In further progress to *Amphioxus* and the vertebrates the old sessile adult stage has been abandoned, and it is the larval type that has initiated the advance.

ment. But it has recently been demonstrated that they are a very ancient group indeed. Paleontologists have long been familiar with a variety of small tubelike structures termed "graptolites" which were abundant in the seas far before the appearance of the oldest vertebrates. Close study shows that these tubes are similar to those which shield the modern ptero-branchs. With this indication of their antiquity, it can be reasonably argued that the pterobranchs are a truly primitive group of chordates (in a broad sense), from whom, by progressive development of a gill-filtering system, the more typical chordates are derived.

VERTEBRATE ANCESTRY

Having followed down the vertebrate pedigree to the simplest and apparently most primitive chordate type, it remains to determine, as far as can be, the relationships of the chordates as a whole with the other great groups of animals, often "lumped" (from an anthropocentric point of view) as invertebrates. In discussing this question we must rely, as in the last section, mainly on the evidence from living forms; for although many invertebrates have hard skeletons or shells capable of fossilization, it seems certain that these have been acquired independently within the various phyla and that the ancient forms which might have been connecting links were surely soft-bodied.

The main stock of the more advanced invertebrates seemingly lies among the coelenterates—multicellular animals with radial symmetry, such as the corals, jellyfish, and sea anemones. From them, or some form a bit more advanced, seem to have arisen two main lines of invertebrate evolution, one leading to the echinoderms, a second to the annelid worms, the mollusks, and the arthropods (or joint-limbed animals), a great group including insects, crustaceans, millipedes, and arachnids, among others.

Except for the mollusks, every phylum mentioned has been put forward for the role of vertebrate ancestors. Some suggest that the coelenterates have been the group from which the vertebrates sprang. This may well have been true in the long run, since these lowly animals presumably were ancestral to all the higher types of life. But may it not be that the vertebrates progressed farther along one of the main lines leading upward from the basal stock before branching out on their own? Can we not find some indications of vertebrate connections in one of the more advanced invertebrate groups?

Annelid worms.—The annelid worms, which include not only the earthworm but a great variety of more highly developed marine types, have been advocated by some as our progenitors. There are a number of common features. The annelids are bilaterally symmetrical, as are vertebrates. They are segmented, each joint of their body repeating the structures of the one ahead; and vertebrates too are segmented, at least as regards backbone, muscles, and nerves, as we have seen.

Annelid worms also have a nerve cord and a good blood system. But here we find marked differences from the vertebrate plan of structure. The

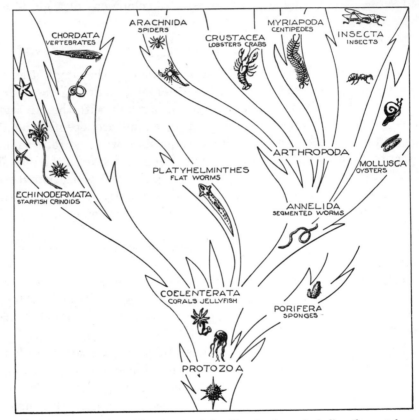

A simplified family tree of the invertebrate phyla. The chordates, including the vertebrates, are probably akin to the echinoderms (starfish, sea urchins, etc.) but only distantly related to other groups.

nerve cord lies on the underside rather than the upper side of the animal, and the blood flows in opposite directions from that in vertebrates along the two aspects of the body. But, as shown in the accompanying figure, these differences may be corrected if the position of the animal is reversed and the worm is assumed to have turned over to become a vertebrate.

It may be that top and bottom mean little to a worm; but the reversal of position raises as many problems as it solves. We must, for example, close the old worm mouth and drill a new one through the former roof of the head, for in both worms and backboned animals the mouth lies on the underside, beneath the brain. Then, again, where in the worm are the notochord and gill slits, those vertebrate structures to which even the lowest of chordates clung tenaciously? Their homologues have been sought in annelids but sought in vain. There is little positive evidence for belief in the origin of vertebrates from segmented worms.

Arachnids.—The arthropods include among their numbers some of the most highly organized of all invertebrates. It is not unreasonable to seek vertebrate ancestors in this phylum. The insects and crustaceans, most

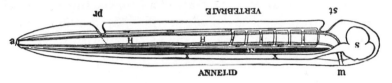

Diagram to illustrate the supposed transformation of an annelid worm into a vertebrate. In normal position this represents the annelid with a "brain" (*s*) at the front end and a nerve cord (*x*) running along the underside of the body. The mouth (*m*) is on the underside of the animal, the anus (*a*) at the end of the tail; the blood stream (indicated by arrows) flows forward on the upper side of the body, back of the underside. Turn the book upside down and now we have the vertebrate, with nerve cord and blood streams reversed. But it is necessary to build a new mouth (*st*) and anus (*pr*) and close the old one; the worm really had no notochord (*nt*) and the supposed change is not as simple as it seems. (From Wilder, *History of the Human Body*, by permission of Henry Holt & Co., publishers.)

numerous of arthropods, have not been seriously considered in this connection, nor have the centipedes and their relatives. A more worthwhile case has been made out for the arachnids.

These forms include not only the spiders, mites, and scorpions but also the horseshoe crab (*Limulus*) of the Atlantic coast and a number of old Paleozoic relatives of this last type, the eurypterids, or water scorpions. The late Dr. Patten, of Dartmouth College, devoted much of his life to the fascinating theory of vertebrate descent from the arachnid stock. The eurypterids, like other arthropods, were armored types, with an external rather than an internal skeleton. But some of the early vertebrates were armored too, as well as possessing deeper-lying skeletal structures. The armor of the water scorpions resembled to a considerable extent that of some early vertebrates. May not armored arachnids have given rise to armored vertebrates?

But there are many objections. Where, for example, have gone the numerous jointed legs of the water scorpion? There is no trace of them in vertebrates. Further, it is the top of the arachnid which is comparable with the top of the vertebrate; but in arachnids (as in worms) the nerve cord is on the underside. We must turn the animal over, and this destroys most of the resemblance. The anatomy of some of the oldest and seemingly most scorpion-like vertebrates is now well known; and there are no features suggestive of arachnid relationship. Interesting as this theory is, it finds little positive support.

Echinoderms.—The echinoderms—starfishes, sea urchins, sea lilies, and the like—seem the most unpromising of all as potential ancestors of the vertebrates. They are radially symmetrical, in contrast to vertebrates; they have no internal skeleton, no trace of any of the three major chordate characters of notochord, nerve cord, or gill slits; in addition, they have many peculiar and complicated organs of their own. At first sight one would be inclined not even to consider them. But here embryology and biochemistry shed an unexpected gleam of light.

The early embryo of the echinoderm is a tiny, free-floating creature, almost beyond the range of vision of the naked eye. Despite the disparity between the adults, the larvae of some echinoderms are so close to the larva of the acorn worm that the worm larva was long thought to pertain to some echinoderm type. Further, in the development of the embryo, the mode of laying down the main structural elements of the body in the echinoderms is radically different from that found in other advanced invertebrate groups—but agrees well with the embryological story seen in such chordates as the acorn worm and *Amphioxus*. Still further, recent advances in biochemistry show that, in such important features as the oxygen-carrying pigments of the blood and the chemistry of muscle action, the echinoderms show significant similarities to the vertebrates and contrast in these features with most other invertebrates.

Here, at last, are some positive facts and from a quarter where, at first sight, one would least expect them. But on further reflection, this suggestion of echinoderm relationship is not as astonishing as it might seem. We think of vertebrates as active animals. But if our reasoning in the last section has any merit, it is probable that the ancestral chordate, from which the vertebrates are descended, was, in contrast, a stalked, sessile animal making its living from tiny food particles gathered by ciliary currents along its outspread "arms." Most living echinoderms are free-moving (although their movement is of a limited and peculiar nature). But the fossil history shows clearly that the ancestral echinoderms, like the ancestral chordates, were sessile forms. The ancestral echinoderm groups are now extinct, but the living crinoids, or sea lilies, are echinoderms of similar habits. A crinoid has a complex skeleton and other specialized features. But basically, it is a "stalked, sessile animal, making its living from tiny food particles gathered by ciliary currents along its outspread 'arms.'"

These are, as you may see, exactly the words used a few lines above to describe a primitive chordate. Actually, a little pterobranch seems to be a type of such simple and unspecialized nature that it may well be close to the humble dwellers in ancient seas from which both chordates and echinoderms, in their very different fashions, took their origin.

CHAPTER 2

Vertebrate Beginnings:
The Evolution of Fishes

We have followed, as far as can be seen from our all too scanty evidence, the trail of vertebrate ancestry upward through small and primitive water dwellers to a point where, in such an animal as *Amphioxus*, a number of the base characters to be expected in the ancestors of backboned animals have already developed. We shall now continue the story onward and upward through the vertebrates themselves. But before doing this in any detail, we may lay out the general plot of the drama and distinguish the main successive types that will be encountered.

All existing (and most fossil) vertebrates can be readily arranged in a series of major groups—technically "classes." The class Mammalia includes the mammals, the familiar warm-blooded, hair-clothed animals, among which man himself is to be included. The birds, class Aves, are readily distinguished by the presence of feathers and wings and by their possession, along with mammals, of a high, controlled body temperature. The class Reptilia, lacking the progressive features of the birds and mammals, represents a lower level of land life, with lizards, snakes, turtles, and crocodiles as living representatives. A fourth group is that of the class Amphibia, including frogs, toads, and salamanders—four-legged animals, but reminiscent of fishes in many respects.

One commonly lumps the remaining lower vertebrates as "fish," and these forms (or most of them) are sometimes included in a single vertebrate class, the attitude being that, after all, they seem to be built on a common plan—water dwellers with gills and with locomotion performed by fins rather than limbs. But this is a rather personal, human viewpoint. An intellectual and indignant codfish could point out that this is no more sensible than putting all land animals in a single class, since, from his point of view, frogs and men, as four-limbed lung-breathers, are much alike. Actually, when we look at the situation objectively, a codfish and a lamprey, at two extremes of the fishy world, are as different structurally as amphibian and mammal. The fishes are perhaps best arranged in four classes of lower vertebrates: class Agnatha for jawless vertebrates, such as the living lampreys and their fossil relatives; class Placodermi for the primitive jawed fishes of the Paleozoic, now extinct; class Chondrichthyes,

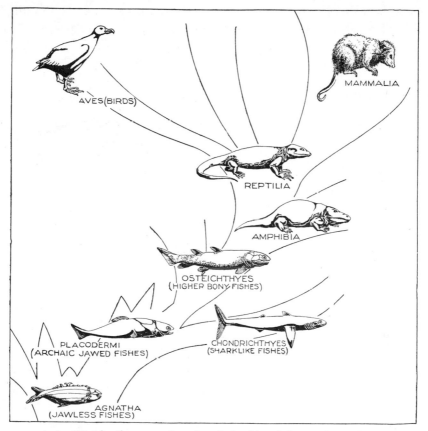

A simplified family tree to show the relationships of the vertebrate classes

cartilaginous fishes, sharks, and their relatives; and class Osteichthyes, the higher bony fishes which today constitute most of the piscine world.

In the above figure these eight classes of vertebrates are arranged in the form of a simple family tree.

PRIMITIVE VERTEBRATES—BACKBONE

What marks the first stage in vertebrate evolution? What sort of creatures were our earliest backboned ancestors?

Lampreys and hagfishes.—A partial answer to these questions may be obtained from a consideration of the lampreys and hagfishes, the living representatives of the class Agnatha, the jawless vertebrates. Because of their round mouths they are usually termed the "cyclostomes." Best known is the marine lamprey, which is common in the colder regions of the northern hemisphere. There are related forms from Chile and Australia. This fish is eel-like in appearance but much more primitive in its structure than true eels, which are highly developed bony fishes. The lamprey is soft-bodied and scaleless and, though having a feeble skeleton of cartilage, lacks bones entirely. There are no traces of paired fins, and, most especially, jaws are totally lacking. The adult lamprey is predaceous,

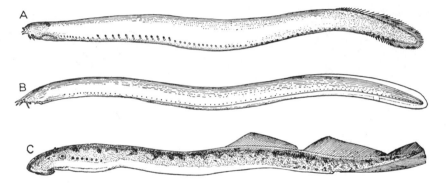

The three main types of cyclostomes. *A*, the slime hag, *Bdellostoma; B*, the hagfish, *Myxine; C*, the lamprey, *Petromyzon*. (From Dean.)

nevertheless; the rounded mouth-cup forms an adhesive disc by which it attaches to the higher types of fishes upon which it preys, and a rasping tonguelike structure within the mouth is a fairly effective substitute for the absent jaws. There is but a single nostril, opening high on top of the head. The gill passages are not slits, as in most fish, but spherical pouches, connected by narrower tubes with gut and body surface. In various less obvious structural characters, the lampreys likewise show a series of features in which they differ from typical fishes—features which appear to be in part primitive, in part aberrant.

The excessively slimy hagfishes are, like the typical lampreys, marine in habit, but differ in a number of ways. The rasping tongue is present, but the mouth is surrounded by short tentacles instead of a sucker. The hags are scavengers rather than active predators, burrowing into the flesh of dead or moribund fishes. The nostril is at the tip of the snout rather than

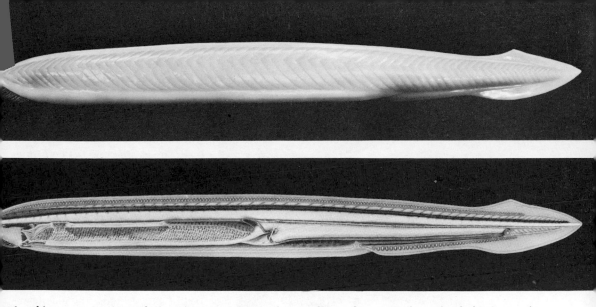

Amphioxus, a chordate relative of the vertebrates, is a small translucent marine animal. At the top is an external view. *Below,* a model cut longitudinally to show the internal anatomy; the "head" end is at the left. The prominent white band running the length of the body is the notochord. Above it (*in black*) is the nerve cord, slightly expanded anteriorly in the position of the vertebrate brain. The striped layer above the nerve cord indicates muscle; the white dots are fin supports.

Below the notochord the digestive tube occupies much of the body, from the mouth at the left to the anus beneath the tail fin. Nearly half the length of the gut is occupied by the gill region (pharynx), with a highly developed lattice-work of gills. There is no distinct stomach; the hind half of the gut is a simple, straight intestine. From this a pocket branches off in front as the liver; this extends forward, mostly concealed in the model by the gill bars.

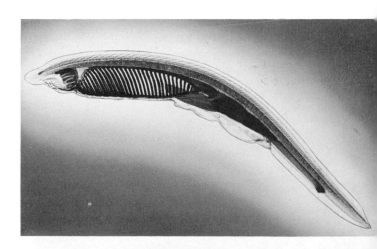

Right, a photograph of a young *Amphioxus* in which most of the same structures are readily seen which are represented in above models. The gill lattice-work, however, is not so highly developed, and the anterior extension of the liver sac is more readily visible. (Photographs above courtesy American Museum of Natural History, New York; larva from photograph copyright by General Biological Supply House, Chicago.)

Brook lampreys spawning. The eggs are laid in a shallow "nest" scooped out in the bed of a brook. The eyes and round gill openings are well shown. (Photograph courtesy New York Zoölogical Society.)

The lamprey attack. A model illustrating the mode of life of the lamprey. The round mouth sucker is applied to the skin of the prey, and the flesh is rasped off by the protrusible toothed "tongue." The catfish illustrated shows scars due to former attacks from lampreys. (Photograph courtesy American Museum of Natural History, New York.)

A Chimera, or ratfish. A representative of a deep-sea group with cartilaginous skeletons. They are distantly related to the sharks but differ in a number of respects, such as the fold of skin covering the gills (compare with the exposed gill slits of the shark below) and the peculiar jaw apparatus, suitable for mollusk-eating. (Photograph courtesy American Museum of Natural History, New York.)

The sand shark (*Carcharias littoralis*). Attached to its belly is the **remora,** or "shark-sucker" (*Echeneis*), which feeds on the shark's food. A second specimen is seen at the bottom. Note the flattened head topped by a sucking device. (Photograph courtesy New York Zoölogical Society.)

atop the head, and the gill pouches in some hagfishes do not open directly to the surface but join to a common external opening on either side.

The hagfish eggs are laid in the sea and the young develop directly there; the marine lamprey, in contrast, has a distinct fresh-water larval stage. Every spring, lampreys run up the streams to spawn (and die), and the developing young spend several years of their lives as little larvae, which lie nearly buried in the mud of brooks and streams. These larvae are not at all predaceous; there is no tongue rasp, no mouth sucker. Instead, they are filter-feeders, which strain food particles much as does *Amphioxus*. A stream of water is brought into the mouth by ciliary action,

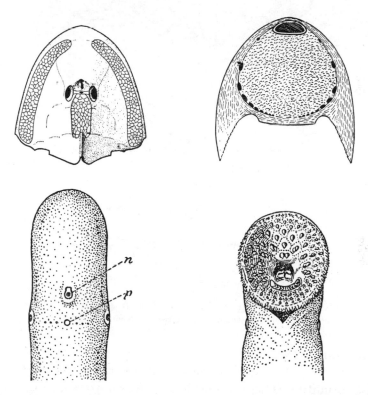

Upper and lower surfaces of the head in a Paleozoic armored jawless vertebrate (*Cephalaspis, above*) and a modern, degenerate type, the lamprey (*below*). In both the nostril is a single opening on the top of the head. Behind the nostril (*n*) is a median eye (*p*) between the paired eyes. On the underside the ancient type had but a small mouth which appears to have sucked in nutritive material which was strained out in its passage to the round gill openings at the side of the throat. In the lamprey, with its carnivorous habits, a large sucking mouth has developed, with a toothed rasping "tongue" in its center (the gill openings, not shown, were farther back on the sides of the throat). (*Cephalaspis* after Stensiö; lamprey from Norman, *A History of Fishes.*)

passes into the throat, and, with food particles strained out, returns to the surface through the gill slits. At the end of the larval period there is a sudden change of structure, and the young lamprey, with adult features fully developed, descends to the sea. It is possible, however, for lampreys as adults to remain in fresh waters—the sea lamprey has successfully invaded the American Great Lakes—and we even find, in certain instances, that small species of lampreys do not take up a predaceous life but reproduce and die in their native streams shortly after assuming adult shape.

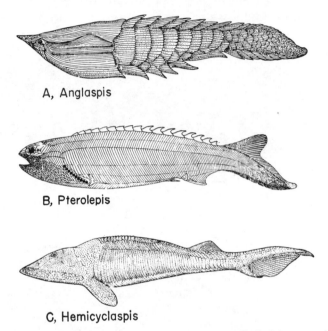

A, Anglaspis

B, Pterolepis

C, Hemicyclaspis

Some ostracoderms, oldest of known vertebrates. These grotesque little fishes were jawless filter-feeders or mud-grubbers in Silurian streams. (After Heintz, Kiaer.)

It is generally agreed that the absence of jaws and, probably, that of fins are primitive features of cyclostomes. Other characters, however, are more dubiously primitive. There is considerable reason to regard the absence of a bony skeleton as a degenerate feature; the predaceous or scavenging habits can hardly have been present in ancestral vertebrates (mutual cannibalism is not, to say the least, advantageous), and the tongue rasp is a cyclostome specialty. Cyclostomes represent a primitive level of vertebrate development; they are not, however, in themselves ancestral vertebrates. We must go further in quest of really primitive types.

Ostracoderms.—Here we may for the first time avail ourselves of the knowledge gained from the fossil record. In the second period of the Paleozoic era, the Ordovician, we find the first faint traces of our oldest known vertebrate kinsmen in the shape of small flakes of bone in rock deposits, apparently formed in a shallow bay in an old sea that once covered much of Colorado and other western states. These oldest fragments, dating from a period in the earth's history perhaps 450,000,000 years ago, are not well enough preserved to give us any idea of the nature or appearance of the forms from which they came. It is not until a whole period later, in the late Silurian, that we first find complete skulls and skeletons of these most ancient predecessors of man.

And here we find ourselves in a quaint, bizarre world. These archaic vertebrates were water-living types, with fishlike bodies and tails. But, whereas modern fishes often have naked skins or are covered in many cases merely by thin scales or denticles, these ancient vertebrates were, one and all, armored with thick plates of bone or bonelike material—scales over the trunk and tail, where motility for swimming was essential, and a solid layer of armor plates over the head region. This armored condition has led to the frequent use of the term "ostracoderms" (shell-skinned) for

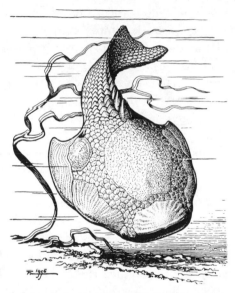

A flat-bodied ostracoderm, *Psammolepis*. (After Mark and Bystrow.)

these oldest vertebrates. In their general appearance they were quite unlike the lampreys and hagfishes, but when these fishes are more closely examined, it is seen that there are real resemblances to the lampreys, indicating true relationship. Like the cyclostomes, these oldest vertebrates lacked any trace of jaws, the mouth being merely a small hole or crosswise slit; and there are no typical limbs or paired fins but, at the most, a pair of flaps behind the head region or small spines at the sides of the body. Even in detail there are many resemblances between ostracoderms and cyclostomes. Many of these old fossils had, as in lampreys, a single nostril high up on top of the head, and the casts of such structures as brains and ears show that we are dealing with forms fundamentally similar to the lamprey type.

Putting together the evidence from both living and fossil forms, it seems certain that the jawless and limbless condition is one to be expected in the oldest vertebrate types. Ostracoderms and lampreys thus represent the most primitive stage in vertebrate evolution. But we have been loath to believe that the armored ostracoderms were actual ancestors of later types; it has been assumed that the true ancestors of later vertebrates were soft-bodied forms which might have escaped preservation and that the ostracoderms were merely an aberrant group having nothing to do with the line of descent of higher fishes or land forms.

Recent work, however, has tended strongly to show that some of the ostracoderms probably are exceedingly close to the line of descent of both cyclostomes and higher fishes. Bone, in the shape of both surface armor and internal skeletal parts, seems to be an extremely old vertebrate character. It is probable that many vertebrates now lacking such structures are degenerate rather than primitive. We know, for example, that such a fish as a sturgeon, with hardly a bone in its body, has descended from forms with a well-ossified skeleton. Armor was probably a character common to the ancestors of all vertebrates, and the ostracoderms are really an ancestral group.

Where did the early vertebrates live?—The old ostracoderms of the Silurian, which we have just described, are in a broad way primitive vertebrates, but all were already specialized in one way or another. Obviously, vertebrates had already been in existence for a considerable period of time in order to produce this variety of forms; but we know nothing of these still earlier fishes except for some scraps from a single formation belonging to the preceding period. Why?

The answer very probably lies in the fact that the ancestral vertebrates lived in fresh waters.

Most invertebrates are sea dwellers; so are all the living lower chordates, and so are the vast majority of living fishes. It is hence but natural to believe that early fishlike vertebrates likewise were marine forms. But some decades ago this began to be questioned, and there are two strong lines of evidence which suggest that the early evolution of fishes took place in fresh waters and that only later did vertebrates re-enter the marine environment of their remote ancestors.

All animals of any high degree of organization have kidney structures of some sort to eliminate waste materials from the blood and body fluids.

Vertebrates, however, have a very distinctive type of kidney, differing markedly from that of even their close marine relative, *Amphioxus*. The kidney is composed of large numbers of tiny tubules. Into each is filtered a quantity of liquid from the blood stream, containing waste materials and forming a very dilute watery urine in almost all fishes and the amphibians. The primitive vertebrate kidney is essentially a pumping system which continually draws off water from the interior of the body. The blood serum and other liquids in the body of any vertebrate contain much the same salts that are to be found in sea water but in a more dilute state. Bathed in this salt solution, the cells of the body thrive; if the salts become either too concentrated or too dilute, the cells—and the animal—die. If a fish or amphibian lives in fresh waters, the internal liquids tend to take in water from outside and become too dilute. The vertebrate kidney, continually pumping out water, is an ideal structure to counteract this tendency and would seem to have evolved for fish living in fresh waters.

But a majority of modern fishes do live in the sea. In such an environment they tend to accumulate too much salt in their blood. And yet most of them have the same sort of kidney that fresh water forms have, pumping water out of the body when they should, rather, conserve it. This is a real handicap. They overcome it, but overcome it in two different fashions in two different groups. The advanced bony fishes, which are marine forms, have evolved special glands which excrete superfluous salts into the water passing through the gills. The sharks have not developed these glands but solve the problem in another fashion. If, in them, the materials in solution in the blood consisted only of salts in proper proportions, their contacts with the more salty ocean waters would tend to increase the internal salts to a point where their cells would die in the resulting brine. But it is a well-known principle of physiology that if two liquids separated by membranes (here the inside and outside of the shark) have the same amount of materials—*any* materials—in solution, they will remain in equilibrium and neither will lose water to the other. The shark neatly applies this principle to solve his dilemma. He has, of salts, only the proper vertebrate amount. But to this he adds a quantity of urea, and this brings the *total* materials in solution in his blood to the same—or even higher—concentration than that in the sea water surrounding him. Retaining urea in the blood in our own case results in a sickly condition. It does not, however, harm the shark, and, thus protected inwardly, he swims through the briny deep without concern about the fact that his kidneys are pumping water from his body just as freely as in a fresh-water fish.

To sum up the argument: The vertebrate kidney is ideally constructed for life in inland waters. Most marine fishes have this type of kidney but counteract its disadvantages by specializations of two very different sorts in two different groups. The only logical explanation is that the ancestral vertebrates, in which this kidney developed, lived in fresh waters.

What does the fossil record show? It is in many cases difficult to interpret the conditions under which a given set of sediments were laid down, and, further, one cannot be sure that the beds in which the remains of a fish are found represent the place where he lived. (A dead river fish, for example, may float out to sea before it sinks.) But on the whole the fossil evidence strongly supports a fresh-water origin. The almost complete lack of records of vertebrates in the very oldest geological periods can be correlated with the fact that nearly all the sediments of those days that have been preserved are of marine origin. It is only in the Silurian period, and especially toward its close, that any fair amount of fresh-water deposits appear in the geological column. And as these rocks appear, vertebrates are met with in numbers for the first time. Following this, the early Devonian has many vertebrates, mainly fresh-water forms; as that period progresses, the percentage of marine forms increases, indicating, it seems, an increasing tendency for fishes to migrate from ancestral fresh waters into the ocean.

Vertebrates versus water scorpions.—But why should the early vertebrates have been armored? An answer may perhaps be found by considering their habitat. These oldest of backboned animals were mainly flat-bodied, comparatively sluggish types whose lives appear to have been spent along the muddy bottoms of fresh-water ponds and streams. The absence of jaws prevented them from becoming predatory, and the only mode of existence open to them was that of grubbers in the mud.

The only reasonable suggestion as to the presence of armor is that it was for protection against enemies. But what enemies had they? Most of the larger invertebrates which one would think of, such as the ancient cephalopod relatives of the squids and octopi and the larger crustaceans, were marine forms, living in quite a different environment from that of our stream-dwelling ancestors. We do, however, find that there was one conspicuous group of invertebrates of Paleozoic days many of whose members inhabited fresh waters. These were the eurypterids, or water scorpions, mentioned before; indeed the two groups are often found in the same deposits and must have dwelt together. These old arachnids were, on the average, much larger than the ostracoderms. The oldest vertebrates were seldom more than six inches to a foot long, while eurypterids were, on the average, much larger—some nearly a dozen feet in length—and were obviously carnivorous forms with biting mouth parts. Apart from vertebrates, there are few animals in these ancient deposits which we can imagine to have afforded a food supply for the old eurypterids; and apart from the water scorpions we know of no enemies against which vertebrates needed defense.

Later vertebrates tended to increase in size, became faster-swimming types, and many fishes migrated into salt waters where they were comparatively free from attack. Correlated with these developments we find, in the fossil records, that the water scorpions dwindled in importance and disappeared. The higher vertebrates, freed by their progressive development, tended in great measure to lose their armor. Portions of it, however,

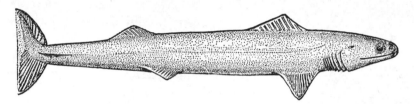

Cladoselache, a fast-swimming, sharklike fish with broad-based paired fins. (After Dean.)

survive in even the highest forms. In our own case we have no bony scales or other superficial armament. But the greater part of the bones of our head are dermal elements, covered, it is true, with thick skin and some muscles but still comparatively superficial in position.

DEVELOPMENT OF JAWS AND LIMBS

To reach the evolutionary stage of typical fishes, two great advances must be made over the structural plan of the ostracoderms or lampreys, in the development of biting jaws and of paired fins. Only with the evolution of jaws and teeth could vertebrates leave the bottom, where food debris accumulates, and become predaceous types, preying on larger living things. With the development of this new mode of life, faster locomotion for the pursuit of prey became necessary, and most true fishes have ceased to be flattened types and have developed rounded bodies with a torpedo-like, streamline construction.

Jaws.—The origin of the jaws is readily surmised by a study of the skeleton of the sharks. There are, in every primitive vertebrate, skeletal bars lying on either side of the throat between the gill openings. These have two principal parts—upper and lower. The jaws lie in line with these elements and, like them, are composed of upper and lower parts, with a joint between. It appears highly probable that the jaws are merely a front pair of gill bars which have enlarged, have rotated their free ends forward, and have been pressed into service to perform a new function.

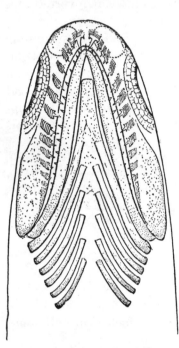

Lower view of the head and throat region of the Devonian shark *Clado-selache* to show the ventral parts of the gill bars and the jaws, seemingly homologous with them. The position of the nostrils (*dotted lines*) and orbits may be seen; teeth are in position along the jaws. (From Dean.)

In our own head and that of most higher forms the upper jaw is fused to the skull. In sharks, however, the upper jaw is usually still a separate element which is only loosely attached to the skull. A detail of construction which we may note here because of its future importance is the fact that the main upper joint (hyomandibular) of the arch just back of the jaws is utilized to prop the jaw joint against the side of the braincase.

The origin of the jaws is fairly obvious in a modern shark but even more so in *Cladoselache*, an ancient Devonian shark which had already become an active salt-water dweller. Frequent finds of this form have been made in concretions in the deposits from the ancient seas which once covered the Cleveland region and have revealed not only skeletal structure but even such details as muscle fibers and kidney tubules. In *Cladoselache* the jaws seem obviously in series with the gill bars behind them. Here, too, we gain a clue as to the origin of teeth. The skin of sharks and other primitive fishes contains small pointed structures, or denticles, which give it a sandpaper-like texture; these denticles are essentially similar to teeth in structure. In *Cladoselache* the teeth which line the jaws are similar in appearance to them, and it is probable that teeth are merely denticles which lie in the skin along the mouth margins and have become enlarged and used in a new way.

Limbs.—The main source of movement of a shark or any typical fish is, we have noted, by sideways undulations of the body and tail, which press the water back and the fish forward. Steering is aided by the various fins.

In all fishes, even the jawless forms, median fins, lying along the midline of the body—the anal and dorsal fins—help in steering, and the caudal fin (also median) is a major propulsive organ. The paired fins, corresponding to our arms and legs, first typically appear in the jaw-bearing fishes as accessory stabilizing organs. In most fishes the paired fins have a narrow base; in *Cladoselache*, however, they are very broad at the body end and similar in structure to the fins of the back. This suggests that they, like the median fins, developed as simple flaps from the body. In an old bottom-dwelling ostracoderm, motion was presumably all in a horizontal plane, and vertical median fins were all that were necessary for steering and turning. But with an active life off bottom, rolling motions became possible and had to be controlled. The paired fins were to become structures of supreme importance in later vertebrates; but in *Cladoselache* and other early fishes these appendages seem to have been merely stabilizing rudders.

Diagram to show the situation of the fins in fishes. The dorsal, caudal, and anal fins all lie in the midline; the pectoral and pelvic fins, however, are paired structures comparable with our arms and legs.

Placoderms.—Members of the sharks or higher bony fish groups, with jaws and fins of "orthodox" construction, were slow to make their appearance; preceding them in the fossil record in the earliest Devonian were a series of "experimental models" in jaw and fin development, all of which became extinct before the close of the Paleozoic. These were mostly grotesque forms, quite unlike any fishes living today. They are currently grouped as a special class of vertebrates, the Placodermi, the name referring to the fact that most were, like ostracoderms, covered to a variable degree by armor-plating.

All had jaws. The jaws of placoderms, however, are frequently of peculiar types and are seemingly primitive or aberrant in build; nature was, so to speak, experimenting with these new structures. Paired fins, too, were developing in connection with the new freedom which fishes were acquiring, but these structures were variable and often oddly designed (from a modern point of view). The early placoderms were essentially fresh-water dwellers, like the ostracoderms before them; but during the Devonian many of them had invaded the seas.

Of placoderms, the most "normal" in appearance were the acanthodians, usually termed "spiny sharks." The general body proportions were sharklike, but the acanthodians were most unsharklike in other respects— particularly the fact that they were fully clad in well-developed bony scales comparable to those found in some of the higher bony fishes. The fins mainly consisted of spines—sometimes of large size—with, apparently, but a small web of skin behind them. Further, these acanthodians had not "decided" how many paired appendages were desirable, for some of them (such as the one illustrated) have several small extra spines and fins be-

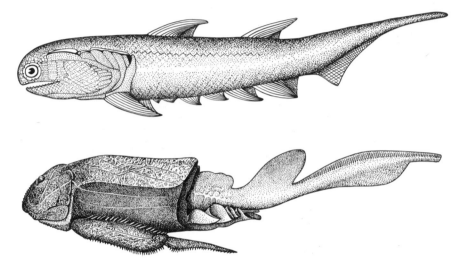

Above, a primitive "spiny shark" (*Climatius*), representing the oldest and most primitive stage in the evolution of jawed vertebrates. These little fishes, only a few inches long, appeared in the Silurian and were abundant in the early Devonian. Jaws were present but in a primitive condition. Limbs have appeared as paired fins supported by stout spines but were variable in development; the form shown sports several extra pairs. (After Watson.) *Below*, an antiarch—an armored placoderm with bony, jointed "arms." (After Patten.)

tween the ones corresponding to the normal two pairs found in orthodox fishes and higher vertebrates.

More common in much of the Devonian were the arthrodires ("jointed-necked fishes"). In these the head and gill region was covered by a great bony shield, and a ring of armor sheathed much of the body, the two sets of armor being connected by a movable joint. Peculiar bony plates served the function of jaws and teeth. The posterior part of the body was quite naked. In some forms there have been found true paired fins, but in the most primitive arthrodires all we find is a pair of enormous, hollow, fixed spines projecting outward from the shoulder region—some sort of holdfast or balancing structures. A third group of placoderms was that of the antiarchs of the Devonian—grotesque little animals which had two sets of armor like the arthrodires, but small heads, tiny nibbling jaw plates, and, for limbs, a pair of jointed "flippers" projecting from the body like bony wings. Still other (but poorly known) placoderms had reduced armor and more normal fin development and rather suggest a transition from armored ancestors to shark types.

Restoration of *Dinichthys*, giant Devonian arthrodire with a length of perhaps thirty feet. It is here shown pursuing the contemporary shark, *Cladoselache*. (From Heintz.)

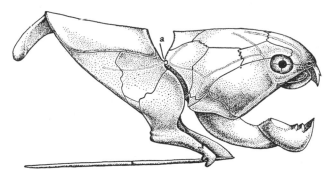

Side view of the armor of the giant arthrodire *Dinichthys*. The head and chest shields moved freely on each other by means of a pair of joints (*a*) at either side of the neck. The chest region was covered below by a broad plate which appears slender as seen in side view. The armored region was about ten and one-half feet long. (After Heintz.)

Most of the placoderms were obviously far off the main lines of vertebrate evolution, and few, if any, of the known types may possibly be regarded as actual ancestors of later vertebrates. As a group, however, they appear to represent nature's first essays in the development of jawed vertebrates. Most of them were not, in the long run, successful; others—poorly known or still unknown—gave rise to the two classes of more advanced fishes.

SHARKLIKE FISHES

Sharks.—While a few of the earliest sharks remained in fresh waters until the end of the Paleozoic, the vast majority of these forms tended rapidly toward a marine existence and also tended rapidly to shed the ancestral armor. *Cladoselache* was far along in this evolutionary process, and in the Carboniferous there appeared the first representatives of the modern shark order. For several periods these forms were none too common, perhaps because of the rarity of suitable food for flesh-eaters in the ancient seas. Later there occurred a vast migration of bony fishes into the seas. These higher fishes have been the main source of food supply of the sharks from that time to this, and from late Mesozoic times to the present shark types have been a conspicuous element, although not numerically a large one, in marine faunas.

The living sharks for the most part adhere fairly closely to a common pattern of body form and function. All swim by undulatory movements of trunk and tail, the latter with an upturned tip and the caudal fin developed mainly beneath it (the common pattern in the earliest bony fishes as well). There are, however, great differences in speed and agility, some forms being very speedy swimmers, others slow and sluggish. The body, in contrast to the skates and rays, is generally rounded in section with little flattening, nor are the paired fins greatly enlarged. There are no bony plates or true scales in the skin, but instead there is present a vast number of little toothlike denticles. The mouth and nostrils are situated well below and back of the projecting "snout." There are typically five normal gill slits, as in most bony fish also, but in contrast to these higher fish (and to chimeras as well) there is no flap of skin or bone covering the gill openings; on this account the sharks, together with their skate and ray cousins are often termed elasmobranchs ("strap-gilled"). In front and above the regular gill slits there is often a small round gill slit of special nature termed the "spiracle"—a feature absent in most advanced bony fishes and in chimeras.

In most sharks the dentition consists of numerous sharply pointed teeth. These are relatively loosely attached to the jaws and are readily and rapidly replaced. As the dentition suggests, sharks are universally eaters of animal food—mainly the ubiquitous teleost fishes, with squids as a second staple in many cases. Sharks are generally aggressively predaceous animals, which will strike at almost any object. It is probably this general characteristic that explains most of the numerous instances of attacks on man, although a few forms—particularly the large white shark—seem to be especially dangerous.

The eggs of primitive vertebrates were probably of modest size, as they are today in such widely varied animals as lampreys, primitive bony fishes, and amphibians. The water-living larva which develops from such an egg must, we have noted, begin to seek its own food at a very early stage and often adapt itself to a mode of life very different from the adult. There are thus many advantages to the production of a large egg rich in yolk which will allow the embryo to grow to considerable size before hatching. We shall see later that the reptiles developed a large-yolked egg, and the sharks (and other cartilaginous fishes) do the same.

Such an egg, if laid unprotected in the water, would be an eagerly sought tid-bit. For protection the shark egg is covered, before being laid, by a horny pillow-shaped shell. But of course this egg must be fertilized before the shell is formed around it. (In other fishes, insemination generally takes place after the female has laid her clutch of eggs.) Here internal fertilization is necessary. To facilitate this, the males in sharks and other cartilaginous fishes have long extensions of the pelvic fins—the claspers—which aid the sperm in entering the female's reproductive tract.

Internal fertilization has led to a further shift in reproductive methods in some sharks, skates, and rays. Since the eggs are fertilized and begin development within the mother's body before being laid, it is possible to retain the egg one step further and bear the young alive. There are very few instances of this method of reproduction—termed "viviparity"—among fish groups other than the sharklike forms. Only those which have developed large-yolked, shelled, eggs and practice internal fertilization show any strong trend toward viviparous habits. As we shall see, the other vertebrates, apart from the sharks and their relatives, which have developed an egg of this sort are the reptiles. Many lizards and snakes have paralleled the sharks in bearing their young alive. The viviparous process has reached its highest peak in the mammals, of reptile descent.

Of the 250 or so known sharks, a very few are thought to be relatively primitive because of the presence of one or two gill slits in addition to the normal five. This is rather dubious, however. A better claim to primitive nature can be made for the Port Jackson shark of the western Pacific. This has a type of dentition with sharp teeth in the front of the mouth, flattened teeth in back. As a result this shark has versatility in food selection, for he can be, at will, a carnivore or a mollusk-crusher. In other features as well as this type of dentition he resembles many early sharks of the late Paleozoic and early Mesozoic.

All the remaining sharks can be divided, for present purposes, into two groups of very unequal size. One, much the larger, is that of typical sharks which are generally fast-swimming types, with fusiform, torpedo-shaped bodies and other features associated with efficient navigation. The second, much smaller, assemblage, consists of forms which suggest in one character or another an approach to the skates and rays, considered in the next section of this chapter.

In the first category, the most active predators and, on the whole, the speediest members of the group are the forms known in general as the mackerel sharks—a name which in itself implies speed, for the true mackerels (which are higher bony fish) have the reputation of being among the speediest fishes of any type. Among the mackerel sharks, the largest and most aggressive and savage is the white shark, which will tackle prey of any size—up to and including man. White sharks frequently attain a length of twenty feet, with thirty-six feet as a record measurement. They are thus the largest predatory animals of the modern oceans. In late Tertiary and Quaternary beds, however, there are teeth of a shark of this type which are more than twice the size of those of the living white shark. We can be thankful that the monster which bore them is not with us today.

From the white shark, typical predaceous sharks range in size down through a large series of forms of rather good size, such as the tiger shark and the blue shark—commonest of sharks on the high seas of the Atlantic —to a host of little fellows termed dogfish, which are content to prey upon tiny fishes and invertebrates. A curious side line is that of the hammerheads—forms of normal shark build in every respect save one. The front part of the head expands widely on either side into the shape of a double-headed hammer, with an eye at the tip of either head.

The largest sharks are not among the rapacious types. The two giants of the tribe, and largest of any living fishes, are the basking shark and whale shark. The two are not closely related but have taken on similar feeding habits and concomitantly assumed large size. The basking shark often

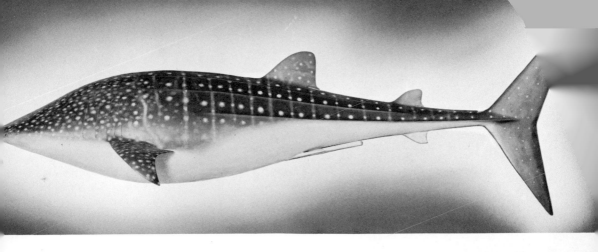

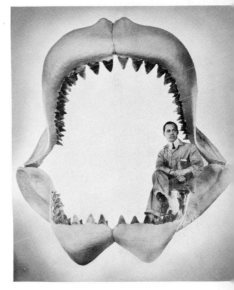

Various sharks. *Above*, the whale shark (*Rhinodon*), largest of all fishes living today, with a length of about 50 feet in the largest known specimens. This form is rare but widespread in the warmer waters of the world. Despite its size, however, this shark is a harmless form, with very small teeth, and makes its living on minute forms of sea life. Some fossil sharks appear to have been larger, although only incompletely known. *Right*, the restored jaws of a Tertiary shark (*Carcharodon*), with a man seated therein to give an idea of the gape. This form was related to the living sand shark on an earlier plate, and with its powerful dentition must have been a terror in Tertiary seas. *Below*, the hammerhead shark (*Sphyrna*), from a painting by Charles R. Knight. In this peculiar creature the eyes are placed in prominent lateral projections from the head. (Photographs courtesy American Museum of Natural History, New York.)

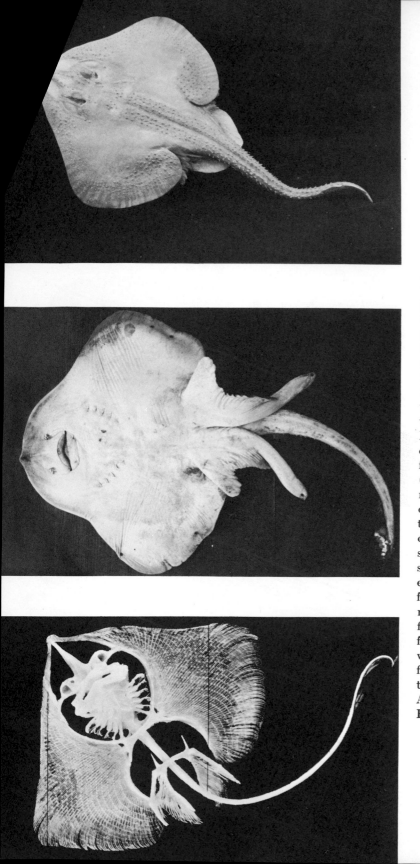

Skates and rays. A rather successful side branch of the shark group are the skates and rays, depressed-bodied forms with the teeth flattened into crushing plates for feeding on mollusks and other invertebrates. These bottom-dwellers have lost the typical body form of their shark relatives. The tail has become a whiplike structure, while the pectoral fins (corresponding to the arms of land vertebrates) are greatly enlarged, as may be seen from the figure of the skeleton, and even meet in front of the nose. Swimming is accomplished by undulations of these fins, which account for the breadth of the animal—six feet in one form. In another ray the fin muscles discharge electrical energy. The form figured is the prickly skate (*Raja scabrata*), common on the northern part of the Atlantic coast. The upper figure shows the dorsal surface. The large opening back of the eyes is the spiracle, a specialized first gill slit through which water is taken in. On the white undersurface can be seen the nostrils, mouth, and gill slits. The upper figure is a female; that shown from the underside is a male, with long claspers on the pelvic fins for internal fertilization of the eggs. (Photographs courtesy American Museum of Natural History, New York.)

reaches forty feet in length and may exceptionally attain fifty feet or a bit less; the whale shark, heavily built, has a known maximum length of forty-five feet and an estimated weight of as much as thirteen tons. The food items of these sluggish and inoffensive giants are minute in nature— tiny invertebrates which are present as "plankton" in quantities of sea water which these sharks take into their throats and sieve through their gills. This is close to being a return to the habits of the little filter-feeding chordate ancestors of the vertebrates.

Now to consider the sharks which show suggestive trends toward the skates and rays. In that group of bottom dwellers the body is flattened, the eyes point upward, the spiracle (often lost in the major shark series) is present and of large size; the pectoral fins are expanded and fused to the body above the ventrally placed gill slits; and the tail is reduced and losing its fins, beginning with loss of the little anal fin. Let us follow down a series of sharks showing an increasing number of these skatelike features.

First, the spiny dogfish and its relatives. These little sharks are very abundant in Atlantic waters, and a number of species of similar character are present throughout the world. Oddly, however, a related type (but one lacking spines) is one of the largest of sharks—the sluggish Greenland shark, very common in northern waters and reaching more than twenty feet in length. In this group there are only faint traces of skatelike traits, such as the presence of a well-developed spiracle and the suggestive loss of an anal fin, detrimental if an animal rests on the bottom.

Second, the saw shark. This form has a forwardly projecting rostrum armed with sharp teeth, much like a skate type termed the sawfish. The present form is not a member of the skate group but shows more indications of a trend in that direction; additional characters seen here are that the spiracle is definitely enlarged, the pectoral fins are somewhat expanded, and the front part of the body is flattened.

Third and last in this series are the angelfishes. These go far in the skate direction, for, in addition to the features already listed, the eyes here look upward, the gills face partly downward rather than in the lateral position of sharks, and the pectoral fins are greatly enlarged, although they are not widely spread at their bases in skate fashion.

Skates and Rays.—The typical sharks, as we have seen, are eaters of animal food which pursue their prey through the water. But there is a rich source of animal food among the sessile or relatively slow-moving invertebrates of the sea floor, particularly the mollusks and crustaceans. Some sharklike fishes became adapted for a mollusk-eating mode of life at an early date in shark history, but the skates and rays, living members of the shark group which live mainly on "shellfish," did not evolve until the recrudescence of shark life in late Mesozoic and Tertiary times.

Typical of skate and ray structure and habits are the common skates, found in shallow as well as deep waters in most parts of the world. Most skates are of modest size although the expressively named "barn-door skate" may have breadth and length dimensions of as much as six feet. We have noted briefly some of their characters in discussing, just above, some of the rather skatelike sharks. In relation to their bottom-feeding habits, the body is very broad and flat. Most of the breadth, however, is accounted for by the very great expansion of the pectoral fins outward from the body proper. These have broad bases which attach to the body all the way forward along the sides of the head to the tip of the "snout." The tail is reduced to a tapering whiplash, and with its reduction the median fins are likewise reduced. These shifts in fin proportions are correlated with the type of locomotion. Unlike sharks, which progress by throwing trunk and tail into a series of sideways curves, the skate progresses along the bottom by undulatory lengthwise rippling motions of the great pectoral fins; the tail is unimportant.

A diagnostic feature of the skate and ray group is that the broad fin attachment passes forward above the gill series and above the mouth as well. This presents a problem in breathing; with gills and mouth both close to or in the sand or mud of the bottom, clear water cannot be taken in through these openings. However, the first gill opening, the spiracle, opens above the fin. This serves as the water intake; it is a greatly enlarged opening not far behind the eye, and in a living skate it can be seen to pulsate rhythmically as it pumps water into the throat to be passed out again through the gill slits below. On the flat underside of the body in front of the gill openings is the broad mouth opening, its bounding jaws set sideways. The teeth are small but numerous and are generally flattened; several rows of teeth are in service at the same time, and the tooth battery as a whole is a fairly effective apparatus for crushing crustacean and mollusk food.

The skates, then, may be taken as typical of the group as a whole, but there are many variants. Some of these are forms which are in many regards transitional between sharks and the more typical skates and rays. For example, the sawfish has as a specialty a long rostrum armed on either side with teeth. We mentioned earlier a saw shark with a similar "saw." In both the saw shark and sawfish the pectoral fins are but moderately expanded, and swimming is accomplished by normal shark movements of the body and a well-developed tail. But in the sawfish—and not in the saw shark—we see a major "trademark" of the skate and ray group in that the pectoral fins run forward above the gills to reach the head. The same general structure is seen in the guitar fishes: they also have a stout tail, swim like sharks, and have the pectoral fins only moderately expanded; but like proper skates and rays they have the gill openings below the forward extension of the fins, with a large spiracle above for breathing purposes.

The torpedos, or electric rays, are more advanced toward typical ray conditions, for the pectoral fins are so expanded that with the front part of the body they form an almost circular disc. The disc, however, is not very flexible, so that what swimming these sluggish forms accomplish is done by the back part of the trunk and the tail. In these forms we meet with the first of several fish types which have developed electric organs. In the torpedo, which appears to use this special development to shock and stun its prey, the organ is formed by modification of part of the large fin muscles. The cells of any muscle are structures which are capable of a sudden release of energy. In normal

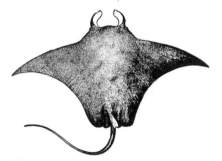

The great devil ray (*Manta*), which may reach a "wing" spread of twenty feet. (After Jordan.)

muscle this energy is spent in contraction; here it produces electricity.

Still more specialized than the forms just considered and the skates are forms to which the term "ray" is often restricted. In these the tail tends to be even more slender and reduced than in the skates, the disc of greater breadth. The sting rays bear one or more sharp serrate spines on the tail; with the spines are associated venom glands. The eagle rays, some of which also bear tail spines, reach good size, with a maximum of seven to eight feet. They are notable for the development of the teeth into a pavement of very stout crushing plates which enables them to deal effectively with the stoutest of mollusk shells. Largest of all rays, reaching a width of twenty-two feet and a weight of a ton and a half, are the devil rays. Their appearance is even more grotesque than that of typical members of the group. The head ends bluntly with a broad terminal mouth; an eye is perched at either side with a peculiar earlike flap extending forward from the eye region. The eagle and devil rays are known to "flap" their fins in a fashion somewhat like a bird's wings rather than progress merely by undulatory movements. And it is curious that (much as in the largest sharks) the very largest rays—the devil rays—live on the smallest of food materials, tiny pelagic animals which are strained out by the gill apparatus.

Chimeras.—An odd side branch of the sharklike fishes is that of chimeras, the rat fishes. These are marine animals, not over a few feet in length, which are mainly inhabitants of the deep sea but in some areas (as the Pacific coast of the United States) may be found in shallow coastal waters. Like the fabulous monster after which they are named, their appearance is bizarre. The head is short and deep with large eyes above and a small nibbling mouth below. There is a large spine behind the head; unlike the sharks, the gills are protected by a large flap of skin; the pectoral fins are large; at the back, the body typically tapers to a whiplike tail. The males have a stiff structure armed with sharp "teeth" projecting from the forehead. Presumably this is associated with breeding habits, of which we know little; it is suggestive of cave-man tactics in chimera courtship. Much of the diet appears to consist of mollusks and other hard invertebrates, and in correlation, most of the teeth are fused into a set of stout toothplates, one in each upper and lower jaw; further, to give a firmer bite, the upper jaws, unlike those of sharks, are firmly fused with the braincase. To this last condition the chimera owes its technical name —the order Holocephali ("fused-heads").

With these grotesque types we end our excursions among the lower sharklike fishes. They have not, on the whole, proved a success; many types are quite extinct, and the surviving groups constitute but a small percentage of living fishes.

HIGHER FISHES—THE SKELETON PROGRESSES, LUNGS BEGIN

Much more important have been the higher bony fishes which constitute the vast majority of the world's present fish population and have occupied this commanding position since the late Paleozoic.

Bone.—The term "bony fishes" (Osteichthyes), which is applied to these types, is not an altogether fortunate one, for, as we have seen, bone is an exceedingly ancient character in vertebrates. But it will serve to distinguish them from the degenerate lampreys and sharks of today, in which bone has been entirely lost. It is only in the bony fishes and the land forms descended from them that this hard skeletal material has been efficiently and permanently used.

The bony fishes appeared in the Devonian, the Age of Fishes, and almost immediately rose to a position of prominence in fresh waters. In ancient bony fish the body was completely inclosed in bony scales, the head and shoulders covered with stout, bony plates. The top and sides of the skull were covered by bones superficially placed in the skin (dermal bones), and further plates protected the gill region, the shoulders, the throat, and the lower jaws. Still others were formed in the skin lining the mouth, fusing the old upper jaws to the skull and supplanting them for the most part. Other bony elements tended in great measure to replace the

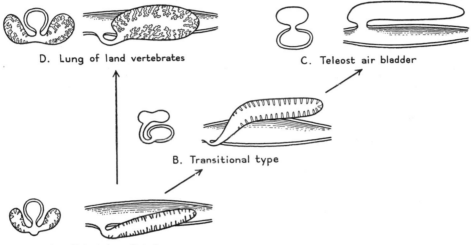

D. Lung of land vertebrates

C. Teleost air bladder

B. Transitional type

A. Primitive fish lung

The evolution of lungs and swim bladder. Lungs appear to have been present in early bony fishes as simple paired sacs opening out from the underside of the throat. In land forms these have persisted but with improved efficiency, owing to foldings of the internal walls of the lungs. In most fishes, however, the structure has been much modified. The sac and its opening have shifted to the top of the throat, and the sac serves not as a lung but as a hydrostatic organ, the air bladder. These evolutionary stages are diagrammed above. In each case there is shown a section across the body and a longitudinal section of gut and lung to show the changes in their relative position. (Based on a figure by Dean.)

shark cartilages not only in the braincase but in the body and limb skeleton as well. It is of interest that the pattern of bones laid down in these old fishes may be traced, with modifications, into almost every later and higher vertebrate type. Almost every element of the human skull can, for example, be directly compared with a corresponding element in the skull of ancient bony fishes. Higher forms have often lost old bones, but only rarely have new ones been added.

Lungs.—A still more characteristic feature of bony fish appears to have been the early development in them of lungs or lunglike structures. We usually think of lungs as attributes of land animals. But in two diverse types of primitive living bony fishes, functioning lungs are still present today, and there is strong evidence suggesting their universal presence in the oldest bony fishes.

The reason for this development of lungs may be found through a consideration of probable Devonian climatic conditions. The Age of Fishes was, it is believed, a time of violent alternations of seasons; much as in certain regions of the tropics today, there were rainy seasons alternating with times of severe drought. If the streams and ponds in which the oldest fishes were living tended to dry up, the water would become stagnant and foul, lacking the necessary oxygen for water-breathing. Such conditions would militate strongly against sharks and other lungless fishes. But if some sort of membranous lung sac were developed in the throat, a fish in such a pool could come to the surface, gulp down air, and breathe atmospheric oxygen in default of oxygen in its native waters.

The primitive lung appears to have been a double sac lying on the underside of the chest region of these fishes. This is essentially the type which has been retained and developed in land animals and is found today among a few tropical fishes which live in regions subject to seasonal drought. But in most later fishes the original lung structure has been much modified. It is obvious that a lung on the underside of the body renders an animal floating in the water top-heavy, and in most living forms the air sacs have been fused into a single bladder-like structure which lies above instead of under the throat.

Further, in times beyond the Devonian, climatic conditions appear to have been less fluctuating and the lung of comparatively little importance as a feature of survival value. This sac has in most modern fishes ceased to function in breathing but has been turned to a new use. By filling the sac with air, or with gas secreted in its lining, specific gravity is reduced, and the fish tends to rise in the water; upon emptying it the fish tends to sink. The old lung has become a hydrostatic organ, analogous to the ballast tanks of a submarine.

FISH ANCESTORS OF LAND VERTEBRATES

At the very beginning of their history the bony fishes had divided into two groups. One, that of the ray-finned fishes, or Actinopterygii, will be treated in a separate chapter. The second group, sometimes called the Choanichthyes but which we shall term the Sarcopterygii, or fleshy-finned fishes, will be discussed here. These forms have not, as fishes, been a success, but they are of immense evolutionary importance as the stock from which all land vertebrates have arisen.

The fleshy-finned fishes in turn are divisible into two orders—the lungfish, or Dipnoi, and the lobe-finned fishes, the Crossopterygii, both almost extinct at the present time. Although the crossopterygians are actually closer to the line of ascent, we shall first describe the lungfish, since their

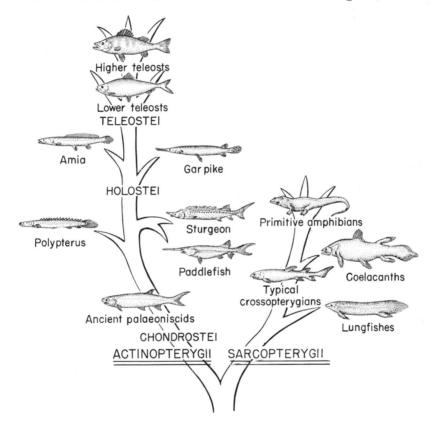

A simplified family tree of the bony fishes, to show their relations to one another and to the amphibians.

living representatives show many of the traits expected in the fish ancestors of land forms.

Lungfish.—The Dipnoi include today but three forms, one of which is present in each of the three southern continents. One is found in rivers in the interior of Australia; a second in the upper part of the Nile Basin of Africa; a third in South America inhabits the swampy region of the Gran

Chaco, in western Paraguay and adjacent areas. It is of interest that these are regions subject to seasonal drought and hence are habitats in which lungs may be highly useful. The lungfish group derives its popular name from the presence of these organs; apart from two archaic ray-finned forms (mentioned later) which dwell under comparable conditions, they are the only living fishes which have retained functional lungs. Lungs are so vital a part of lungfish economy that they are apparently necessary for normal life as well as during drought. An American scientist, Homer Smith, in his book *Kamongo*, tells how for a long time his efforts to keep African lungfish alive failed because he thoughtlessly kept them in shallow tubs of water. Under such conditions they were unable to tilt their bodies sufficiently to get the mouth above water and take in air; the poor fish had drowned.

In most fishes the nostrils are merely pouches, containing water, in which the olfactory sense functions. In the lungfish, however, the jaw construction has been so modified that there is an opening from the nasal sac to the inside of the mouth as well as to the exterior. This opening is potentially useful for allowing an animal to breathe air at the surface without opening the mouth and thus "shipping water," although the lungfish themselves appear not to use it in this fashion. Internal nostrils of this sort are present in all land vertebrates but are found in fish only in the lungfish and certain of their crossopterygian cousins.

Apart from nostrils, lungfishes and their lobe-finned relatives have many common features which show that together they form a natural group. We may cite the significant fact that in both groups the paired fins have prominent fleshy lobes; they contrast strongly with those of the actinopterygians, which are supported only by horny rays. The soft anatomy of the lungfishes reveals many structures in the nervous system, blood vessels, etc., which are highly comparable to those found in primitive land dwellers, and in the early stages of their development a lungfish and a frog show almost identical conditions. Lungfishes are thus close to the ancestry of land animals. But there are certain specializations which show that they are slightly off the direct line. The lungfish are, in part, eaters of small mollusks and crustaceans; their jaws are peculiarly modified for this diet, and the specialized teeth are fused into fan-shaped crushing plates. The lungfish are, so to speak, not the ancestors but the uncles of land dwellers.

As our figures show, the oldest fossil lungfish were similar in many regards to their crossopterygian cousins in general body proportions and fin structure. The paired fins were long, stout, leaf-shaped organs, scale-

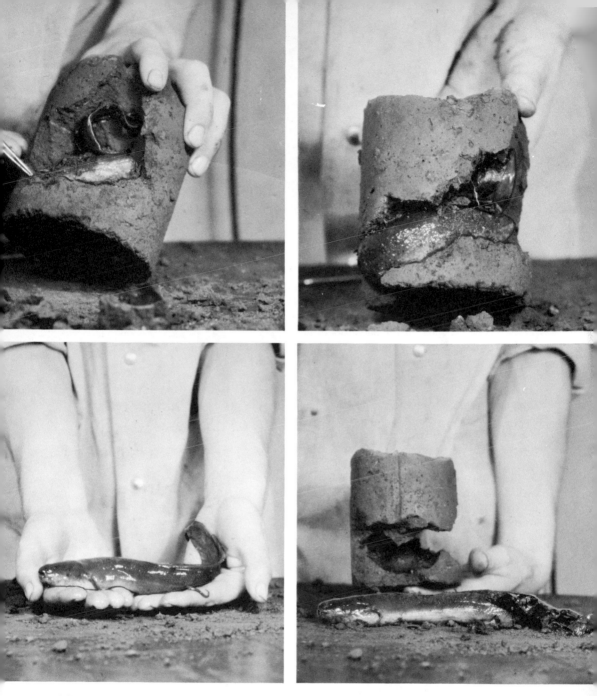

Canned lungfish. The African and South American lungfish have the ability to burrow into the bottom and pass the season of drought in a mud "cocoon." It has been found that the African form (*Protopterus*), illustrated here, can be persuaded to form its cocoon in a tin can and thus be readily transported. When desired, the can may be opened and the fish removed; on placing it in the water it soon resumes its normal activities. This potentiality of passing a hot or dry season in a condition of reduced body activity is termed aestivation and corresponds to the hibernation of many animals in cold climates. Above are shown stages in the opening-up of a canned lungfish. On the next plate is shown a specimen in an aquarium shortly after removal. (Photographs copyright General Biological Supply House, Chicago.)

Fish relatives of land animals. *Above*, the African lungfish (*Protopterus*). Although the living lungfish are quite specialized in a number of regards, they retain many features to be expected in the ancestors of land types. Closer, however, to this ancestry are the lobe-finned fishes (Order Crossopterygii), two examples of which are illustrated. The group is nearly extinct. *Right*, a restoration of *Eusthenopteron*, a fossil Devonian form frequently found in American deposits. Its fins appear to have been stout enough to support it, as shown here, on a stream bank. *Below*, *Latimeria*, the only surviving lobefin, representing a side branch of that group long thought to be extinct but recently found alive in the sea off the Comoro Islands. (*Protopterus* courtesy New York Zoölogical Society; *Eusthenopteron* from a painting by F. L. Jaques under the direction of W. K. Gregory, courtesy American Museum of Natural History, New York; *Latimeria* after Millot.)

covered, with an internal structure that surely included fin muscles as well as good skeletal supports. In the median fins, the tail fin was originally, as in all early bony fishes, an upturned sharklike affair. Above, there were (as in crossopterygians) two median fins; below, a small median fin lay between tail and anus. If we follow through a series of successive fossil forms toward and to the living types, we see the tail fin gradually straightening out (as it did in many crossopterygians) and a backward movement of the other median fins, so that by modern times all have fused into a single symmetrical structure surrounding the hind end of the body. The African and South American forms have become still further modified. The body has become elongate and rather eel-like, and the paired fins have become slender structures of relatively little importance. In the South American form the pelvic fins have become accessory water-breathing organs; with a rich blood supply to the feathered-out margins, they function much like a pair of gills.

The presence of lungfish in the tropics of the southern continents without representatives in the intervening areas is a type of discontinuous distribution which we shall encounter time after time in other groups of freshwater or terrestrial animals. In this case, as in most, it is explained by the fact that in earlier geological periods the ancestors were widespread in northern continents but have become extinct there—frequently, it would seem because of the unfavorably cold climates which developed in the north during Cenozoic times.

The Australian lungfish is found only in a few rivers in Queensland; known locally as the "barramunda," it is a large fish, reaching six feet in length. In times of drought its lungs enable it to survive in the stagnant water of drying pools. If the pool dries completely, however, the barramunda is in difficulty, although it may stay alive for a time by wallowing in the mud of the pool bottom.

The African and South American lungfishes are smaller animals of eel-like shape. They are closely related and of similar habits. They are more highly adapted to drought than the Australian form, for they can successfully survive the complete disappearance of the waters in which they normally live. In the Gran Chaco of South America, for example, there are vast areas which for much of the year are shallow lagoons. But during the southern winter, in June and July, they dry up completely. At the onset of these conditions the lungfish digs a burrow in the mud of the pool bottom and, curling within it, lives in a state of partly suspended animation until the rains begin and the waters return. That this habit is no recent devel-

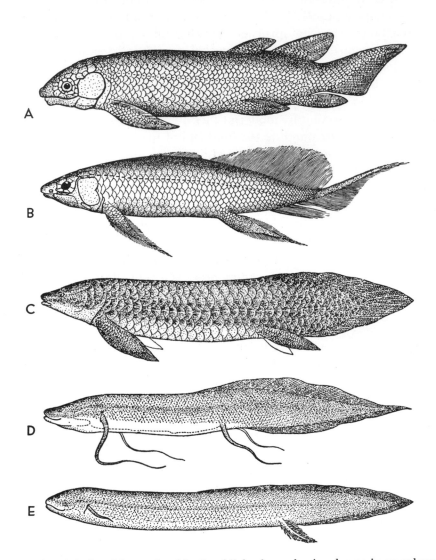

The evolution of the lungfish; a series of fossil and living forms showing changes in general proportions and fin structure. *A*, the earliest fossil lungfish (*Dipterus*). The median fins are all separate,
and the tail tilts strongly upward; the whole appearance is similar to that seen in the related lobe-
finned fishes. *B*, a somewhat later fossil type (*Scaumenacia*); the fins are tending to concentrate at the
back end of the body. *C*, the living Australian lungfish (*Epiceratodus*); the tail is straight, and the
median fins have fused into a single structure. *D*, the African lungfish (*Protopterus*), and *E*, the
South American form (*Lepidosiren*), show a transformation into an eel-like shape and reduction of
the paired fins. (*A*, after Traquair; *B*, after Hussakof; *C–E*, from Norman, *A History of Fishes.*)

opment is shown by the fact that there have been recently discovered in late Paleozoic sediments in Texas ancient lungfish burrows—occasionally containing the remains of a fish which failed to survive and revive successfully. Many animals in temperate zones tend to "hole up" over the winter, a process termed "hibernation"; this comparable trick of the lungfishes and some other tropical animals of remaining quiescent during an inhospitable summer season is reasonably termed "aestivation."

Crossopterygians.—The lobe-finned fishes, or crossopterygians, are the group which we believe to have included the ancestors of land vertebrates. Lobefins were abundant in the Devonian, like the related lungfish; in later

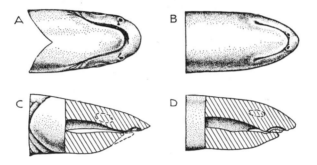

A, B, the underside of the head in a shark and in a lobe-finned fish to show the development of the internal nostrils by a forward growth of the jaws into the nostril region; *C, D,* the same heads sectioned in side view. In the shark the nasal pockets lie in front of the mouth; a forward growth of the jaws to the extent indicated by the dotted line would bring one of the nasal openings into the mouth cavity.

periods, however, they dwindled rapidly, and no typical member of the group survived the end of the Paleozoic. In consequence, our knowledge of them is more restricted than is the case with the lungfish. From the fossils we can gather much information about other organ systems as well as the skeleton, and we feel confident that in the typical older lobefins there were present many of the features, such as internal nostrils, lungs, etc., in which the lungfish resemble land types. The old crossopterygians had avoided the specialized diet and related dental peculiarities of the lungfish, and in almost every respect the skeleton can be closely compared with that of primitive amphibians. These fish lack typical legs, of course. But their fins are of exactly the sort one would expect to find in an ancestor of the amphibians. As in lungfish, the fins contain a fleshy lobe, within which are bony skeletal supports. Of these bones only one attaches to the shoulder skeleton. At its far end are found two elements forming a second fin joint, while beyond this there is an irregular branching series of bones. This is a pattern basically comparable to that of the leg of a land animal.

Typical lobefins were fresh-water fishes which, as we have said, became extinct early. Much longer-lived were members of a side branch of the lobefins termed the coelacanths, in which the head was shortened, the body became stubby, and there were various other departures from the typical crossopterygian pattern. The coelacanths early migrated into the seas and therefore departed widely in habits and, presumably, in various

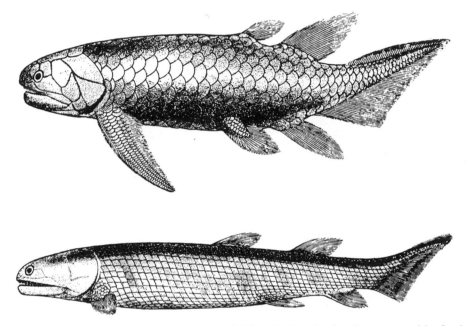

Typical Devonian crossopterygians, or lobe-finned fishes, closely related to the ancestry of land animals. *Above, Holoptychius*, a form which is closely comparable in general contours to early lungfish seen on p. 62. *Below, Osteolepis*. (From Traquair and Smith Woodward.)

functions and internal structures in relation to this environmental shift. Fossils of marine coelacanths have been discovered in many formations of late Paleozoic and Mesozoic age, but there is no trace of coelacanths in any deposits of the Cenozoic. It thus seemed that they had become extinct some 70 million or so years ago, and in consequence it used to be flatly stated that crossopterygians of all sorts, both typical and specialized, were long since extinct.

But this statement was of course based on negative evidence only; and any one piece of positive evidence is sufficient to refute any belief based on negative grounds. Such positive evidence we now have, and those of us who used to be so emphatic about crossopterygian extinction have now eaten our words—with great pleasure.

In 1939 a commercial fisherman, trawling off East London, South Africa, brought up a type of fish he had never seen before—a big, deep-bodied, five-foot fish with large, bluish scales. He brought it ashore as a

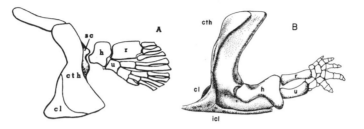

Fish fin *vs.* land limb. The shoulder girdle and limb of a lobe-finned fish (*A*) and of a primitive amphibian (*B*), placed in a comparable pose. The main elements of the limb—humerus, radius, and ulna (*h, r, u*) are clearly homologous in the two cases, and the irregular branches of the distal part of the fish fin have settled down to the pattern of toes of a land animal.

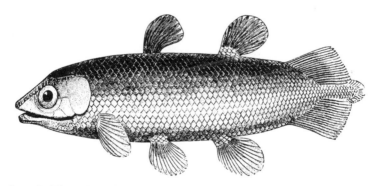

A fossil coelacanth. The coelacanths are crossopterygians, related to the ancestors of the land vertebrates. They have been long known as fossil forms, and the old figure shown here was based on material from Mesozoic rocks. That the restoration was essentially correct is proved by the discovery of a living coelacanth shown on an accompanying plate. (From Traquair.)

curiosity, and, since its oily body was decaying, it was skinned, stuffed, and mounted. When a zoölogist from a nearby college was consulted, he received a surprise which would have been little greater had he seen a dinosaur walking down the street. For this unknown was a coelacanth, a lobefin belonging to a group supposedly extinct since the days of the dinosaurs.

Little of this animal remained for scientific study except the skin and outer skull bones. But surely this specimen represented a stray from some region off the East African coast where presumably a colony of these supposedly extinct coelacanths still survived. The outbreak of World War II prevented immediate search for coelacanths. Once the war ended, however, the scientist concerned, Professor J. L. B. Smith, circulated descriptions of the animal widely up and down the coast. Presently the coelacanth home was discovered—in deep waters off the shores of the Comoro Islands, north of Madagascar. Already a number of specimens have been recovered and are being intensively studied by Professor Millot of Paris and his colleagues. As this is written, these studies are far from complete, but already many facts have come to light which show that, as expected, the coelacanths combine specializations associated with their long life in marine waters with many primitive features retained from the early days of crossopterygian existence. As a striking example: although the coelacanths have been for many millions of years marine fishes to whom lungs are useless for breathing purposes, they still retain them as a pair of ventral outgrowths from the throat, much in the fashion of their ancestors of Paleozoic fresh waters.

CHAPTER 3

Modern Fishes

While the fleshy-finned fishes, the crossopterygians and lung-fishes, have been of major importance as ancestors of higher vertebrates, they have played but a minor role in the drama of fish history. Here the important roles have been played by members of the other major group of bony fishes—the Actinopterygii, or ray-finned fish.

As has been indicated, the group derives its name from the structure of the paired fins. Exceptional cases apart, these are composed only of a web of skin supported by horny rays. Flesh and bone do not invade the fin to any extent and are confined to its base alone. There are, however, many other differences between these forms and their cousins, the Sarcopterygii, including marked differences in the structure of the scales, different patterns of the bones of the head, lack of internal nostrils, etc.

What the factors are that have been responsible for the success of the group, it is difficult to say. One item of interest has to do with sense organs. Like most vertebrates, the Sarcopterygii appear to have been largely dependent on the sense of smell for their knowledge of the external world, and the eyes were not markedly developed. In the ray-finned group the olfactory organs are present but relatively unimportant. The eyes, on the other hand, tend to be large and apparently are the dominant sense organs; this contrast is reflected in the brain organization. Eyes may thus have been important in actinopterygian history.

The air bladder, too, may have played a part in their success. Only the most primitive of rayfins retained a lung. In this group alone is found the transformation of this structure into an air bladder, a hydrostatic organ which, as suggested in the last chapter, may have proved of great utility, particularly in the seas which have been the home of most later actinopterygians.

Still more important in the rise of the rayfins may have been the mere force of numbers. In other groups of fishes there tend to be but relatively few eggs laid—from a few dozens to a few hundreds. Here, however, the eggs, although small in size, may number thousands or even millions for every female. Other fish, such as the Sarcopterygii, may have gone in for conservation of the individual, laying fewer eggs, furnished with a more abundant food supply of yolk. But quantity not quality has been the actinopterygian motto. Individuals may die in shoals, but the fecundity of the race will fill the ranks again in a short time.

PRIMITIVE RAY-FINNED FISHES

The oldest rayfins (the palaeoniscids) appeared in the Middle Devonian. They were, however, not particularly abundant at that time and were greatly outnumbered by the contemporary lobe-finned fishes and lungfish.

Two living African fish represent the earliest ray-finned stage, although in somewhat modified form. The bichir, or *Polypterus* ("many-finned"), is the better known of the two. Like the oldest fossils, the fish is covered

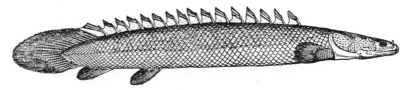

Polypterus, an African fish which represents a persistent (if somewhat specialized) descendant of the paleoniscids of the Paleozoic; the most primitive of living ray-finned fishes. (After Agassiz, modified.)

with thick and shiny bony scales, and there are many primitive features. Most interesting is the fact that we find here alone, among living rayfins, well-developed and functional lungs opening in primitive fashion from the underside of the throat. This is to be correlated with the fish's habitat—the upper Nile and other tropical African areas where seasonal droughts occur. The fins have changed somewhat, however, from the primitive type, not only in the division of the back fin into the series of sail-like structures which give the fish its name, but also, most unusually, in the fact that the pectoral fins have developed a fleshy lobe. This last feature caused *Polypterus* to be included among the crossopterygians for many decades before its proper place was realized.

Two other types of fish, both present in American waters, also represent survivors from an early stage of evolution of the rayfins—the sturgeon and the paddlefish, or spoonbill, of the Mississippi. These types represent a somewhat more advanced condition, for they possess a typical air bladder rather than a lung. On the other hand, the fins have remained more primitive than in *Polypterus;* the tail fin, for example, is still essentially shark-

Four primitive ray-finned fishes.
Most living rayfins are teleosts—
the highest subdivision of the series. Of the few survivors of more
primitive actinopterygians, a majority are found in the United
States. *Top*, the paddlefish, or
"spoonbill cat" (*Polyodon*), of the
Mississippi, which may reach 200
pounds in weight. It is degenerate
in many ways and has become a
bottom-dwelling mud-grubber with
a long, spoon-shaped snout. It still
retains a sharklike tail. *Next below:*
the sturgeon (*Acipenser*) also rather
degenerate; note the peculiar "feel-
ers" at the mouth. Little cherished
except for caviar, the sturgeons are
found in various regions of the
northern continents, including the
Mississippi Valley and Great Lakes.
Third, the bowfin, or "fresh-water
dogfish" (*Amia*), a form not far below the teleost level. It is a common
lake fish of the Middle West and
South. The gar pike (*Lepidosteus*) is
a Mississippi Valley inhabitant, notable as one of the few living fishes
which have retained the thick
ganoid scales of early fishes. The
form shown is the giant alligator
gar of the southern states. (*Polyodon* and *Lepidosteus* photographs
courtesy American Museum of Natural History; *Acipenser* and *Amia*
courtesy New York Zoölogical Society.)

The most primitive ray-finned fish. *Above*, a restoration of *Cheirolepis* of the Middle Devonian, oldest actinopterygian. (From a painting by F. L. Jaques, under the direction of W. K. Gregory.)

An Australian sea horse. *Phyllopteryx, below*, differs from the ordinary sea horse in peculiar outgrowths like seaweed. (Photographs courtesy American Museum of Natural History.)

Color change in fishes. *Above*, two catfish kept for a day in dishes with light and dark backgrounds, respectively, show color change. *Below*, two photographs of the same specimen of flounder placed on different backgrounds; in this form there is a tendency for change not merely in color but in pattern as well. Such changes are initiated by vision, for if the animal is blinded, the skin becomes permanently dark, and there is no response to the environment. The color changes are effected chemically by neurohumors, substances given off either by nearby skin nerves or by the pituitary. (Catfish courtesy G. H. Parker; flounders after S. O. Mast.)

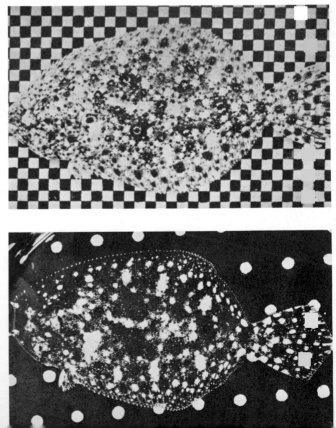

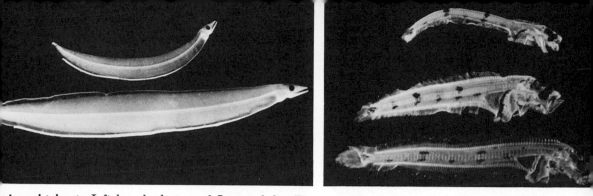

Larval teleosts. *Left*, larval eels, termed *Leptocephalus*. These are thin, translucent, and leaf shaped. The larva of the European eel is 3 inches long before the "leaf" turns into a smaller cylindrical baby eel. *Right*, tiny flounders. In contrast to the pancake-shaped adult, the young are slender. Note that both eyes (*seen as black dots*) are already on this upper surface, the left eye having migrated around the upper edge of the head. (Photographs courtesy H. C. Bigelow.)

Unusual teleosts. Many of the modern bony fishes are familiar from the dinner table. At the right and overpage are figured some of the more unusual types:

Deep-sea fishes. In the lightless depths of the ocean are numbers of teleosts which have become adapted to conditions there. In this scene (from a museum exhibit) a carnivore with an enormous mouth (*Chauliodus*) is seen pursuing a flock of "big-heads" (Melamphids). (Photograph courtesy American Museum of Natural History, New York.)

The four-eyed fish (*Anableps*). A small fish of the American tropics which swims at the surface. Each eye is divided into two parts externally, the upper part adapted for vision in air, the lower for water vision.

The sea robins (*Prionotus*) are common fishes of the Atlantic coast, usually found on the bottom in shallow waters; when disturbed they may bury themselves to the eyes in sand. The front rays of the pectoral fins move separately and can be used as digging organs. (Photographs courtesy New York Zoölogical Society.)

The mudskipper (*Periophthalmus*), which lives along the coasts of the Old World tropics. Unlike ordinary teleosts, the fish has muscular fins by means of which it is able to leave the water and crawl over the mudflats and the roots and lower branches of mangroves near the water.

The electric eel. This fish (*Gymnotus*) of the Orinoco and Amazon regions is eel-like in shape but is actually a relative of the carp and catfish. Its peculiarity (shared with several other fishes) is that parts of its muscular system are so modified as to have become a powerful electric battery. The voltage averages about 550 volts (about five times that of ordinary house current); it can be drawn on for enough power to light more than a dozen ordinary lamp bulbs—enough for an apartment.

The archer fish (*Toxotes*) of the East Indies derives its name from the fact that it can accurately shoot drops of water to capture insects or spiders near the surface. (All photographs courtesy New York Zoölogical Society.)

like, with the tip of the body tilting up into the upper part of the two-lobed tail. Again we find degenerate features, for the scales have been in great measure lost, and these fishes are long-snouted mud-grubbers with feeble jaws.

In the Mesozoic era the rayfins were the dominant fishes and were progressing toward modern conditions. The only survivors today of these intermediate members of the group are two common American fishes—the long-snouted, shiny-scaled garpike, common in rivers of the Mississippi system, and the bowfin, or fresh-water dogfish (*Amia*), a typical lake fish of the Middle West and South. These fishes are fresh-water forms, but most of the Mesozoic actinopterygians had migrated into salt water, and the seas were the centers of later evolution of the group.

TELEOSTS—THE DOMINANT MODERN TYPE

Toward the middle of the Age of Reptiles there arose the teleosts, end products of actinopterygian evolution. No sharp line of demarcation marks them off from their ancestors, although technically they can be characterized by numerous details such as the assumption of a superficially symmetrical tail fin, thinning of the scales, etc. The teleosts form an overwhelming majority of all living fishes, including every fish familiar to us except those already mentioned—every food and game fish, every fishy inhabitant of the fresh waters of the northern continents (except their four primitive relatives cited above), and the vast majority of marine fish, numbering perhaps 20,000 species.

These fishes constitute about half of all living vertebrates. How should they be treated here? To give them their just due, half of this book should be devoted to teleosts. But if that were done, I suspect, good reader, that (unless you are an ardent fisherman) you would give up at this point (and I would be inclined to do so myself). On the other hand, you might say that there are just too many of these creatures; that fish are fish; and let us skip them and go on to some more interesting topic. But that would be equally unfortunate. I propose to compromise. The accompanying figures give a series of fairly representative teleosts, and I shall, referring to the illustrations by number, give a brief account of the groups which they represent. The order in which they are arranged is to some degree arbitrary but so are all existing attempts by learned ichthyologists, none of whom has ever evolved a generally satisfactory systematic arrangement of the components of this overwhelmingly large group.

The first teleosts appeared at about the middle of the Mesozoic era and by the end of the Age of Reptiles were already the dominant fish types. Broadly viewed, most teleosts may be arranged in two categories, both with central types and both with aberrant side groups, some so far removed that their relationships are in doubt. To pick obvious differences between the two, readily seen in surface views (or pictures): the primitive forms have the fin webs supported by relatively soft and pliable rays, and the paired pelvic fins tend to remain in their primitive position, well back along the ventral side of the body; in the advanced forms, part of the fin supports are stout spines, and the pelvic fins move forward to the shoulder region or beyond.

Small teleosts of a primitive nature, which would have appeared when alive something like herrings, are the oldest of known teleosts. The herrings (1) and their relatives, such as the sardine, alewife, shad, anchovies, menhaden, and that great game fish, the tarpon, are a very primitive and generalized series of teleosts, which have departed little from the primitive pattern.

A bit more advanced are the members of the salmon group, such as the Atlantic salmon (2), the Pacific salmons, the trout, whitefish, smelts, grayling, and capelin; many are readily identifiable by the presence of a small fatty fin above the base of the tail. Typical members of the group are confined to northern temperate and Arctic regions. There are southern temperate zone river fishes of troutlike appearance and habits but of dubious relationships. Primitive teleosts were probably evolved in marine waters, but many members of the salmon group breed in fresh waters, and a number are permanent fresh-water dwellers. On the other hand, a number of forms assigned to the group inhabit the deep sea.

Beginning now to depart from primitive stocks among primarily marine forms, we may mention the flying fishes (3) and needlefishes (4). The two are common members of a teleost order, bound together by certain technical characters, but they are exceedingly different in all obvious regards. The flying fishes are common, traveling in schools on the high seas in the warmer regions of the ocean. The jaws are short, and the pectoral fins greatly enlarged. Their flight is due to alarm or for escape from enemies. They shoot rapidly to the surface, "taxi" a distance to get up speed, and set off on a "flight" (actually a glide) which may last half a minute or so, may extend several hundred yards, and may take them well above the water. Some members of the flying-fish family have smaller fins and longer jaws and are transitional to the slender, small-finned, and long-beaked needlefishes—voracious forms, some of which run to five or six feet in length. A curious variant on the needlefish pattern is that of the half-beaks, in which the upper jaw is short, the lower one long. This would seem to be an unsatisfactory eating arrangement, but the animals thrive, nevertheless, and it is of interest that in at least one of the needlefish, the young are seen to develop a lower beak before the upper jaw elongates.

Deep-sea fishes, living in twilight or utter darkness in depths where food, friends, and enemies are rare and hard to find, have developed in a num-

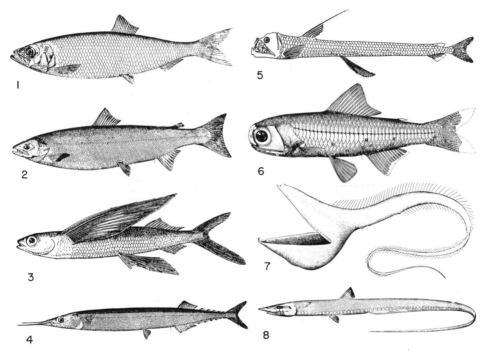

An album of teleosts. In this and the three illustrations that follow are shown a small sample of the varied members of this group. The figures are mainly from Bigelow and Schroeder or Goode and Bean; a few are from Jordan and Evermann. *1, Clupea,* the herring. *2, Salmo,* the Atlantic salmon. *3, Clypselurus,* a flying fish. *4, Scombresox,* the needlefish. *5–8,* deep-sea fishes for which, in general, there are no popular names: *Chauliodus, Diaphorus, Eurypharynx, Aldrovandia.*

ber of teleost groups. Figures 5–8 show a few of these forms which have split off from some of the more primitive stocks. We have mentioned that some salmonoids have descended into the depths; the little but voracious viper-fish (Fig. 5; also shown in a plate) with its large mouth and fanglike teeth can be identified as a salmonid by its little fatty fin at the tail base. Also presumably of salmonoid origin are the lantern fishes, so-called because they bear along the flanks a series of phosphorescent organs with the appearance of a row of lighted portholes. The one pictured, the headlight fish (6), gets its name from the fact that it has an additional large "light" at the tip of the snout. Some of the most grotesque of all deep-sea fishes are the gulpers (7), possibly related to the eels. A good proportion of the whole animal consists of an enormous mouth and throat; food seldom descends to the depths where the gulpers live, and when it does, the fish must be prepared to make the most of it. A final deep-sea

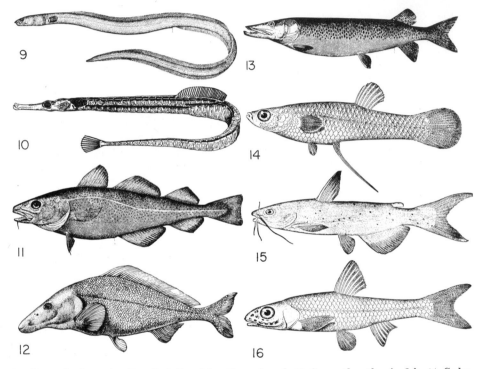

An album of teleosts (*continued*). *9, Omochelys*, the snake-eel. *10, Syngnathus*, the pipefish. *11, Gadus*, the codfish. *12, Mormyrus. 13, Esox*, a pike. *14, Gambusia*, a cyprinodont. *15, Ictalurus*, the channel catfish. *16, Ericymba*, the silver-jaw minnow.

form pictured (8) is one of the halosaurs—there is no popular name—long, slender, soft-bodied fishes, black in color with a long series of "lighted portholes." These four fishes are only samples of the many odd types inhabiting the deeper waters and rarely seen except when dredged up by exploring vessels.

Coming back from the depths, we may mention further marine offshoots of the more primitive teleost stocks. The eels, such as the snake-eel pictured (9), form an abundant series of fishes, most of which have a common body pattern. All are long and slender with very long dorsal and anal fins and without a distinct tail fin; the pelvic fins are lost, and the gill opening—a prominent slit in most teleosts—is reduced to a small rounded opening. A fraction of them live in fresh water although (as far as known) breeding in the ocean; most are purely marine. Typical eels are extremely voracious. The pipefish (10), one of a group of little fishes (of which the sea horse is better known), has a body incased in armor and a tiny nibbling mouth. More primitive members of this group, which tend to connect these creatures with more "orthodox" fish types, are the sticklebacks, some of which (unlike the pipefish or sea horse) are found in fresh-water streams. Here the mouth and body shapes are of more normal type, but there is a variable development of armor plates on the flanks. As the name suggests, the sticklebacks have some development of fin spines, and these peculiar fishes are probably fairly advanced, if aberrant, forms.

The cod (11) and its relatives, such as the haddock, pollock, and hake, form a very distinctive group but are of somewhat problematical relationships; they are primitive in the absence of fin spines, but the pelvic fins have moved forward to a point below the shoulder, as in the spiny-finned fishes seen in later figures. Typical members of the group feed near the bottom on northern marine banks of moderate depth. One group of codlike fishes, the grenadiers, or rat-tails, is confined to the oceanic depths in the North Atlantic and North Pacific, but one form, the burbot, is found in fresh waters.

So far, we have made mention of few but marine teleosts. A number of fresh-water types may now be passed in review. In the tropics of South America, Africa, the East Indies, and Australia are to be found members of families of fresh-water fishes which appear to be primitive offshoots of the early teleost stock. None will be familiar to most readers; I have pictured the mormyr (12), a common Nile fish belonging to one of these families. The presence in these several southern regions of fishes with common characteristics has (as usual) led to the idea of former land connections; the explanation (as usual) is that they probably existed in intervening northern regions in happier climatic times. Much more advanced in some structural features are those common fishes of northern waters in both North America and Eurasia, the pikes (13), fast-swimming, long-jawed forms that are some of the most voracious of fresh-water fishes. Equally advanced, but very different in nature, are the small fishes known technically as the cyprinodonts, including the killifishes, guppies, the top minnows (14), and so forth; the largest of the group reaches a foot in length, but many are very tiny, and this family includes many favorites for home aquaria. The four-eyed fish, shown in a plate figure, is a cyprinodont. Some venture into salt waters in river mouths and along shores, but most are inland dwellers; they are particularly abundant from Mexico south into the American tropics. In the cyprinodonts are found the only teleosts which bear the young alive; as in the form shown, the male may have a greatly elongated anal fin, which aids in internal fertilization.

While the fresh-water forms so far mentioned are either offshoots of marine groups or may reasonably be suspected of being descended from ocean-dwelling forms, we come at last to one group of teleosts which includes the vast majority of living river and lake fishes and whose history, as far as known geologically, is one of continuous existence in fresh waters. Unfortunately there is no popular name for the group as a whole, which is known technically by the awkward term Ostariophysii. Equally unfortunately, they are not tied together by any readily observable surface features, although they are clearly united by the presence in all of them of a series of bones connecting air bladder and internal ear which gives them better hearing power than ordinary fishes. Included are the catfishes (15), the characins, the European carp, and a host of varied and extremely common little fishes most of which the ordinary person lumps as "minnows" (16). The catfishes, readily recognizable by their scaleless nature and peculiar barbels, are extremely varied and widespread; there are about one thousand species, mainly in the tropics of South America and Africa. One of the few fishes to develop electric organs is an African catfish. In South America a number of catfishes have compensated—or overcompensated—for the loss of scales by the development of bony armor which makes them look somewhat like modern editions of the ancient placoderms.

The characins are of normal appearance, many resembling the trout or the pike; they are found in Africa and in South and Central America (another example of southern-continent distribution). Some are ferocious—notably the piranha of South America, a small fish but one whose bite has been compared to a razor cut; attacking in droves, they can literally cut a swimmer into bits. A curious form is a further example of a fish with a developed electric organ—the so-called "electric eel" of South America, shown in a photograph. The final group of this great assemblage includes few fish of any size except the carp, native to Europe; the smaller forms, the various "minnows," loaches, and suckers, are present in abundance in Eurasia and North America and have been claimed to comprise close to 2,000 species; almost none, however, are present south of the Northern Temperate Zone.

A very considerable proportion of the teleosts belong to an advanced structural type termed the "spiny-finned teleosts" (technically Acanthopterygii). In typical members of this progressive group there is a great development of hard spines rather than soft rays in the supports of the fins. The dorsal fins frequently form a double structure, the front part having strong spines. A further easily seen feature is that the pelvic fins have

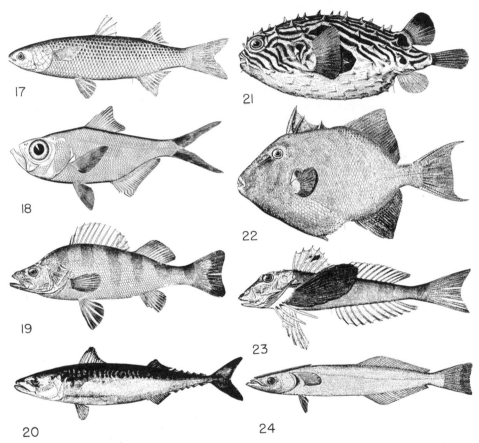

An album of teleosts (*continued*). *17, Mugil*, the mullet. *18, Beryx*, a primitive spiny-finned form. *19, Perca*, the perch. *20, Scomber*, the mackerel. *21, Chilomycteris*, a puffer. *22, Balistes*, a triggerfish. *23, Prionotus*, a sea robin. *24, Remora*, the shark-sucker.

moved forward, so that they lie beneath the pectorals in the shoulder region or even farther forward. A number of minor fish groups show features transitional between lower types and the spiny-finned forms, for example, the mullets (17), rather small fishes of coastal waters. They show little spine development, but there is the beginning of a forward shift of the pelvic fins. Related are the silversides, smaller minnow-like fishes of fresh waters and estuaries, and the barracudas. The last are slender-bodied, fast-swimming, voracious fishes of warm sea waters with long knifelike teeth. The barracudas are one of the few marine teleosts which are definitely known to attack a swimming man; cases are particularly numerous in the Caribbean region. *Beryx* (18) is a small marine fish in which, again, there is little spine development, but the pelvic fins have moved well forward, and the body is somewhat deepened—a feature found in many of the typical spiny-finned forms.

The perch (19) are excellent examples of generalized spiny-finned forms. The body may be somewhat deepened; the pelvic fins have shifted forward along the belly to the shoulder region; fin spines are well developed, notably in the presence of a prominent and separate spiny dorsal fin. The true perches and some of their close relatives are fresh-water forms, these including the bass and "crappie" and the American sunfishes (not to be confused with the great oceanic sunfish). However, most of the perchlike fishes (and, indeed the vast majority of all the remaining teleosts to be noted) are marine, and a large proportion are native to warm tropical waters. To list a few of the more common and better known forms—the sea perches, sea bass, jewfishes, groupers, "dolphins," croakers, drumfish, weakfish, snappers, porgies, scup, tautog, and parrot fishes, and the deep-bodied angelfishes and butterfly fishes, many of whom frequent the still waters of coral reefs; the little archer fish, shown on one of the plates, is a deep-bodied perch type.

A group of spiny-finned forms very progressive in many regards is that typified by the mackerel (20). These are fast-swimming forms, with a slender, beautifully streamlined body and a forked tail fin set on a slender base. Related are the bluefish and albacore; in this group, too, belong some of the most spectacular of game fishes, notable for the development of long-spiked upper jaws as well as for great size—the sailfish, the marlins, and the great swordfish, which may reach twenty feet in length.

From the more typical spiny-finned teleosts there have arisen many aberrant side branches. The next two fishes illustrated are members of a group which have developed a tiny mouth but one armed with heavy teeth and strong jaws, set at the tip of the "snout"; the body is deep and rounded in side view and the pelvic fins have been lost. In the trunkfish, puffers (21), and porcupine fishes, the body assumes a nearly spherical shape, capable of great distension in the puffers and often incased in spines or armor. In this group also are the triggerfishes (22) and filefishes; most of the development of the spiny dorsal fin is in the form of a large erect spine, which may be movable. These forms are essentially circular in side view but flattened from side to side. A much larger flattened member of the group but one without spines is the enormous but lethargic oceanic sun-fish, or head fish (illustrated in our first chapter), which may reach nearly ten feet in body diameter and a ton in weight.

A further group clearly derivable from the perchlike forms is that of the mail-cheeked fishes. The name derives from the fact that there tends to develop from the cheek bones a set of bony plates which may expand to cover most of the head. Little of this, however, is noticeable in superficial view, and some of the more primitive forms, such as the rock "cod," are very similar in appearance to the perches. There is, however, a strong trend toward development of large pectoral fins and an exaggerated spininess, which reaches its extreme in such types as the sculpins of Northern Temperate Zone waters and the gurnards, such as the sea robin (23). Still another group of aberrant nature but clearly derivable from the perchlike fishes is the remora, or shark-sucker (24). Its notable specialization is a large flat sucking disc atop the head with which it can fasten firmly to the body of a shark or other large fish or sea turtle and obtain a free ride (as well as a possible share in its host's food). This, oddly, is a modification of the spiny front dorsal fin of its perchlike ancestors. The mud-skipper, shown in a plate figure, is a representative of the gobies—small Tropical-to-Temperate Zone forms which are found in shallow waters in the tidal zone or in nearby ponds and swamps, frequently burrowing into muddy shores. The mud-skipper, in strong contrast with most fishes, can use its fins to walk or skip about the shores in search of food. A small goby from the Philippines measures only about half an inch long when fully grown and is probably the smallest of all vertebrates.

The curious-looking toadfishes (25) may possibly be transitional from normal spiny-finned teleosts to the repulsive anglers, of which the goose-fish, or "fishing frog" (26), is a common shallow-water representative. The body is broad and flat, and—quite in contrast with ordinary teleosts—the pectoral fins form great fleshy flaps extending out on either side of the body. The goosefish is predaceous; there is a very broad mouth armed with sharp teeth; above it, an enlarged front spine with an expanded tip dangles like a baited hook. There are a number of variants on the goosefish pattern. Some anglers are deep-sea forms. In some the "bait" is made attractive by a luminous organ at the tip. We will comment later on the "marriage" customs of deep-sea anglers (p. 81).

A considerable number of other variants from the primitive spiny-finned pattern might be described if space (and the reader's patience) permitted. I shall restrain myself to two further examples. The rarely seen ocean fish variously called "the king of the herrings," ribbon fish, or oarfish (27) is a spectacular creature, silvery in color with long crimson-hued fins and a very long body, flattened from side to side, which may reach a length of

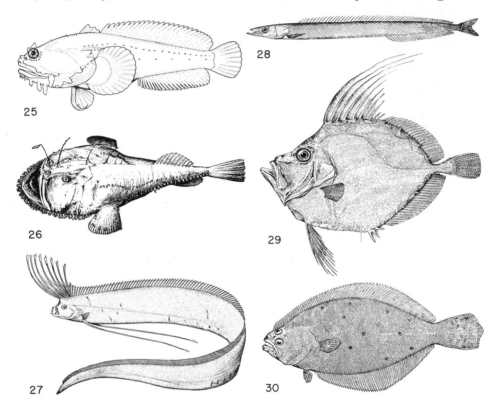

An album of teleosts (*concluded*). 25, *Opsanus*, the toadfish. 26, *Lophius*, the goosefish. 27, *Regalecus*, the oarfish. 28, *Ammodytes*, the sand lance. 29, *Zeus*, the John Dory. 30, *Paralichthys*, a flounder.

twenty feet; a series of long slender rays form a head crest, and there are long slender trailing pelvic fins. It is uncertain whether this type is to be considered a degenerate and aberrant descendant of the spiny-finned group or has its roots further back in teleost history. The little sand lances (28), coast dwellers often burrowing in the sand beneath the breakers, may be related to the blennies, larger fishes which are, like the sand lances, long-bodied forms with long median fins.

A final and important group of teleosts to be noted is that of the "flat-fish"—the flounders (30), soles, plaice, turbot, and their giant relative, the halibut. These are fishes which live on or close to the sea bottom where they feed on bottom-dwelling invertebrates. In mode of life and in general shape they resemble the skates and rays; but actually they are constructed in diametrically opposite fashion. They are greatly flattened, not a top-to-bottom flattening, but a side-to-side one. The fish is very deep-bodied and thin. It begins larval life swimming in normal fashion but then turns over to rest upon one side—right or left, generally according to the species. During growth (as shown in a plate figure) the eye from the prospective lower side migrates over the top of the head to join its mate. The flat fishes lack spines on their fins, but this may be due to secondary loss, and it is possible that some deep-bodied, perchlike fishes, such as the John Dory (29), point out the ancestry of the flounders and their kin.

Distribution.—Although it is probable that the ray-finned fishes originated in fresh waters, it is possible, as suggested earlier, that the major center of teleost evolution lay in the sea and that hence many if not all of the present fresh-water teleosts are forms which have secondarily returned to inland waters. The one possible major exception, as mentioned above, is that of the important carp-catfish group (the Ostariophysii) whose members make up a large fraction of the world's fresh-water fish fauna. We mentioned in the opening chapter the difficulties which a purely fresh-water type of fish must meet in becoming distributed over a wide territory. It is thus not unexpected that there are considerable differences between the teleost populations in the different continents. There are a number of types, such as the pike and perch, which are widely distributed over the Holarctic area, from Europe across northern Asia to North America. But even here there are numerous contrasts; catfishes, for example, are numerous and varied in North America, rare in Europe and northern Asia. The South American fresh-water teleost fauna is in strong contrast with that of North America and consists in great measure of ostariophysians—characins and a host of catfishes. This is presumably to be correlated with the fact that the two continents were separated for many millions of years during the Cenozoic. As was said earlier there are striking resemblances between other types of South American teleosts and those of Africa and the Orient, which have been interpreted by some—probably incorrectly—as due to former direct land connections. Australia has been separated from the other continents since the days of the dinosaurs, and hence it is not surprising to find that it has almost no members of the normal fresh-water teleost types of other continents. Most of the fishes in its streams are "stray" species, belonging to groups normally marine, which have, here and there, ventured from the sea into streams otherwise barren of fish life.

In addition to purely fresh-water forms we find, of course, a number of teleosts which (like the sticklebacks) seem to be at home in either salt or fresh waters, some which migrate during the life cycle from one environment to the other, and in the case of the salmon-trout-whitefish group some types which live happily all their lives in fresh waters, having dropped the marine phase from their life cycle.

The general discussion in the first chapter of the horizontal and vertical distribution of marine life applies particularly to the teleosts. The great majority of marine teleosts are dwellers in the shallow waters of the continental shelf, and a smaller number are surface or near-surface dwellers on the high seas. Since it is the surface waters which are most susceptible to zonal climatic influences, shallow-water teleosts often show considerable restriction in north-south range; tropical shallow-water fishes make up an extremely large percentage of the total number of teleost species, although the less varied dwellers in colder waters, such as cod or herring, may be extremely abundant in numbers of individuals. Deeper-water denizens tend to have broader ranges; such waters are cold in any latitude. Down to perhaps 250 fathoms the deep-water populations consist mainly of small silvery teleosts with large eyes that can make the most of such light as penetrates to that depth. Still deeper, the fish are mainly blackish in color, and their eyes are small to vestigial; vision is of little use in the depths.

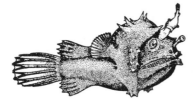

A female of a grotesque deep-sea angler fish, with the tiny parasitic male attached to her forehead. (After Norman.)

Breeding.—Among marine teleosts breeding habits in general are of a relatively simple nature. Fertilization of the eggs is external; males and females gather in swarms, and the shedding of eggs and sperm takes place, so to speak, as a community project. Only rarely is there individual mating in marine forms. However in deep-sea forms, where life is sparse and light feeble, breeding is more of a problem. Presumably the luminiscence of the lantern fishes has developed in great measure as a means of recognition of their own kind. The deep-sea anglers have solved the problem of finding a mate in unique fashion—the female carries her mate with her. The male, of tiny size, attaches himself in a haphazard way to some part of the female's body. He is thus always at hand for mating-time; all his structures degenerate except his reproductive system, and, as a parasite, he obtains his nourishment by connection of his blood stream with that of his host-mate.

Many vertebrate groups lay but a modest number of eggs and care for them in some fashion or other, but most marine teleosts lay vast numbers of tiny eggs, which float near the surface. Few teleost eggs exceed one-sixteenth of an inch in diameter; many are much smaller; usually they are buoyed up by a tiny oil droplet inclosed within the membrane surrounding each egg. A small herring may lay 20,000 to 40,000 eggs; females of the flounder group produce a quarter of a million to a million; a good-sized cod up to 6 or 7 million; a specimen of ling—a cod relative—has been estimated as having had more than 28 million eggs in its body. In a few cases some care is taken of their eggs by salt-water forms; in pipefish and sea horses, for example, the male has a brood-pouch on his abdomen in which the young develop. Generally, however, the eggs are abandoned as soon as laid and fertilized. Most, of course, never reach maturity but are eaten, along with the rest of the tiny animal and vegetable life of the sea, by various invertebrates or small fishes. But if, on the average, only two of the many millions of eggs laid during a female's entire life develop to maturity, the survival of the species is assured.

A relatively small number of marine teleosts lay eggs of a different type —eggs which are heavier than water, sink to the bottom and, usually having a sticky surface, tend to adhere to rocks or weeds. In these forms the eggs are generally of somewhat larger size and fewer in number.

Eggs of this latter type are, as might be expected, almost universal among fresh-water fishes; floating eggs laid in streams would be at the mercy of the currents. Here, too, some further attention is paid to the eggs in many cases, and hence paired matings are common, sometimes with elaborate courtships, as in the sticklebacks. Nests, varying from a simple hole in the bottom scooped out by the tail fin to much more elaborate structures, may be prepared for reception of the eggs, and the nest in some cases is guarded by one or both parents. Nearly all teleosts have external fertilization; exceptions to the rule, however, are many of the little guppies and their allies, in which, as we have noted, fertilization is internal and the young are born alive.

Of interest are various teleosts which breed in an environment different from that in which they spend most of their lives. Not uncommon are forms which are marine as adults but breed in fresh waters. These are mainly found among relatively primitive teleosts of the herring and salmon groups. The Atlantic salmon is for the most part a marine fish. It develops, however, in fresh-water streams and spends some time there before descending to the sea. After four years or so, on the average, it enters the mouths of rivers in the cooler parts of the Temperate Zone in western Europe and eastern North America. At one time hordes of salmon entered such rivers as the Hudson, the Connecticut, and the Thames, but with stream pollution and the blocking of rivers by power dams, the salmon runs are now mainly along more northerly streams in more sparsely settled country. Arriving in small and shallow streams at the headwaters of such rivers, the salmon mate, and the females scoop out shallow trenches in which the eggs are laid. The increase in size of reproductive organs in preparation for breeding, the journey upstream, and the actual breeding process are all exhausting to the fish. Few of the males survive, but many of the females drift back downstream in a spent condition and, after regaining good condition in the sea, may return to spawn several more times during an average life of about nine years.

The several species of Pacific salmon have somewhat similar life histories. But breeding occurs only once. After attaining maturity in the ocean, they return upstream; in the case of the Yukon, the breeding grounds may be over two thousand miles from the ocean. In the headwater streams, after a journey that is often long and difficult, males and females mate and spawn for the first and last time and, spent, die.

Most interesting in many ways of all life-histories is that of the fresh-water eels. These animals abound in a number of regions, notably those bordering both shores of the North Atlantic. We shall confine ourselves to the European species, since its history is better known than that of the American form. Eels are common inhabitants of the streams of western Europe but have never been seen to breed, and eggs and larval forms are unknown there. It was, therefore, not unnatural that the ancients believed that they arose spontaneously out of the mud in which they dwelt.

However, a century or so ago, it became obvious that eels migrated from, and returned to, some salt-water breeding-grounds. Every fall mature eels migrate downstream to the river mouths, and every spring tiny young eels, called elvers, ascend into fresh water. The eel remains in inland waters for a period of from eight to a dozen years and constantly increases in size, the adult males averaging a foot and a half in length, the females a yard or so. Eventually sexual maturity arrives, and the eel, now silvery in color, joins the fall migration down to the sea.

Eels migrate in enormous numbers. One writer has estimated that, from a single lagoon on the Mediterranean coast of France, about 100,000 silver eels enter the sea each fall and that the whole number leaving the European shore in a single season may be on the order of 25 billion. Where do they go? It was at first thought that they bred in the waters not far from the river mouths they had left and that the elvers ascending the rivers the next spring were the product of these activities. But if this were true these well-fished coastal waters should yield plenty of breeding eels, eggs, and small larvae. This is the reverse of the truth. Of the billions of eels which leave the land each fall, hardly a one has ever been seen after leaving the shore. They simply disappear. Where do they go?

There have long been known in the oceans small fishes termed *Leptocephalus* ("thin heads"). Not merely the head but the whole body is thin, and these translucent little creatures, a few inches long, have very much the shape of a tree leaf. A first clue to the major problem was obtained a few decades ago when an Italian scientist kept some of these little fishes alive in his laboratory. To his astonishment they underwent a complete metamorphosis, or change of shape; the leaflike body contracted into a thin cyclinder, and the *Leptocephalus* became a baby eel!

Leptocephalus was known to be present in many areas of the North Atlantic Ocean. With this clue to go on, a Danish oceanographer, Johannes Schmidt, collected specimens from various parts of this great body of water. When the data as to their size, the time of year, and locality were plotted on charts, the eel story finally became clear.

Every spring, in April, eel eggs and tiny larvae may be found in a circumscribed area of the Atlantic, a few hundred miles in extent, southeast of Bermuda. It is to the depths beneath this spot that the adult eels travel to breed. The adults have never been found there; presumably they perish once their reproductive duty is accomplished. The young, however, do not remain on the spot or drift with the currents; they set a course for Europe. By the following spring many of them have reached mid-Atlantic and have grown to two inches or so in length. Another year finds the surviving larvae fully grown to three inches and arriving off the margins of the European continent. During the following summer and winter metamorphosis occurs, and by the following spring—three years after their birth—the elvers migrate upward into the inland waters.

The American eel has a similar but simpler history. Its breeding-ground lies in the same general region but more to the southwest; and since the distances are shorter, the larvae take but a year to reach the coast.

We have, thus, the major facts of the eel story. But why this curious life-history? And how do adults and young find their way in these long journeys through trackless seas?

Fisheries.—Apart from the flesh and milk of the common hoofed mammals, man's greatest utilization of animals in support of his economy is in the food supply furnished by marine fishes. It is estimated that currently 30 million tons of fishes—primarily teleosts—are gathered annually for food. Some fresh-water fishes are included in this total. For example, the American Great Lakes fisheries were important before the advent of lampreys, and carp are "farmed" commercially in ponds in Europe. The overwhelming majority of the tonnage, however, comes from the sea. The three most important fishing nations are, in order, Japan, Great Britain, and the United States. The amount of fishing depends upon the economic needs of the population and the abundance of fish in nearby waters.

Most of the world's fisheries are conducted in the shallow coastal waters of the continental shelf. The reasons are twofold: First, fish tend to deteriorate rapidly, and although from ancient times fish have been salted, smoked, and pickled to preserve them, and although in recent years effective refrigeration has entered the picture, nearness to market is extremely important. Second—and more important—it is in such shallows that fish life is most abundant. If one looks at a map of the world on which ocean depths are marked, it will be seen that such shallows are very unevenly distributed. Around most of Africa, for example, the depths increase rapidly offshore, so that there is little possibility of extensive fisheries developing there. Directly off the coast of Japan the shallows are restricted in area, so that Japanese fishermen must range widely to gather the sea food vitally important for the large population of that small country. In contrast, both sides of the North Atlantic have extensive fishing banks—on the European side, extending from the western coast of France around the British Isles to the North Sea, and on the American shores, from the great Newfoundland banks south to the New England coast. It is fortunate that these major sources of fish food are close to two of the world's major concentrations of population. And, on the other hand, it seems certain that the availability of this source of food has been important in the development of these flourishing areas. Even today the sacred cod (stuffed) hangs from the wall of the representatives' chamber in the

Massachusetts State House, commemorating its importance in the early survival of the colonies.

The wealth of fish in the continental shallows is due to the nature of the food cycle in the sea. Larger fishes may eat smaller ones; fishes of all sorts may feed upon invertebrate animals; but all the animals, large or small, depend ultimately upon the supply of small plants in the ocean—principally the tiny but immensely abundant diatoms. Such plants alone can create the organic materials upon which the whole marine community subsists. Their process of manufacture (photosynthesis) requires sunlight, and hence it is only in shallow water that new food materials are created. Living matter is composed largely of three chemical elements—hydrogen, oxygen, and carbon. The diatoms can obtain these simply from sea water and the carbon dioxide contained in solution in it. But other elements are needed as well for the building-up of organic materials—notably nitrogen and tiny but vital amounts of phosphorus. Nitrates and phosphates released by the decay of dead animals and plants in the shallows can be reused to supply these missing elements, and in some areas, particularly in northern latitudes, rising currents of cold water may bring up to the shallow banks quantities of such matter which had sunk to deep waters and would otherwise have been lost (a situation which makes northern fishing banks more productive than comparable areas in the tropics). And an additional factor which makes near-shore surface waters more productive of life in general—and hence of fishes—than the high seas is that here the basic supply of food-building substances is constantly renewed by organic materials brought down by rivers.

Fishing had its origins well back in human history; for example, the oldest population of northern and western Europe after the end of the Ice Age seems to have been in great measure a race of shore dwellers who ate fish as well as mollusks and other invertebrates gathered from the tidal flats and shallows. Today, however, fishing from the shore or from small boats, except in the case of the Pacific salmon industry, is an unimportant factor in the fisheries picture. Most operations are conducted from larger vessels—formerly sailing ships, today powered by steam or motor—which may make voyages of considerable duration before returning to port. Surface types, such as the herring-like fishes and mackerel, are generally caught by the use of long nets in which these fishes are enmeshed. Most important is trawling—dragging bag-shaped nets over the shallow bottom to trap such abundant and valuable bottom dwellers as the flatfishes of the flounder-sole-halibut group. Cod and haddock, which swim near the bottom, may also be caught by trawling, but a more common method uses long lines carrying innumerable baited hooks.

Relatively few types of teleosts make up the bulk of the catch. Of fishes swimming close to the surface, the primitive forms represented by the herrings and sardines, smelts, and so forth are small in individual size but bulk large in the total volume. Of the salmonids, the Pacific salmon is notable, and the related whitefish is economically important. A very large fraction of the trawler's catch consists of flatfishes. The cod and its relatives, such as the haddock, are particularly abundant on the Newfoundland banks. The great majority of genera and species of teleosts are of the advanced, spiny-finned type; although many are eaten regionally, the mackerel is the only member of the group notable in the total world catch.

CHAPTER 4

Toward Land Life:
The Amphibians

Our story of vertebrates has so far been that of primitive water dwellers—fish and fishlike forms. This evolutionary history has not been unmarked by progressive features, for we have witnessed the appearance of jaws and paired fins, structures whose development enabled our earliest mud-grubbing ancestors to leave their sluggish life on the bottoms and become active, aggressive forms. But the greatest adventure of all still lay ahead for the backboned animals—the conquest of the land, a feat which led to the development of higher groups of four-footed terrestrial vertebrates.

The change from water to land life was first initiated in the Devonian by the early amphibians and completed by the reptiles in the late Paleozoic; later, from reptilian types, were developed the warm-blooded birds and mammals. Once the water was left behind, there took place wide and repeated radiations of four-footed types. These land dwellers have adapted themselves to almost every conceivable mode of life on the surface of the earth and have, aided by the advances made during land life, taken to the air and invaded the seas. Snakes, birds, men, and whales are but a few examples of the widely varied types which have evolved from the primitive and ancient forms which in Devonian and Carboniferous times first left the streams and pools to walk upon the land.

Progressive fish development.—It was not all at once, or in a single evolutionary stage, that there occurred all the many changes necessary for a fish's descendants to become good land dwellers. Some of the essentials for land life had already developed in the bony fish and, particularly, the lobe-finned forms from which the tetrapods sprang. Lungs, a prime essential for land animals, we have seen already present in the primitive bony fish. These forms too (in contrast with sharks, for example) had a well-developed bony skeleton without which life on land would not have been pos-

sible; the problems of bodily support are much intensified on land, and a
backbone or limbs of cartilage would not stand the strain of terrestrial life
without bending or breaking. Further, in the lobe-finned fish, we have seen
developed a stout fleshy and bony lobe in the paired fins, which gave the
possibilities of development into land limbs. Potential land adaptations
were thus already being initiated in the fishes; and, on the other hand, land
adaptations, as we shall see, are only partial in the amphibians, first of
four-footed animals.

Life-history of amphibians.—The amphibians are the most primitive
and earliest known of four-footed animals; they are, as a group, the basal
stock from which the remaining land vertebrates have been derived. Living amphibians include but three comparatively unimportant orders: the

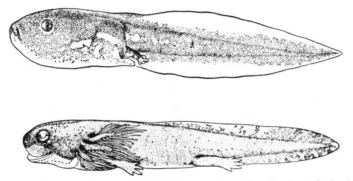

Two amphibian larval types. *Above*, a tadpole, *below*, a salamander. In the tadpole the gills are
covered by a flap of skin; in the salamander the external gills are projecting feathery structures.
(After Noble.)

frogs and toads, the salamanders and newts, and some rare wormlike
forms. All are highly specialized and have departed far from the first land
forms. The life-history of the more familiar frogs or toads, however, shows
many of the essential characteristics of the class. The eggs are rather small
and, except for a jelly-like covering, are without the protective membranes
or shell found in reptiles or higher types. Normally the frog lays its eggs in
the water, just as has been the case throughout fish history, whereas the
reptilian egg is laid on land. The embryo hatches out as a tadpole while
still very small (for there is little yolk in the egg) and must find its own
food as it grows in the water. There is a well-developed tail for swimming;
breathing is done by a peculiar type of gills concealed beneath a hood of
skin over the future neck region. The tadpole, unlike the adult, eats water
vegetation, and both jaws and intestines, in relation to this, are very dif-

ferent from those of the adult. Later, with approaching maturity, metamorphosis occurs, and the structure of the body changes radically: the gills
disappear; lungs and limbs develop rapidly; and the pollywog becomes a
frog; the animal becomes an air-breather and a potential land dweller instead of a water type. In salamanders the story is essentially similar. The
egg usually hatches into a larva which is not as specialized as the frog tadpole, for the body shape resembles that of the adult, and limbs bud out at
an early stage. The animal, however, is a water-breather. Gill slits develop,
but the actual breathing is done by means of feathery external gills like
those of larval lungfishes. Further, the larva differs from the adult in other
features, such as a thinner skin and absence of eyelids.

But even after metamorphosis the amphibian is not entirely freed from
an aquatic environment, from the need of living the "double life" to
which the name of the group refers, for at the breeding season the frog
must return to the streams or ponds. A complete adaptation to land life, a
decisive break with water-living, is impossible as long as this old-fashioned
type of reproduction is retained.

Various devices have been developed by amphibians, which tend to
avoid, to some extent, this necessity for a double mode of life and a double
set of adaptations. In many instances the eggs are laid in moist burrows
rather than in the water. Some tropical tree frogs never descend to the
streams but lay their eggs in rain-filled hollows in the trees; some toads
carry their eggs about in pockets on their back or twined in strings about
their legs. In many cases the larval stage is shortened by an initial development away from the water, the larva only entering the pond or stream
when part of its growth has been attained. This is made possible by increasing the amount of yolk in the egg; and with a still larger amount of
yolk present in a large egg, we find that in many instances the tadpole
stage is entirely eliminated and the young hatch from the egg as full-
fledged, if tiny, frogs or salamanders. There are, in some of these cases of
direct development, outgrowths which at least partly cover and protect
the developing young and air-breathing organs to supply oxygen to the
embryo before the lungs develop. In fact, except for a hard shell, we find in
one amphibian or another structures comparable to those found in the land
type of egg characteristic of reptiles, which we will discuss in a later chapter.

What selective processes have led so many modern amphibians to develop, to a variable degree, embryological processes similar to those "in-
vented" by the ancestral reptiles many millions of years ago? Not, rather

surely, any "urge" toward a strictly terrestrial existence, for the adults are generally quite happy in an amphibious mode of life. The reasons appear to be two. One—probably of minor importance for modern amphibians—is that elimination of a tadpole stage eliminates the danger of drought, of the pond drying up, with consequent death of the larvae during a dry spell. More important, it would seem, is the fact that eggs laid in a pool are an attractive source of food for a variety of enemies ranging from insects to other vertebrates—a sort of amphibian caviar. With the eggs kept away from the water and hidden or carried about, the chances of survival are greatly increased.

Amphibian locomotion.—Among living amphibians the salamanders and newts approach closely in body form and general appearance the most ancient land dwellers. The body and tail are elongate; median fish fins have disappeared, but the tail is often much flattened and is still an effective swimming organ. The paired fins of the ancestral fish have been transformed into land limbs. These limbs in salamanders are quite small and feeble as compared with the legs of other higher types but are large compared with fish appendages. Salamanders can move their legs freely; but the body is still thrown into sinuous curves which push it forward on the legs supporting it, much as the fish pushes forward by pressure on the water which surrounds it.

Primitive amphibians.—But while the salamanders may resemble in superficial fashion the ancestral amphibians, such modern types are quite specialized and degenerate; the primitive four-footed animals from which all land forms have descended were quite different in many of their structural features.

Among the oldest and most primitive of known amphibians was a group (the labyrinthodonts) abundantly represented in deposits laid down in the Carboniferous swamps from which most of our coal deposits have come. One of these (*Diplovertebron*) is illustrated. This form appears to have been, at its largest, about two feet in length, but some related types were as large as modern crocodiles, which they may have resembled somewhat in appearance. Much of the life of these ancient amphibians was still spent in the water, and small fishes which abounded in the pools of the coal-measures swamps seem to have been their main source of food. In their long, slim bodies, well-developed tails, and many internal features these early amphibians were still not far from the lobe-finned fishes from which they sprang, but the presence of limbs capable of locomotion on land is an obvious and striking difference. The skulls of these old forms were completely covered by an armor of bone, just as in their fish ancestors, and the older amphibians are very often called stegocephalians, or "roof-headed" amphibians, because of this fact. (The very considerable reduction in skull bones in modern amphibians is one of many features which show that they are degenerate types.) Back of the head in the fishes was a bony covering for the gill region; but the gills have disappeared in every truly adult amphibian, and the gill covering has gone with them. In fishes, a slit in the side of the skull just in front of the gill plates contained the small first gill opening or spiracle. This slit is still present in the skull of primitive land animals and still lodges this gill pouch which, as we shall see, now serves a very different function.

In fishes there was no special mechanism for transmitting sounds to the internal ear, which lay deeply buried within the braincase, but on land the problem of hearing is a very different one. Vibrations in the air are (except for something on the order of an explosion) too feeble to set up vibrations in the animal's body and reach the hearing organ in this fashion. For the reception of air waves, tetrapods, from the early amphibians up, have established an amplifying mechanism. Across the tube of the old spiracle, and primitively in the notch mentioned above, is a membrane, the eardrum, which picks up the sound waves. Between this membrane and an opening in the side of the braincase beneath it, which communicates with the internal part of the ear, stretches in lower land types a small bone called the stapes, or stirrup; this is a modification with a new use of the hyomandibular bone which, in fishes, helps prop up the jaw joint.

The most striking contrast between the early amphibians and their fish ancestors is seen in the limbs. These were quite small in most of the early amphibians but already showed the pattern of the bony structures seen in land forms in general (p. 92). In the front limb of tetrapods there is always one large proximal element termed the humerus running from shoulder to elbow. A second long segment of the limb contains two elements—the radius lying on the inner (thumb) side and the ulna on the outer side, capped at the top by the "funny bone." Beyond these two bones is a series of small bony elements making up the wrist, or carpus. Beyond this, again, is a series of long bones lying in the palm and each terminating in the free joints of a toe. In the hind leg there is a similar development, but the names are different. The first segment is the thigh bone or femur. The second segment contains two bones, the tibia, or shin bone, and the smaller fibula (needle bone) on the outside. Beyond these are the small elements making up the ankle or tarsus, followed by the bones of the sole of the foot and toes much as in the hand. In a majority of unspecialized land forms, just as in man, there are five digits in both front and hind limbs. Modern amphibians never have more than four front toes, and the same is true of many fossil members of the class. However, some primitive fossil forms had five, and there is even some reason to believe that there may have been as many as seven in some primitive land types. There is no a priori reason why five, particularly, should have been selected as the proper number; very likely there may have been considerable variation among early types before higher tetrapods "settled down" to the orthodox five toes.

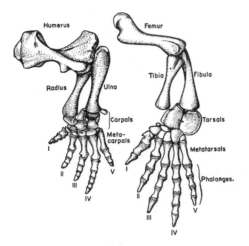

The front and hind legs of a primitive reptile, to show the basic pattern of the legs of land vertebrates. Except for differences in proportions and in the loss of a few bones in hand and foot, even such a mammal as ourselves has preserved this ancient series of structures little changed. In both front and hind legs there is a single bone in the upper joint, two bones in the forearm or shin, a series of wrist or ankle bones (carpus, tarsus), bones in the palm or sole (metacarpals, metatarsals), and bones of the fingers or toes (phalanges).

Problems of air-breathing. — We have little knowledge of changes which must have been going on in the softer parts of the body of these ancient amphibians, but we can draw many inferences from existing types as to the evolutionary processes under way and the new problems encountered in the development of land life. Chief among the difficulties in the transition between water and land life were those which had to do with oxygen supply. The fish lung is generally developed to a greater extent in an air-breather but never becomes a very efficient organ in modern amphibians. Much of the breathing is done through the moist skin, and some small salamanders rely entirely on this latter means of gaining oxygen and have lost their lungs. Reptiles and higher types use the ribs and chest muscles to fill their lungs. Amphibians have never acquired this method of breathing and use their throat muscles to swallow air just as their fish ancestors used those same muscles for pumping water to be utilized in the gills.

A further difficulty in lung-breathing is the problem of efficiently circulating the oxygen absorbed in the lungs. In fishes the heart receives only the impure blood from the veins and transports it all to the gills; from the gills only aerated blood is passed back through the arteries to the body organs. But from the lungs, now introduced into the circuit, the pure blood mixes in the heart with the impure venous blood, and this mixed fluid flows alike to lungs and body. The separation of the two blood streams has not been completely brought about in amphibians, and their inefficient circulatory system has probably been one of the causes of the failure of amphibians to become a successful group.

We have earlier discussed the question of water regulation in the blood and tissues of a vertebrate. A fresh-water fish needs to pump out water, in a dilute urine, to keep up a proper internal salt concentration. When a vertebrate becomes a truly land animal, conditions are reversed; in the air, water tends to be lost by evaporation through the skin rather than absorbed, and reptiles, birds, and mammals have evolved kidney modifications to conserve water and prevent their internal liquids from becoming too briny. This problem does not arise, however, in most amphibians, since typically they spend much of their lives in the water or in moist areas near by, where evaporation is not too great.

Amphibian ancestry.—Many features in the anatomy of the early amphibians point definitely to the lobe-finned fish as the ancestors of all land forms. In the skull pattern and even in such details as the minute structure of the teeth, there is very close agreement between the two. The lungfish of today are quite similar to the amphibians in their mode of development and in many of their internal organs. But this merely means that the lungfish are related to the crossopterygians; were primitive lobe-finned fish still in existence, we would probably find them to be even more similar to the living amphibians in their soft parts, and the surviving *Latimeria*, despite the specializations due to its living in the deep sea, shows a number of comparable features. The one really conspicuous difference between the two types lies in the limbs; the fins of these fish were much smaller than the legs of the amphibians. But in some cases, at least, we have noted that the fins were essentially similar in structure to the land type of limb. A primitive amphibian was, in essence, only a lobe-finned fish in which limbs capable of progression on land had been developed.

Why land life?—The most primitive of known amphibians were, as we have said, inhabitants of fresh-water pools and streams in Carboniferous and Devonian times. Alongside them lived representatives of the ancestral crossopterygians, forms similar to them in food habits and in many structural features and differing mainly in the lesser developments of the paired limbs. Why should the amphibians have developed these limbs and become potential land dwellers? Not to breathe air, for that could be done by merely coming to the surface of the pool. Not because they were driven out in search of food, for they were fish-eating types for which there was little food to be had on land. Not to escape enemies, for they were among the largest animals of the streams and pools of that day.

The development of limbs and the consequent ability to live on land seem, paradoxically, to have been adaptations for remaining in the water, and true land life seems to have been, so to speak, only the result of a happy accident.

Let us consider the situation of these two types—lobe-finned fishes and amphibians living in the streams and pools of the late Paleozoic. As long as the water supply was adequate, the crossopterygian was probably the better off of the two, for he was obviously the better swimmer; legs were in the way. The Devonian, the period in which the amphibians originated, was a time of seasonal droughts. At times the streams would cease to flow, and the water in the remaining pools into which the fish and ancestral amphibians were crowded must have been foul and stagnant. Even so, the lobe-finned fish, since he possessed lungs, was at no disadvantage, for a short time at any rate, for he could come to the surface and breathe air as well as the amphibians.

If, however, crowded conditions continued for a period of time, the local food supplies would be exhausted, and the situation would be a desperate one. Still worse, the water might dry up completely. Under such circumstances the crossopterygian would be helpless and must die. But the amphibian, with his newly-developed land limbs, could crawl out of the shrunken or dried-up pool, walk up or down the stream bed or overland, and reach another pool where he might take up his aquatic existence again. Land limbs were developed to reach the water, not to leave it.

Once this development of limbs had taken place, however, it is not hard to imagine how true land life eventually resulted. Instead of immediately taking to the water again, the amphibian might have learned to linger about the drying pools and devour stranded fish. Insects were present by coal-swamp days and would have afforded the beginnings of a diet for a land form. Later, plants were taken up as a source of food supply, while (as is usually the case) the larger forms on land probably took to eating their smaller or more harmless relatives. Finally, through these various developments, a land fauna would have been established.

Older amphibian types.—Until recently the oldest-known remains of amphibians dated from the Carboniferous, the time of formation of the great coal deposits. But already at that period amphibians were quite diversified, and we felt sure that their origin must have taken place further back in the Devonian, when the development of early fish groups was at its height, although no amphibians had ever been reported from rocks of that age. In recent years, however, this gap in our knowledge of the ancestry of higher vertebrates has been filled in by the discovery of very primi-

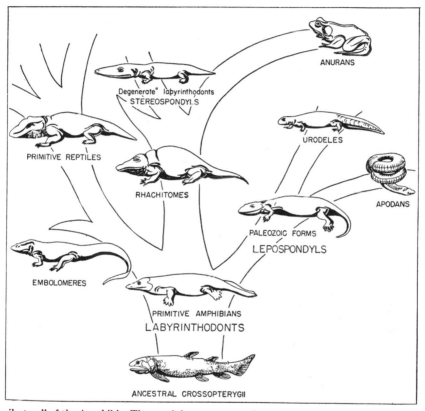

A "family tree" of the Amphibia. The surviving groups are shown at the upper right. Urodeles, limbless amphibians (and Apodans) and their fossil relatives are frequently grouped as lepospondyls, as contrasted with the labyrinthodonts, from which the frogs appear to have descended, as have the reptiles. (From Romer, *The Vertebrate Body*, W. B. Saunders Company.)

tive amphibians from the late Devonian of Greenland. Their skulls resemble those of lobe-finned fishes even more than do those of the Coal Measures labyrinthodonts, and there is even a somewhat fishlike tail fin. The limbs are very short and stubby but with the elements found in later terrestrial forms already developed. Here, then, are forms which continued to live in fishlike fashion, although they already possessed limbs, the advantages of which eventually led to the evolution of land forms.

In the late Paleozoic—the Coal Measures and on into the Permian period—amphibians formed an exceedingly prominent part of the vertebrate life-picture. Many were members of a great group of amphibians which in a broad sense constitute the basic stock of most, perhaps all, land vertebrates. The term "labyrinthodonts" refers to the minor point that their teeth are grooved and infolded like those of their crossopterygian ancestors; more basic are technical features of the skeleton, such as the fact that each joint of the backbone consisted of several discrete blocks of bone. Most of the old labyrinthodonts continued to spend much of their lives in

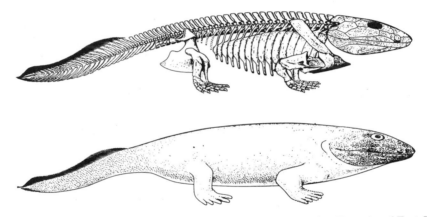

A restoration of one of the most ancient of known amphibians, from the late Devonian of East Greenland. Limbs, although very short, are already developed, but there is a remnant of the old fish tail fin. (After Jarvik, modified.)

the water; but a few developing stout limbs, tended to become land dwellers. Many grew to large size. Fairly typical and moderately advanced was *Eryops*, whose skeleton is to be seen in a number of museums; this is an animal of six to eight feet in length, with sturdy if clumsy-looking limbs, which may have spent much of his life in swampy environments but was perfectly able to stroll about the shore. Alongside the labyrinthodonts there existed numerous members of a second, less important group of ancient amphibians, termed the lepospondyls ("scale vertebra"), so called because the main part of each vertebra consisted of a thin sheath of bone surrounding the notochord. The lepospondyls were less progressive in nature and appear to have persisted in an almost purely aquatic life. Most were of small size; the limbs were generally feeble and sometimes lost entirely; and there was in many a strong trend (which we shall see repeated in some modern salamanders) toward body elongation and an eel-like shape.

Before the Carboniferous had closed, the reptiles, a group much better adapted to land life, had already sprung from the primitive amphibian stock, and toward the end of the Paleozoic amphibians rapidly decreased in numbers. A few large degenerate remnants of the old amphibian groups survived into the Triassic but then became extinct. Beyond that time the only known amphibians are members of the three living orders, which constitute but a small and unimportant part of the vertebrate life of modern times. Two of the three are represented by fossils from the latter part of the Mesozoic era onward, but except for one Triassic fossil form which is an archaic frog ancestor, the pedigree of the modern orders is quite unknown.

A restoration of *Eryops*, a large common early Permian labyrinthodont of fairly advanced type

Salamanders.—As we have noted, the living salamanders and their relatives are not dissimilar in general appearance and proportions to the ancient types, but in the skull and skeleton there are many degenerate features. As in modern amphibians generally, the old bony scales of the fish have disappeared; the skin is soft and moist and acts as an accessory breathing organ. There are numerous skin glands; in some salamanders (and in various frogs and toads as well), some of these glands secrete a toxic substance which acts as a deterrent to would-be eaters of these otherwise defenseless animals. In many salamanders and frogs color changes, such as also occur in fishes, may take place in the skin; change is controlled by the pituitary gland or by direct action of sunlight; there is no nervous control. Salamanders are unable to exist under dry conditions and when not in the water are usually found in moist wooded areas or buried under logs and stones. They are mainly dwellers in the more humid regions of the North Temperate Zones, where they are widespread, and are not adapted to life in the tropics. None are present in Africa south of the

Mediterranean coast; none are found in Australia; almost none in South America. Adult salamanders and frogs are eaters of animal food, but because of their typically modest size, their diets are largely restricted to insects and other small invertebrates.

The salamanders, or urodeles, are customarily divided into half a dozen or more families, most of which show the common pattern in structure and habits. The more familiar small salamanders and newts are members of a family common to Europe and North America. In North America a second family of salamanders of somewhat larger size includes the spotted and tiger salamanders which are more terrestrial in habit than most urodeles; typically their coloration is a combination of a black background with orange to white stripes or spots. (There is a related group in eastern Asia.) The most abundant family of salamanders, for which no general popular name exists, is that of the Plethodontidae, mainly North American forms but also found in Europe and (alone of urodeles) in South America. These are mostly small slender forms, seldom seen by the ordinary person. An interesting specialization is that lungs are not developed in members of the family, although gills are absent and they are air-breathers. Due in part to their small size, they are able to take in sufficient oxygen through their moist and highly vascular skin.

The forms already mentioned are all typical salamanders, of small size and, to anyone but the specialist, rather uniform in characters and habits. We may conclude our catalogue of urodeles, however, with mention of a number of types which depart from the normal pattern—and also tend to be purely aquatic in habits. First, salamanders which run to large size. The "mud-puppy" of the American Midwest—essentially the Ohio River drainage—runs up to two feet in adult length and among other characters retains external gills and water-breathing throughout life. Some salamanders have taken up life in underground streams or caves; one such, *Proteus*, a relative of the mud-puppy, is blind and without pigment in its skin. Larger is the "hell-bender" of the Allegheny River, a drab-looking form with a greatly flattened head and body. The giant of the order, however, is a close relative of the hell-bender which is found in Japan. This giant salamander (relatively a giant, anyway) may exceed five feet in length. Incidentally, a fossil form of this type from the upper Rhine region was pictured in one of the early editions of the Bible as the remains of a poor sinner drowned in the flood!

A second series of divergent types (also purely aquatic) includes several salamanders having in common a trend to an elongate, eel-like body ac-

Primitive four-footed animals. Restoration of early amphibians (labyrinthodonts, genus *Diplovertebron*) of Carboniferous days. These ancient forms spent much of their lives in the water but nevertheless possessed the ability to walk on land. (Painting by F. L. Jaques under the direction of W. K. Gregory; photograph courtesy American Museum of Natural History.)

Walking salamander—a newt of the genus *Pseudotriton* photographed from above. Much of the progression of the body is accomplished, as in the ancestral fish, by throwing the body into sinuous curves. (Photograph by Sherman C. Bishop.)

Skeleton of a fossil amphibian, *Eryops*, one of the largest of primitive Paleozoic amphibians, seen from above. This figure illustrates well the sprawled pose of the limbs in early land types, a pose which is retained in the salamander shown above. (Photograph courtesy American Museum of Natural History.)

Tiger salamanders. *Left,* the common American orange and black variety (*Amblystoma tigrinum*). In the next figure, immersed in water, is the axolotl, a salamander from Mexico and the Rockies, a water dweller, breathing by means of gills. The axolotl is a variety of tiger salamander which tends to remain throughout life in a larval condition. However, various stimuli (including thyroid) may cause it to change into the true adult form. (*A. tigrinum* copyright General Biological Supply House, Chicago; axolotl courtesy New York Zoölogical Society.)

The mud puppy (*third figure*) is a large salamander, well over a foot in length, found in many parts of the Mississippi Valley. Like the axolotl, the mud puppy is a permanent larva, for it never leaves the water. In this case, however, no stimulus has brought about the adult condition; it has, so to speak, forgotten how to grow up. (*Necturus,* courtesy New York Zoölogical Society.)

A cave salamander. In several instances salamanders have become adapted to underground life. Directly below is *Lyphlotriton*, photographed in an Ozark cave. It is blind and nearly colorless. (Photograph by G. K. Noble, courtesy American Museum of Natural History, New York.)

The "Congo eel," which is not an eel or a native of the Congo but a salamander (*Amphiuma*) inhabiting swamps and sluggish streams of the southern states. It is a degenerate water dweller in which the limbs are reduced to tiny vestiges. (Photograph courtesy American Museum of Natural History, New York.)

More Salamanders. *Above*, two specimens of the common newt of North America (*Triturus*). In the adult "water phase" shown here, most of the body is olive green; the "red eft" is an earlier, land-dwelling phase in which the body is red.

"Degenerate" types which tend to remain in the water and retain larval characters. *Siren* (*left*) of the southern United States has reduced limbs. The flat-bodied hellbender (*Cryptobranchus*) (*below*) of the Allegheny River reaches nearly 2 feet in length and is the largest of living amphibians except for a Japanese relative. (*Triturus* photograph by Harold V. Green, Montreal; *Siren* and *Cryptobranchus* by Isabelle Hunt Conant.)

Metamorphosis of a common frog (*Rana clamitans*) from a typical tadpole at the lower left to a mature frog at the upper right. Development of hind legs, appearance of front legs, and resorption of tail occur in order. (Photograph copyright General Biological Supply House, Chicago.)

The spring peeper (tree toad, genus *Hyla*). Sound production is aided by the inflated vocal sac. (Photograph by Dr. Frank Overton, courtesy American Museum of Natural History, New York.)

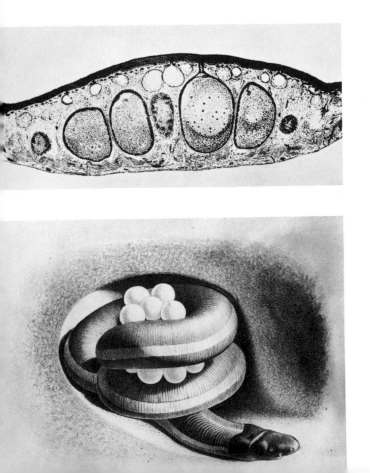

Frog skin; a section, highly magnified. *Above*, the thin outer skin (epidermis) unprotected by scales as in fish, a feature regained by reptiles; *below*, the thicker dermis. In the latter are numerous globular glands. The smaller are mucous glands; the larger secrete a mildly poisonous material. (Photomicrograph copyright General Biological Supply House, Chicago.)

A burrowing, limbless amphibian belonging to the order Apoda and only distantly related to the salamanders and frogs. These burrowers are small and wormlike, with rudimentary eyes but with sensory tentacles between eye and nostril. The Apoda are tropical; the form illustrated (*Ichthyophis*) is from the East Indies. The eggs are kept in a moist burrow and protected by the female, a definite advance toward the reptiles and complete land living. (From the Sarasins.)

companied by a reduction of limbs. They are inhabitants of the south-eastern United States. The "siren" (with looks very different from sirens of mythology) has tiny front legs and none behind. Another southern form, a swamp dweller, is the "Congo eel," or "Conger," which is very eel-like in appearance, with the limbs represented by small flaps.

Neoteny.—Mention of these aberrant aquatic salamanders brings up the topics of neoteny and paedogenesis, previously discussed. Some salamanders which live their entire life in the water have seen no necessity, so to speak, of gaining adult characters for a terrestrial existence which they will never reach and retain certain of their larval characters throughout life. This situation may be found in some of the more normal salamanders as well as the "freaks." For example, over most of its range in America the spotted salamander mentioned above grows from the larval stage into a typical adult form. But in certain highland regions of the West, notably near Mexico City, this is not the case; the animal remains all its life a gill-breathing water dweller and breeds in this larval condition. It was once thought that this highland form was a distinct species, called in Mexico the axolotl. But if experimentally treated in various ways, such as being fed an extract of the thyroid gland (whose product plays a major part in normal metamorphosis), the axolotl changes markedly, loses gills, develops lungs, and comes onto land as a mature spotted slamander.

Some of the more purely water-dwelling types mentioned toward the end of our list of salamanders show as breeding adults one or more larval characters and hence are likewise to be regarded as neotenous to some degree. The Japanese giant salamander and the hell-bender, for example, have nearly all adult characters, but both lack eyelids, as do larval urodeles, and the hell-bender also retains an open gill slit. The Conger also shows general larval features throughout its life. The siren and mud-puppy remain still more larval in character and, for example, bear prominent external gills throughout life. In the axolotl, we have noted, stimulation will cause metamorphosis to occur. Not so, however, in the case of the neotenous forms just listed. For example, scientific workers have fed mud-puppies quantities of thyroid extract without causing any loss of the gills or other signs of metamorphosis. They have lost from their developmental processes the power of ever reattaining the adult characters which they lack.

Frogs and toads.—A much more flourishing group of living amphibians is that of the tail-less forms, the frogs and toads forming the order Anura. Since they are amphibians, and since the amphibians are the lowest of terrestrial groups, it is often assumed that frogs and toads are, *ipso facto*, primitive land vertebrates. But this is very far from being the case; the frogs (described in detail in the next chapter) are in many respects among the most highly specialized of backboned animals. The specializations have to do mainly with their hopping method of locomotion. The back is extremely short; there may be as few as eight vertebrae, whereas a primitive land form would probably have had nearly thirty joints in the backbone, not including the tail. The hind legs are excessively long and highly specialized, and the front legs, although more normally proportioned, are also much modified with adaptations of the shoulder bones which break the shock of landing. It is quite probably their peculiar mode of locomotion which has enabled the frogs and toads, alone of amphibians, to remain a fairly flourishing group.

Members of the group adhere quite closely to a common structural pattern, but there are considerable variations in anatomical details and habits, and they are frequently divided into a dozen or so families, only a few of which will be noted here. The anurans are far more widespread than the salamanders, for they are abundant in the tropics as well as being common inhabitants of more temperate regions. So familiar as to hardly need description are the typical frogs, many of which are included in the single genus *Rana*, which extends over nearly the whole world. Contrasting considerably in habits to these water-lovers are the toads, where the common genus *Bufo* is likewise widespread, but fails to reach Australia. These forms wander far from the water and, in correlation, have a more horny skin which keeps down water loss. A third familiar group is that of the tree frogs, with *Hyla* as the characteristic genus. These are mainly American types but *Hyla* itself is nearly world wide. The tree frogs are typically small forms, rather more related to the toads than to the true frogs; as aids in their arboreal mode of life, their toes generally have expanded pads and (an unusual condition in amphibians) clawlike toe tips. Incidentally, some members of the true frog family have paralleled the tree frogs in mode of life and adaptations.

There remains for further consideration among amphibians only a group including a few inconspicuous tropical forms, the Apoda, or limbless amphibians. Caecilians is an alternative name. These are small, almost blind, burrowing animals, which look very much like large worms and have no vestiges of limbs. These very degenerate creatures are not at all closely related to the groups considered above, and their fossil history is quite unknown.

The failure of the amphibians.—The amphibians are a defeated group. They were the first of vertebrates to emerge from the waters onto the lands; but they were not destined to complete the conquest, and, at first abundant, they have shrunken into insignificance among four-footed vertebrates. Only by the reptiles, their descendants, was the land truly won. The reason for amphibian failure and reptilian success is not hard to find: it lies in the mode of development. The typical amphibian is still chained to the water. In the water it is born; to the water it must periodically return. We have noted various devices among living amphibians which have enabled them to circumvent this difficulty to some extent. But these makeshifts have not been particularly successful. The amphibian is conservative in its basic developmental processes. It is, in many respects, little more than a peculiar type of fish which is capable of walking on land.

CHAPTER 5

The Frog

We here turn aside from the general story of vertebrate evolution to consider in greater detail a common amphibian—the frog. In many elementary courses in biology this animal is dissected as a representative vertebrate. This account is inserted to accompany such laboratory work; in consequence, it is more replete with technical terms than are other sections of the book, and those interested only in more general viewpoints will be well advised to pass this chapter by.

The choice of the frog as a favorite laboratory animal is due to several factors. One item (not a minor one) is the fact that it is common, readily available, and hence inexpensive. There are, however, good scientific grounds for its selection. As an amphibian it represents a group halfway up the family tree from fish to mammal—an "average" vertebrate, better suited for use than either extreme type if but one form is to be studied.

It must be pointed out, of course, that even this animal has its disadvantages. We have noted earlier that modern amphibians are somewhat degenerate in their skeletal system, and the frog is no exception to this condition. Further, the jumping habits of the frogs have caused great modifications of the limbs, so that these appendages are to be considered as highly specialized structures. However, apart from features of the skeleton and muscles and a general "shortening up" of various organs to fit the compact body shape, the frog appears to have adhered rather closely to the general pattern found in primitive land-dwelling vertebrates.

THE LIFE OF THE FROG

The forms most frequently used in the laboratory are members of the typical frog genus *Rana*. The bullfrog, *Rana catesbiana*, is an exceptionally large form; *R. pipiens*, the leopard or grass frog, is the commonest of American species. The bullfrog tends to spend much of its life immersed. The leopard frog is more of a land dweller; in damp weather it may wander far from the water but is commonly found along stream banks, ready to leap in and submerge at an instant's notice.

Food.—Adult frogs are purely carnivorous in a broad sense of the term —eaters, that is, of animal food. Since they are of modest size, the food supply mainly consists of invertebrates. Earthworms, insects, insect larvae, and spiders are favorite foods. However, they have no compunctions against eating other vertebrates, and minnows, tadpoles, and even smaller frogs may form additions to the diet. The frog tends to snap instinctively at any small moving object near by (such as a bit of red flannel!). The teeth, however, are relatively feeble; an important method of obtaining food lies in the protrusibility of the tongue. This organ is highly developed, attached anteriorly, and normally lies with its free tip turned backward in the mouth. From this position it may be flipped out suddenly to gain contact with a fly or other desired titbit. The tip is sticky, so that the object touched adheres to it. A reverse flip, the food is in the mouth, and is swallowed without further ado. The protrusion of the tongue is brought about by a peculiar mechanism. A large sac in the floor of the mouth suddenly fills with lymph and pushes the tongue upward and forward.

Enemies.—The frog has, in turn, numerous enemies. Man, either in pursuit of frogs' legs or in search of your laboratory animal, is a major factor in frog destruction. Frogs offer a substantial and favored food supply for snakes; various birds and some mammals and even fishes prey upon them. The tadpoles are, of course, much more susceptible to attack, and their enemies include not only vertebrates but a number of aquatic insects; under ordinary conditions only a small percentage of the larval frogs ever reach the adult stage. Still other enemies are parasites of various sorts. Leeches suck their blood, and there are numerous internal parasites such as flukes, roundworms, a number of protozoan species, and even parasitic plants which invade their tissues.

Locomotion.—The locomotor abilities of the frog are highly useful in escaping from major enemies, for the frog is an accomplished jumper and swimmer. The front legs are of use mainly to support the head and chest and break the force of the fall on landing; the hop is accomplished by a sudden straightening (extension) of the hind legs, which in resting pose are flexed in readiness. A frog jump under good conditions is well short of a yard (Calaveras County jumping contests not considered). In swimming, the front legs are little used; propulsion is accomplished by alternate kicks of the hind legs, which push the webbed toes against the water. A favored resting pose is one in which the frog floats in the water in a sprawled position with only nose and eyes protruding; the level at which it floats may be regulated by filling or emptying the lungs and thus altering the specific gravity of the animal as a whole. Diving from the bank begins as an ordinary leap, followed by a vigorous downward swim. To dive from the floating position, the frog first gives itself a vigorous push back and down to "submerge," then tilts the body downward to begin its swim to cover.

Besides flight, other factors help it escape from enemies. Some of the skin glands secrete a mildly poisonous material which makes some animals avoid the frog. By puffing itself full of air, a frog may become a round and slippery object, difficult for its enemy to swallow.

Annual life-cycle.—The activities of frogs vary greatly with the seasons, particularly in temperate climates. The animal is a cold-blooded form, that is, it is unable to maintain a constant body temperature; hence in the winter it must hibernate, and so it burrows in the mud to avoid freezing temperatures. During this season the internal activities of the body go on at a much reduced rate, drawing for fuel upon food materials stored in the body, particularly in the liver and muscles. With the coming of spring the frog becomes very active. The sex organs develop greatly, the body of the female fills with eggs, and the breeding season arrives. Once the eggs are laid, life goes on at a slower tempo. During summer and fall new stores of food are laid up against the approach of cold weather.

Breeding.—In the grass frog the breeding season commonly occurs some time during April, depending upon the climate of the region concerned and the nature of the season—the warmer the sooner. Frogs are essentially unsociable at other seasons and solitary in their habits. At the breeding season, however, they become highly gregarious and congregate in large numbers in shallow bodies of water. Fertilization is external, the male clasping the female and discharging sperm which fertilize the eggs as they emerge. The clasping movement is a readily excitable reflex action of the male; typically, male and female remain clasped for several days before the eggs are laid.

GENERAL FEATURES OF THE FROG BODY

Body regions.—In its contours the frog body shows a division into three regions—head, neck, and trunk—found in all higher vertebrates. In the fish there was no distinctive neck. In the frog, with the disappearance of the gills, which once lay in this area, a neck appears. It is, however, shorter than in reptiles or higher groups, and there is in consequence relatively little freedom of motion of the head on the body. The trunk is much shorter than in typical vertebrates. A marked specialization of the frog is the reduction of the tail; only a rudimentary stump remains, concealed in the general contours of the body. The hind legs are developed to an unusual degree, equaled elsewhere only by man and some other bipeds. These various specializations of the frog are to be associated with its peculiar leaping habits.

Orientation.—In dissecting or describing a vertebrate, attention must be paid to terms used to describe the relative position of various parts and organs. The frog is, of course, bilaterally symmetrical, the right and left sides fundamentally mirror images of each other. The direction in which the animal moves (here the head end) is termed "anterior," the opposite end "posterior." Upper and lower surfaces are called "dorsal" and "ventral" (Latin for "back" and "belly"). An annoying feature in the attempt to compare any four-footed animal with man lies in the fact that man's upright posture causes changes in directional terms. Since anterior means the direction in which an animal moves, this term in man becomes the same as ventral, and posterior the same as dorsal. We thus need new scientific terms for the up-and-down directions in man; these are supplied by using "superior" and "inferior." The confusion that can be (and is) caused by this shift in posture can be readily imagined.

It must be remembered that most diagrams of body organs are drawn from the ventral side, as they are seen in the dissection of the abdomen. The right side of the animal is thus to the reader's left, and vice versa.

Superficial features.—Most of the frog body is covered with a soft, moist skin, interrupted by a number of openings or other topographic markers. At the front is the widely gaping mouth; just above its anterior end are the external openings of the nostrils (or nares). Farther back on the head are prominent eyes and the large eardrums. At the posterior end of the body is found the opening of the cloaca, a pocket into which open not only the digestive tract but also the tubes carrying urinary and reproductive products.

Body cavities.—When the dissection of the internal organs begins, it will be found that most of them lie packed in compact fashion in a large cavity—the body cavity, or coelom. Here are found most of the digestive organs—stomach, intestines, liver, and pancreas—as well as the reproductive organs and spleen, while the kidneys are exposed on the back wall. In mammals the lungs lie in separate compartments, separated from the abdomen by a muscular partition, the diaphragm; but in the frog the lungs extend freely into the general cavity. As in every vertebrate, however, the heart occupies a separate pericardial cavity.

The coelom is lined by a thin but continuous membrane known as the peritoneum. This covers not merely the outer walls but also all the contained organs. These do not "float" freely in the body cavity but are attached to the walls by folds of the membrane, called mesenteries. The most important mesentery descends from the midline of the back wall of the abdominal cavity to anchor the stomach and intestine in place.

SKIN

In the frog, as in every vertebrate, the skin consists of two layers. The more superficial is the epidermis, consisting of a sheet of cells which forms a moist and thin but continuous outer covering. Deeper lies the dermis, mainly a feltlike mass of connective-tissue fibers but containing blood vessels, nerves, and sensory structures. In its skin the typical modern amphibian differs markedly from both its fish ancestors and the reptiles. The fish was covered by thick, bony scales placed in the dermis; the frog has lost all trace of these structures. On the other hand, typical reptiles have acquired superficial scales, or scutes, formed by deposits of horny material in the epidermis. Such scales are lacking in amphibians, although slight deposits of horn may form in the skin in some cases. The frog is thus left without the protection of either type of scales, and its soft skin makes it necessary for the animal to stay in damp environments to prevent excessive drying. Nevertheless, this type of skin has one advantage. It is a moist membrane which is capable of absorbing oxygen and giving off carbon dioxide; richly supplied with blood vessels, it acts as an accessory lung.

The skin contains numerous glands of simple construction, essentially globular pockets of the epidermis. Most of these secrete a mucous material which keeps the skin moist. Relatively few in number are larger glands which produce an acrid fluid thought to be poisonous in nature. While this appears to be relatively ineffective in frogs, certain toads secrete definitely poisonous materials.

Just beneath the epidermis is a layer of cells containing pigments of at least two types—one dark, one yellow—and crystalline granules. Combinations of these elements in different proportions and positions give the green, brown, and other frog colors. Further, the colors in many cases are not fixed but may show changes, owing to expansion or contraction of pigment-bearing cells or to changes in their relative positions.

NERVOUS SYSTEM

In the nature of its nervous tissues in general—the spinal cord, the peripheral nerves, and the autonomic nervous system—the frog and man are essentially similar. In fact, apart from the brain, the only noteworthy difference between the two is that in the frog the body is so shortened that there are but ten pairs of spinal nerves in contrast to three times that number in man.

Even in the brain many features are closely comparable with those seen in man. At the back end of the frog brain is the medulla oblongata, which appears in general structure to be little more than an expanded portion of the spinal cord with which it is continuous. In the medulla are carried out many of the more automatic nervous activities of the frog body, and to this region of the brain attach most of the cranial nerves. A projection above the front end of the medulla is the cerebellum, associated with posture and muscular co-ordination. This structure is much smaller in the frog than in most vertebrates.

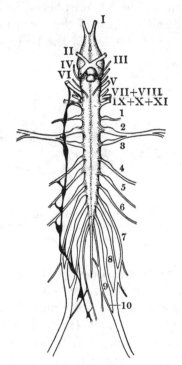

Farther forward, the only notable feature of the midbrain region is the pair of dorsal swellings, the optic lobes to which in the frog (although not in man) run most of the nerve fibers from the eyes. A bit farther forward, in a region anatomically considered to belong to the forebrain, in a general sense, is a pronounced ventral swelling in the neighborhood of the optic nerves —the infundibulum. Just below and behind it is a rounded (readily detached) structure—the hypophysis or pituitary, a vitally important gland of internal secretion.

It is only in its most anterior portion that the frog brain, persistently primitive, differs radically from that of man. Here, in the frog, are found small, paired swellings, each somewhat constricted at midlength. Into the front halves (olfactory lobes) run the nerves from the nostrils; the back portions are the cerebral hemispheres. The frog hemispheres appear to ex-

The nervous system of the frog seen from the ventral side. At the top, the ventral surface of the brain, continuing into the spinal cord. On the left side of the figure the ganglia and trunk of the autonomic ("sympathetic") nervous system are shown in black. The cranial nerves are labeled in Roman numerals, the ten spinal nerves in Arabic. Spinal nerves *2* and *3* form the brachial plexus for the arm, and components of nerves *7–10* form a similar plexus innervating the hind leg. (After Ecker and Wiedersheim.)

ercise some slight control over the animal's activities, particularly in causing response to stimulus received through the sense of smell. However, if the hemispheres are carefully removed, without injury to other parts of the brain, and the animal allowed to recover from shock, it is found that in almost every respect the frog is capable of carrying on its normal life almost exactly as before. Actually, it is the large optic lobes (which are very much reduced in mammals) which form the dominant brain center in the frog and control much of its activities.

Far different, of course, is the situation in man. As we will note in chapter 11, the human brain is notable for the enormous expansion of these same cerebral hemispheres, which have grown so as to exceed greatly in bulk all the rest of the brain together and have come to be the directing centers for much of the body's activities—the seat of consciousness and memory.

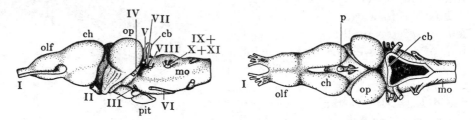

Lateral and dorsal views of the frog brain. The cranial nerves are labeled in Roman numerals in the lateral view; *cb*, cerebellum; *ch*, cerebral hemispheres; *mo*, medulla oblongata; *olf*, olfactory lobes; *op*, optic lobes of midbrain; *p*, pineal; *pit*, pituitary. (After Wiedersheim.)

The frog possesses ten cranial nerves, comparable in major features to the first ten found in man. In man there is a well-developed twelfth nerve, absent in the existing amphibians, but found in reptiles, birds, and other mammals as well. Its absence in the frog has often been thought to be a primitive character. However, there is considerable evidence that it was present in primitive amphibians; here, as in other features, the frog is a bit degenerate.

SENSE ORGANS

Nose.—Since the internal openings of the nostrils lie almost immediately below the external ones, there is no opportunity for the development of large nasal passages in the frog. Nevertheless there are small, well-developed olfactory sacs folded in a complicated fashion, which appear to furnish the frog with important sensory information.

Eye.—The fundamental pattern of the eye is similar in frog and man, but there are numerous differences in details of construction. Lids are present but are poorly developed; they cannot close of themselves, and the eye is shut by pulling the eyeball back into its socket. On closure, a thin membrane attached to the lower lid and called a "nictitating membrane" covers over much of the surface. The lens is nearly spherical and of a "fixed focus" type and cannot change either position or shape to accommodate for near or far vision. The optical properties are such that the frog is nearsighted on land, farsighted with the eye immersed in water. As in all typical vertebrates, the retina contains both rod and cone cells. In man

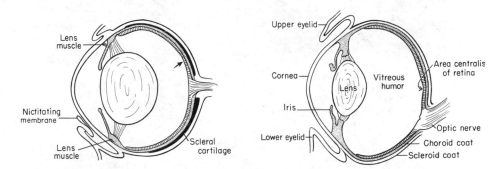

The eye of a frog (*left*) and of a mammal (*right*), cut in section. The part of the eyeball in front of the lens is filled with a watery liquid, termed the "aqueous humor"; the area behind the lens is filled with the jelly-like "vitreous humor." The eyeball has a triple lining; within is the sensory retina; outside of this the choroid coat, rich in blood vessels and pigmented; the external layer, the scleroid coat is a stout fibrous covering, continuous in front with the transparent eyeball surface, the cornea. In frogs the back of the eyeball is stiffened by cartilage. In mammals there is generally a distinct central area in which details of vision are clearest; this is only faintly indicated in a frog (*arrow*). (After Rochon-Duvigneaud.)

the latter, which make for color vision and clear perception of detail, are concentrated in a central area; in the frog the two types are scattered throughout, and hence we may assume that visual acuity is less pronounced. Further, the frog lacks the stereoscopic depth-effects possible in man and many other mammals. This type of vision is rendered possible in mammals by a sorting-out of the nerve fibers from the eyes as they enter the brain, so that the sensory pictures received from the same objects by both eyes are superimposed on one another in specific brain areas to give an impression of depth. This does not occur in the frog.

The frog is, when at rest, popeyed, the eyeballs projecting prominently from their sockets but readily withdrawn. The withdrawal is accomplished by the development of two extra muscles in addition to the six which in all typical vertebrates perform the usual rotary motions of the eyeball.

Primitive vertebrates possessed a third, median eye between the members of the normal pair; this is lost in most living forms, but the structure often remains, as in man, in the form of a pineal body attached to the

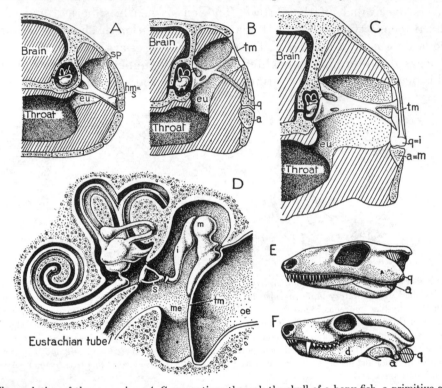

The evolution of the ear region. *A–C* are sections through the skull of a bony fish, a primitive amphibian, and a reptile, at the level of the ear and hind end of the jaw. In fish the hyomandibular bone, propping the jaws, is braced against the braincase, outside the point where the sacs and canals of the internal ear are situated; close beside it is the spiracle, the first gill slit. In primitive land animals the hyomandibular is generally transformed into a sound-conducting ossicle, the stapes (or columella), and the spiracular pouch becomes the middle ear cavity, connected with the throat by the Eustachian tube and closed externally by the eardrum, or tympanic membrane. In reptiles the situation is similar. *D*, the ear region of a mammal in a more detailed view; two former jaw elements have been taken over as auditory ossicles, the malleus and incus. *E*, side view of the skull of an ancient amphibian; the eardrum lay in a notch at the side of the head, in the same position as the fish spiracle. *F*, typical land vertebrates; the eardrum has moved farther backward and in mammals is inclosed within a new growth, the auditory bulla. (*a, q,* bones of the jaw joint of a lower vertebrate [articular, quadrate] which are transformed into the malleus and incus in mammals; *d,* the dentary, principal bone of the lower jaw [and the only bone in mammals]; *eu,* Eustachian tube; *hm,* hyomandibular; *i,* incus; *m,* malleus; *me,* middle ear cavity; *oe,* external ear tube of mammals; *s,* stapes; *sp,* spiracle; *tm,* tympanic membrane.)

brain. In the frog the pineal body has almost completely disappeared externally but may be sometimes identified as a tiny pigmented spot in the skin between the eyes, almost completely separated from the brain.

Ear.—The frog lacks the external tube of the ear seen in mammals, and the prominent eardrum is exposed on the surface. Within the drum is a middle-ear cavity communicating with the mouth (as in man) by a Eustachian tube. A single bone crosses this cavity to carry vibrations from the drum to the inner ear, buried in the braincase. In man three ossicles are present here. That of the frog corresponds to the innermost of the three,

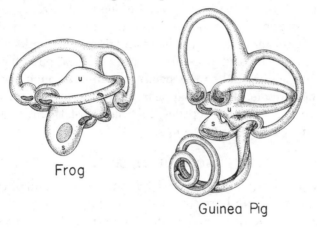

Frog

Guinea Pig

The left internal ear of a frog and that of a mammal. This is a lateral view of the sacs and canals, which are filled with a liquid, the endolymph; surrounding the endolymph system, but within the walls of the ear capsule, is a second fluid, the perilymph. (Sensory areas in the walls of the endolymph vessels are stippled.) The three semicircular canals present in each case register movement of the head in space; the sensory areas in the sacs, the utriculus (u) and sacculus (s), in the main register the static position of the head but also have some association with hearing. In mammals hearing is concentrated in a long, coiled outgrowth, the cochlea.

the stapes, and hence may be so termed (although usually designated as "the columella"). The other two ear ossicles of the mammal, the malleus and the incus, are present in the frog, as in all vertebrates except mammals, as portions of the lower and upper jaws which articulate with each other.

The basic portion of the ear system consists of the inner-ear structures, lying within the braincase. Here there is a system of canals and sacs filled with a fluid (endolymph) and containing areas with sensory cells. The primary function of the vertebrate ear is the sense of balance, of equilibrium; hearing is a secondary function. Three little canals—the semicircular canals—are present, the three at right angles to one another. These register the movement of the head and the frog as a whole. Below the canals lie two sacs (utriculus and sacculus) with sensory areas that register the static position of the head. These structures are similar to those found in man and vertebrates generally. In man and other mammals, however, hearing is registered in a long, coiled, liquid-filled tube which may be larger than all the other parts of the ear put together. This is absent in the frog; hearing reception takes place only in a small patch of tissue in the sacculus. This suggests that, although the frog can hear, it lacks the ability to discriminate between tones of various pitch, as is the case in animals with a cochlea. Bass and soprano may be one to the frog.

Lateral-line organs.—Fish possess a system of sensory organs arranged in a pattern of pits and lines on the head and down the flanks which appear to register movement and pressure in the water about them and thus afford valuable sensory aid to the swimming animal. Such organs are still present in the tadpole. They disappear, however, at metamorphosis and are absent in the adult frog; reptiles, too, have lost them, and they never reappear in higher vertebrate groups.

DIGESTIVE SYSTEM

The frog's mouth functions primarily as the anterior end of the digestive tract. Teeth are feeble and are confined to the upper jaws, where there is a row along the jaw margins and a small patch on the front part of the palate. The tongue was noted in the discussion of frog habits. Mouth glands are poorly developed. Beyond the mouth the alimentary canal passes back through the short pharynx, or throat region, and the distensible esophagus to reach the stomach. Pharynx and esophagus are lined with cilia, tiny hairlike structures, which, beating rhythmically, aid in passing food particles downward. The stomach is a simple pouch not dissimilar to that of man in shape and functions but rather less curved. Here food is stored, manipulated mechanically by stomach movements, moistened by mucus, and reduced to a pulp. Here, too, part of the chemical process of digestion takes place, for carbohydrate materials are partly broken down by the action of pepsin and hydrochloric acid secreted by stomach glands. The stomach terminates at a constriction, the pylorus ("gatekeeper"), where the food passes to the small intestine.

The pancreas, as in vertebrates generally, secretes the major digestive enzymes which break down food materials into simple substances capable of being absorbed into the body. It is rather diffuse in shape and lies along the course of the bile duct, into which its secretions pass to reach the intestine. The liver typically has two major lobes, left and right; between them is a smaller lobe, while the gall bladder occupies a median position on the dorsal surface of the liver. In the liver occur major chemical processes of altering food materials, storing carbohydrates as the complex sugar glycogen, and sending off waste products as bile. The liver undergoes marked seasonal changes. In the summer it becomes large and light colored and is filled with food materials. During hibernation this food store is consumed, and the liver becomes a relatively small dark structure.

The small intestine of the frog has functions similar to that of man; here the breakdown of all food types —carbohydrates, fats, and proteins —is completed, and the simpler molecules to which it is reduced are absorbed into the body through the intestinal walls. The small intestine is relatively short, with but few coils. However, the tadpole, which eats vegetable rather than animal food, has a much longer intestine, coiled like a watch-spring. The large intestine is a short straight structure; there is no caecum or appendix. Its narrower, distal portion, the rectum, opens out into the cloaca, a pocket into which urinary and genital products also empty. The cloaca is a structure common to most lower vertebrates but absent in man and most mammals, where the intestine opens separately to the surface. The outer opening of the cloaca is closed by a sphincter muscle.

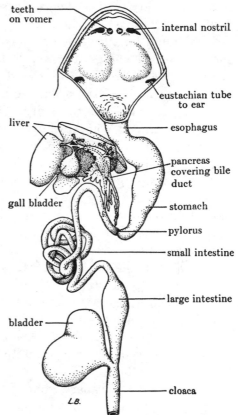

RESPIRATORY SYSTEM

The adult frog is an air-breather and a possessor of lungs. But even in the adult there are accessory breathing structures, while the tadpole is a water-dwelling user of gills.

The digestive system of the frog seen from the ventral side. At the front the lower jaw has been cut away to expose the roof of the mouth. Much of the liver has been cut away and the remainder represented as pushed upward to expose the gall bladder and ducts. (Modified from Wiedersheim.)

Air in passage to the lungs enters the mouth through the nostrils. The internal openings (choanae) are near the front of the mouth, for there is no development of a secondary palate such as is seen in mammals. The entrance to the lungs is the slitlike glottis far back in the floor of the mouth. Below the glottis is a small chamber, the larynx, which is partially inclosed by cartilages belonging to the hyoid apparatus; these cartilages are remnants of the old gill-bar system of the fish. The frog has no development of a long trachea such as is found in man, for the neck is short; beyond the larynx the air tube divides into two passages which almost immediately arrive at the lungs. These are relatively simple (and relatively inefficient) sacs lacking the complicated folds seen in the human lung.

A mammal breathes by movements of the ribs and diaphragm which suck air into the lungs. The frog lacks both of these structures and must use other methods. There are two steps in filling the lung. First, with the mouth shut and nostrils open, the floor of the mouth is depressed. This sucks air into the mouth cavity. Then the nostrils are closed and the mouth floor raised, compressing the imprisoned air, which is thus pumped through the glottis to the lungs. However, the lungs are filled only at intervals, and mouth-breathing is a more frequent practice. As may be observed in a resting frog, the floor of the mouth rises and falls in a gentle, regular rhythm. Every now and then there is a more violent movement, indicating the filling of the lungs. The rest of the time there is merely a flow of air in and out of the mouth cavity. This orifice is lined with moist skin and acts as a breathing organ.

Still further, the skin of the frog is a moist and usually soft membrane. It can and does function in breathing and is richly supplied with blood vessels which utilize its potentialities as a respiratory organ. It has been found by careful measurement that even under normal conditions more carbon dioxide is given off by the frog's skin than by the lungs; and a frog submerged in cool water can obtain enough oxygen through the skin to live for several days.

Most vertebrates, apart from birds and mammals, are voiceless; the frogs, however, have evolved a pair of vocal cords which are fairly comparable to those of man and consist of a pair of elastic bands running across the larynx. Sounds are produced by the passage of air over these cords. In many frog species the males possess vocal sacs on either side of the throat which open into the floor of the back part of the mouth. These serve as resonators to reinforce the sounds produced by the cords.

EXCRETORY SYSTEM

In all vertebrates the basis of the excretory system consists of numbers of tiny tubules which filter from the blood water and waste matters, excreted as urine—very dilute in frogs. The mass of tissue containing the thousands of tubules constitutes the kidney. Despite the fundamental similarity of the units composing this structure, there are, in the various vertebrate types, marked differences in the shape and position of the kidneys and the drains passing the urine to the surface.

In the frog the kidneys form a pair of oval strips of dark-red tissue imbedded in the back wall of the body cavity, whereas in man the kidney is much shortened and thickened to form a large bean-shaped structure, more posteriorly placed. In the frog, as in lower vertebrates generally, the urine passes into a duct (technically known as the "Wolffian duct") which, in the male, also carries the sperm from the testis. In all vertebrates above the amphibian level—reptiles, birds, and mammals—this tube has been given over to reproductive functions exclusively, and the urine leaves the kidney through a newly developed duct—the ureter. This name is sometimes applied to the functional kidney duct of the frog, but improperly so, for the two are not at all homologous. In the frog the urine-carrying ducts pass directly to the cloaca, and a bilobed bladder for urine storage is found in the floor of the cloacal pouch. In man, on the other hand, the urinary bladder is more internally situated, and the two ducts lead directly to it.

REPRODUCTIVE ORGANS

Female structures.—In the female frog the ovaries, the primary sex organs, are paired, more or less lobulate bodies lying in the dorsal part of the body cavity. They undergo great seasonal changes. When breeding is over in the spring, they are reduced to tiny wrinkled bodies. During the summer they increase in size and by autumn may fill much of the abdomen. With the coming of the breeding season, the numerous eggs, which have been maturing meanwhile, burst out of the ovaries to fill the body cavity.

From this cavity the eggs pass to the surface through the oviducts, which also vary greatly from season to season and are much enlarged at breeding time. These tubes run most of the length of the abdomen. Anteriorly, near the base of the lung, is a wide, funnel-shaped mouth into which the eggs pass from the body cavity. Much of the length of the tube is highly convoluted. The inner surface of the duct is much wrinkled at breeding time. The ridges are covered with ciliated cells which are instrumental in passing the eggs down the duct. Between the folds are glands which secrete a gelatinous material to coat the egg. Near the exit to the cloaca there is a thin-walled portion of each tube—the uterus. In these the eggs collect before they are extruded, and the uteri may become greatly distended. These structures are not, of course, particularly comparable to the true uteri of mammals, in which development of the young takes place.

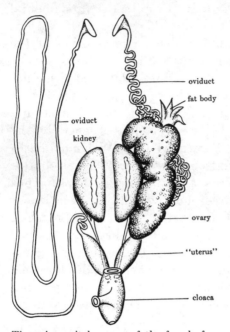

The urinogenital organs of the female frog seen from the ventral aspect (semidiagrammatic). The urinary structures are similar to those of the male, but the sex products do not utilize the urinary duct; instead they pass through the highly coiled oviduct (which may be present in rudimentary form in the male also). On the left side of the figure the ovary has been removed and the oviduct artificially straightened to show its true length. The genital system is shown in a quiescent stage; as the reproductive period approaches the ovaries grow enormously, and subsequently the "uteri" are distended with eggs.

The oviduct does not function in the male, but in some species of frogs, including the leopard frog, it is present in a rudimentary state, lying lateral to the kidney duct. This is illustrative of the fact that in early development the typical vertebrate does not, so to speak, "know" which sex it is destined to be, and the beginnings of the typical organs of both sexes may be present. Later in development the organs of one sex or the other dominate; those of the opposite sex may disappear but may (as in this case) persist in a rudimentary, nonfunctional state.

Male organs.—The primary sex organs of the male, the testes, are rounded bodies lying ventral to the kidneys and bound by membranes to the dorsal lining of the body cavity. The frog lacks the coiled epididymis which in mammals adjoins the testis. Instead, a number of efferent ducts pass across into the front part of the kidney. Crossing this structure, the sperm at breeding season pass down to the cloaca through the Wolffian duct, which, as noted above, also carries the urine. This situation, in which the testes "impose on" the kidneys to obtain an outlet, is apparently a fairly primitive vertebrate condition. In mammals, the two types of products follow separate courses most of the way to the exterior. This is accomplished through the kidney evolving a new duct (ureter) for its exclusive use and politely turning the older structure over to the reproductive system as the sperm duct (ductus deferens). The front part of the kidney has been abandoned to help form the epididymis.

In both sexes we find, just in front of the gonads, a large yellowish organ, the fat body, with branching finger-like processes. This body serves as a storehouse for nutriment, waxing large in the summer and decreasing in size during the breeding season; apparently the stored-up nutritive material is used by the sex organs.

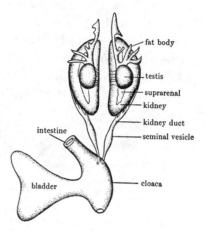

CIRCULATORY SYSTEM

Blood.—All vertebrates have a major liquid component, the plasma, mainly water containing salts in solution as well as complex proteins specific for the blood stream and varying from form to form. Carried along in this fluid are the blood cells, red and white. In the frog the red cells are large and oval in shape, in contrast with the rounder cells of man, for example; further, they tend to retain the nucleus, although at certain seasons a large proportion may be enucleated, as in mammals. White cells abound, although the types present (and shown in the accompanying figure) vary somewhat from those of man, which are usually cited in descriptions of vertebrate blood. All ver-

The urinogenital organs of the male frog seen from the ventral surface (semidiagrammatic). The testes have been pushed slightly laterally to show the slender vessels passing from them into the medial side of the kidneys. The suprarenals are thin masses of tissue lying on the ventral surface of the kidneys. The kidney ducts (also termed mesonephric or Wolffian ducts) drain not only the kidneys but the testes as well. The reproductive system is represented as in a quiescent stage. At the time of sexual activity the testis is enlarged, and the seminal vesicle may be greatly distended with sperm.

tebrates possess some sort of tiny structures associated with the necessary function of blood clotting after injury; in the frog these are known as "spindle cells."

Arteries.—In the fish the blood in the arteries flows from heart to gills before passing on to the body capillaries. In the frog tadpole this primitive situation still persists. In the adult, however, the gills and their circulation are eliminated. The blood from the heart flows forward and upward around the sides of the throat in vessels representing (as in man) three of the paired channels which in fishes lay between the gill slits.

Most anterior of these channels are the carotid arches, which carry blood to the head. Back of them are a pair of much larger vessels, which carry the blood to all parts of the trunk and limbs. These are the aortic arches. They run back above the body cavity and fuse to form a large median vessel, the dorsal aorta, which runs back close to the backbone. A notable difference between frogs and the most advanced vertebrates, the birds and mammals, is that the frog preserves both members of the pair while in mammals that of the right side has been abandoned, and the dorsal aorta is continuous with the left arch; in birds the left arch is aban-

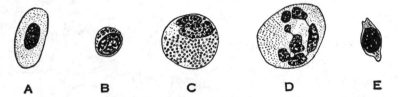

A   B   C   D   E

The principal types of blood cells in the leopard frog. *A*, a normal red blood cell (erythrocyte). *B–D*, white blood corpuscles: *B*, a small lymphocyte; *C*, a large white cell filled with granules taking acid stains; *D*, a large white cell with subdivided ("polymorphic") nucleus. *E*, a spindle cell, associated with blood clotting. (From Noble, after Jordan.)

doned and the right only is retained. Most posterior of the arteries arising from the heart are those which lead to the lungs, the pulmonary arteries.

Veins.—Into the frog heart empty veins carrying blood from the body organs. A pair of pulmonary veins drains the lungs. Two large vessels, anterior cardinals or venae cavae, bring blood from the head, front legs, and skin. These are compared with the superior vena cava of man and other mammals, which has similar functions but has become a single rather than a paired structure. From the back part of the body comes the posterior vena cava, draining kidneys and liver. This large vein is closely comparable with the inferior vena cava of man, the shift in name being, of course, related to the changed posture of the body.

Blood from the legs and gut, however, does not pass directly to the heart; instead, the veins from these regions lead to capillary systems in other organs—liver and kidneys—through which they drain before entering the main venous circulation. Such systems of veins are termed portal systems. One, the hepatic portal system, present in lungfishes and all land vertebrates, receives blood from the gut and sends it to the liver, where part of the food content may be stored. In the frog this system also receives part of the blood from the hind legs via a vein along the abdomen. In the frog we find also a renal portal system, present in most lower vertebrates but abandoned in birds and mammals. This receives the remaining blood from the legs and sends it to circulate in the kidneys, thus assuring a sufficient flow of blood in those organs.

Heart.—In the typical fish the heart is a simple structure with simple functions. In it is received, in a single stream, the blood from all the or-

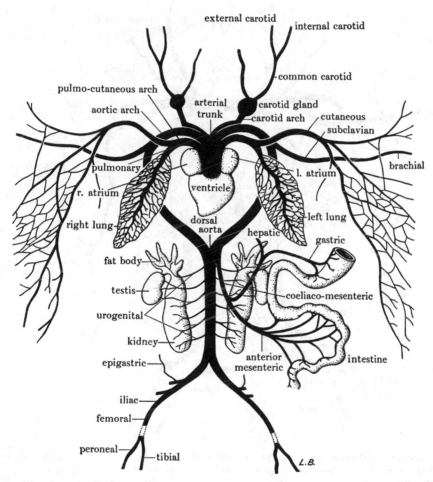

Arterial system of the toad seen from the ventral side. The blood leaves the heart through three pairs of channels. The most posterior (pulmo-cutaneous arches), which are mainly filled by deoxygenated blood from the body, pass to the lungs and to the skin (which is an accessory breathing structure). The main aortic arches carry blood to the body and limbs; the most anterior (carotids) supply the head. (After Jammes.)

gans. This is "spent" blood, deprived of its oxygen in its passage through the body, and thus all of it is ready to pass to the breathing organs, the gills, to receive a new supply of that necessary element. We find, thus, that the fish heart is basically a single-tube structure pumping all the blood it receives forward in a single stream to the gills.

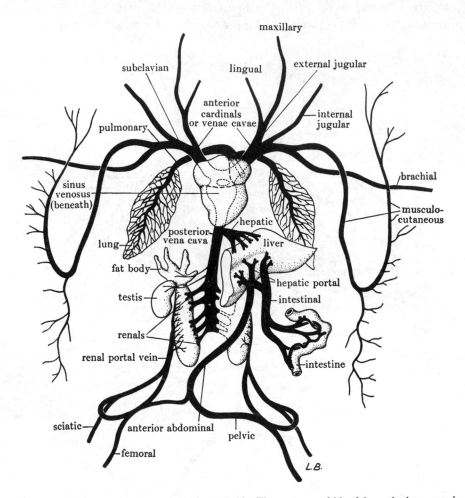

Venous system of the toad seen from the ventral side. The oxygenated blood from the lungs reaches the left atrium via the pulmonary veins. All other blood enters the right atrium by way of the sinus venosus (lying behind the heart and shown in broken lines). Three main channels enter the sinus: paired anterior cardinal veins (or venae cavae) and a single posterior vena cava. Blood from the hind legs may pass through the kidney (renal portal system) before reaching the posterior vena cava; blood from the intestine and some from the hind legs passes through the liver (hepatic portal system). (After Jammes.)

In lung-breathers, however, a major complication arises. Two blood streams are present. Fresh blood from the lungs should pass to the major body structures, while spent blood from the body should be pumped to the lungs. In man and other mammals this "problem" has been adequately solved. Both major parts of the heart—atrium (or auricle) and ventricle—have been divided down the middle, a double-barreled pump has been created, and the two blood streams are completely separated, although

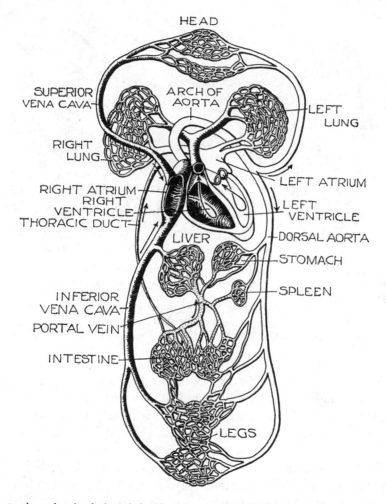

Diagram to show the circulation of the blood in man. Vessels carrying oxygenated blood light in color; those with "impure" blood, dark. In contrast with the frog, the two blood streams are completely separated in mammals.

passing simultaneously through the one organ. In the frog the problem is, so to speak, in process of solution. The heart is incompletely divided; but, nevertheless, the two streams are fairly well separated.

The atrium, single in the fish, here consists of two chambers. The spent blood from the body first enters a thin-walled sac, the sinus venosus, on the upper surface of the heart and thence passes to the right atrium. The fresh blood from the lungs, on the other hand, passes into the left atrium. The two blood streams thus enter the heart separately.

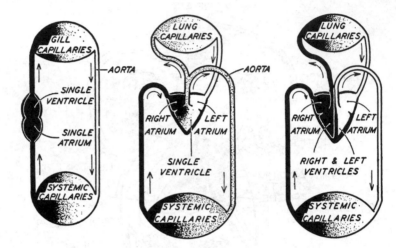

A much-simplified diagram to show the evolution of the double circulation through the heart due to the substitution of lungs for gills in higher vertebrates. In the fish the blood from the breathing organ (gills) flowed directly, via the arteries, to the body (systemic capillaries), and hence the heart was a simple pump. However, in lung-bearing fish and in amphibians the lungs return their blood to the heart, where it tends to mix with "spent" blood from the body. Crosswise subdivision of the heart in birds and mammals separates the two streams. (From Carlson and Johnson, *The Machinery of the Body.*)

Beyond this point, however, structural separation does not occur. The ventricle is a single structure and so is a final region, the conus arteriosus, out of which lead all the major arteries. It would seem that the two streams would become hopelessly mixed in their further passage through the heart.

But, in fact, little mixture does take place. It will be noted that the right atrium opens into the ventricle farther forward than the left. When the atria contract, this topographic situation results in a ventricle filled with blood of both sorts but with the spent blood in advance of that from the lungs. Beyond the ventricle the arterial cone contains a peculiar spiral value which tends to direct blood more readily to the pulmonary arteries than to the other sets of vessels. When the ventricle contracts, the spent blood thus tends properly to pass to the lungs; the blood behind, fresh from the lungs, finds the passages leading back to those structures full and hence flows, as it should, to the arches reaching body and head. The frog system of separation of blood streams seems a poor makeshift; but it works fairly effectively.

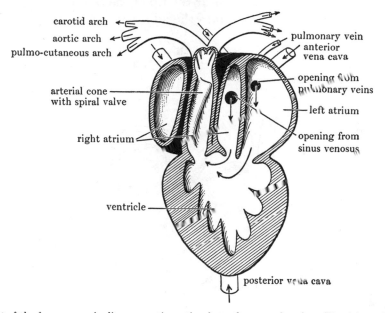

The heart of the frog as seen in diagrammatic section from the ventral surface. Blood from the lungs enters the left atrium through the pulmonary veins and thence passes to the ventricle. Deoxygenated blood from the body first passes via the venae cavae into the sinus venosus, which lies on the dorsal surface of the heart and is not seen in this figure. Thence it enters the right atrium (which is exposed in the figure on either side of the arterial cone) and from this chamber into the ventricle. From the ventricle the blood passes upward in the conus arteriosus. A spiral valve here assists in sorting out fresh and deoxygenated blood so that the latter tends to enter the pulmo-cutaneous arch, while the fresh blood tends to pass to the carotid and aortic arches.

Lymphatics.—Higher fishes and all land vertebrates have, in addition to the veins, a second system of vessels returning fluid from the tissues to the heart. Unlike the veins, these do not connect with the arteries via the capillaries and hence do not carry red blood cells. Except for a duct running along the top of the abdominal cavity which collects fats absorbed by the intestine, these vessels, the lymphatics, are of little importance in reptiles, birds, and mammals. In the frog, however, the lymphatics are numerous. They mostly occur in the form of large, thin-walled sacs of irregular shape, between which there is a sluggish flow of liquid. A lymphatic "cistern" of this sort occupies a large area above the walls of the body cavity and below the backbone; still more important is a whole series of lymph spaces which lie beneath the skin. Circulation of this liquid is brought about through the presence of two pairs of small pumping structures, the lymph hearts. One pair lies at the sides of the third segment of the backbone; the other pair is near the tip of the tail bone. They pulsate regularly and drain the lymph into the adjacent veins.

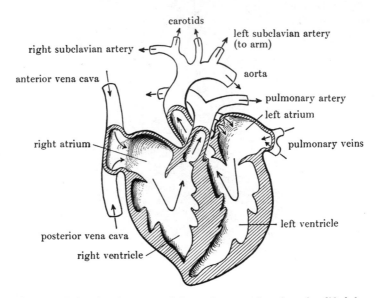

The human heart with its chambers opened from the ventral surface (modified from Jammes; semidiagrammatic). Blood from the lungs enters the left atrium (auricle) through the pulmonary veins, passes through the left ventricle, and leaves the heart via the aorta with its various branches. Blood from the remainder of the body reaches the right atrium through the two venae cavae, passes through the right ventricle, and thence reaches the lungs by the pulmonary arteries.

GLANDS OF INTERNAL SECRETION

In all vertebrates we find, in one part of the body or another, important glands regulating body functions by secreting chemical "messengers," or hormones, into the blood stream. Typical glands of this sort are present in the frog. Since, however, much of the work on endocrines has been done from the point of view of medicine with emphasis on man, our knowledge of amphibian secretions is not so complete.

We have noted the presence of a well-developed pituitary in the frog.

Here, as in man, this gland is of great importance not only in its direct effects but also through its influence on other endocrines. An interesting item to the biologist lies in its influence on the gonads. Many biological problems are studied through investigation on amphibian eggs and larvae. It was formerly necessary to wait until the spring breeding season to obtain eggs for such work. Now it has been discovered that an extra "shot" of pituitary extract may cause the laying of frogs' eggs at almost any season desired. The pituitary also has an influence on the coloration of frogs. If it is removed, the darkly pigmented cells of the skin contract, and the frog becomes pallid; injections of pituitary extract restore the normal color.

The frog thyroid is similar in function to that of man; it forms by a budding-out of tissue from the floor of the throat and secretes a substance (containing iodine) which has an important effect on growth and metabolism. Because of its importance in growth, it is but natural that it has been found to be of great importance at metamorphosis—the time of transformation of the tadpole into the adult frog shape. If thyroid extract is fed to immature tadpoles, they change promptly into tiny adults. If, on the other hand, the thyroid be removed from a young tadpole, it may continue to grow and may even reach sexual maturity but never change from the tadpole body form.

A number of small structures arise from the margins of the gill pouches in all land vertebrates. In mammals they include the thymus, a small mass of throat tissue whose functions are poorly known, and the parathyroids, small nuggets of tissue which become embedded in the thyroid and are of great importance in the use of the element calcium in the body. The frog has a somewhat comparable set of glands; little, however, is known of the possible functions of these bodies. In the frog, as in vertebrates generally, the pancreas contains "islands" of tissue which do not secrete enzymes into the intestine but are the source of insulin. This is a material essential for the regulation of sugar metabolism; if deficient, it produces the condition known in man as diabetes. In all higher vertebrates there are, near the kidneys, endocrine glands termed the adrenals, with a double function. One series of products, while imperfectly known, is of vital importance in regulating the composition of body fluids and kidney function; a very different product is adrenalin, a chemical which rouses the body to activity in times of stress. The frog adrenals consist of bands of yellowish tissue extending along the undersides of the kidneys. In these bands are mixed masses of cells producing these two types of secretions.

SKELETON AND MUSCLES

It is in the skeletal system that we find the most marked specializations of the frog. The hopping gait is obviously responsible for the peculiar construction of the legs and limb girdles and undoubtedly has had much to do with the great shortening of the body. Even the skull is rather specialized and degenerate. Except for the expanded braincase, the human skeleton, "advanced" as we think it to be, has a better claim to be regarded as primitive than has that of the frog. The study of the frog skeleton is not

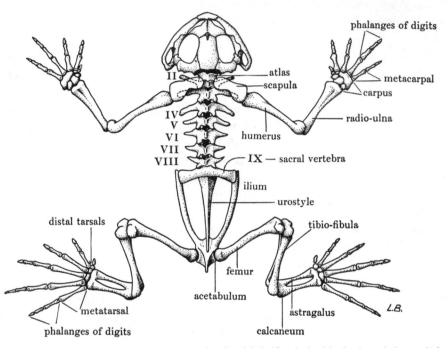

Skeleton of a common toad. Except in minor details, this is identical with the frog skeleton. (After Jammes.)

without interest; but we must not delude ourselves into believing that we are dealing with a "generalized" structure.

The muscles too are highly specialized in many ways, particularly those of the limbs. In the course of study of the frog there is frequently included a dissection of some part of the musculature, such as the thigh. This is of use as an introduction to the methods of muscle dissection but is not to be regarded as of importance for purposes of comparative study. The muscles of this region form a complex system. To them are given names such as "sartorius," "gracilis," etc., which suggest that we are dealing with the muscles known by the same terms in man. Comparative studies, however, show that for the most part this is not the case. The primitive land animal, with a clumsy gait, appears to have had but a small number of relatively massive muscles in its limbs. Both frog and man have become agile types in which these muscle masses have eventually become subdivided into an intricate system. But the subdivisions have taken place independently in the two types, and there are few instances in which we can be sure that frog and human leg muscles are actually identical.

Both cartilage and bone are present in the frog; cartilage is rather more common, however, in the amphibian than in higher vertebrates. Much of the skull fails to ossify. In the limbs of mammals the terminal portions form separate bony epiphyses, with "growing points" of cartilage between them and the shaft. In the amphibians there are no such structures, and the ends of the bones frequently remain in a cartilaginous condition.

Trunk skeleton.—In the frog the length of the body and, in consequence, the number of segments in the backbone are much reduced. In a majority of vertebrates (although not in man and his ape relatives) there is a long tail containing half-a-hundred or more vertebrae in many cases. In the frog the tail no longer projects beyond the body contours, and of the tail vertebrae there remains only a spikelike structure termed the urostyle, or "tail pillar."

Equally remarkable is the reduction in the more anterior part of the vertebral column. In most land forms the neck and trunk together contain about two dozen vertebrae, frequently more. In common frogs there are only nine. A typical vertebra consists of two parts, neural arch and centrum, although the two ossify in the frog as a single bone. The arch incloses the spinal cord, extends upward as a neural spine, and sends out on either side an elongate transverse process. Both anterior and posterior margins of the arch bear a pair of processes termed zygapophyses, by means of which the successive vertebrae articulate with one another. The articular surfaces on the anterior zygapophyses face upward and inward, the posterior ones downward and outward. The centrum is oval in section, concave in front and convex behind, each centrum receiving the projecting posterior end of the one ahead. The first vertebra has, anteriorly, a pair of concave oval surfaces which articulate with the skull; the last of the series, the sacral vertebra, has unusually heavy transverse processes which connect with the pelvic girdle. In nearly all vertebrates ribs are present, articulating with trunk vertebrae; the frogs are notable for the entire absence of these structures.

Head skeleton.—The major skeletal structure of the head region is the skull, inclosing the brain, sheltering the sense organs, and forming the upper jaws. The skull in the frog, as in other living amphibians, is much flattened, and a considerable portion remains in an embryonic cartilaginous condition. As in other living vertebrates above the shark level, two types of bones are present—those which replace cartilage and the more superficial dermal elements. Few replacement bones, however, are present in this degenerate type; of bones seen in the diagrams only three—sphenethmoid, pro-otic, and exoccipital—are of this nature. The remainder are dermal elements. But here, too, many bones primitively present have been

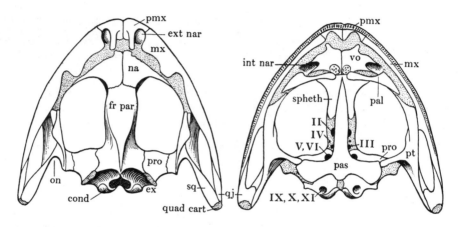

The skull of the bullfrog seen from dorsal and ventral surfaces. Cartilaginous areas are stippled. Certain of the nerve exits are marked in Roman numerals. *cond*, occipital condyle; *ex*, exoccipital; *ext nar*, external nares; *fr par*, fused frontal and parietal bones; *int nar*, internal nares; *mx*, maxilla; *na*, nasal; *on*, otic notch partially surrounding eardrum; *pal*, palatine; *pas*, parasphenoid; *pmx*, premaxilla; *pro*, pro-otic; *pt*, pterygoid; *qj*, quadratojugal; *quad cart*, quadrate cartilage; *spheth*, sphenethmoid; *sq*, squamosal; *vo*, vomer.

lost. At the back of the skull, below and on either side of the foramen magnum (the "big hole" through which the spinal cord emerges), are paired projections, the condyles, which articulate with the first vertebra. Mammals, including man, also have two condyles, whereas reptiles and birds have but one. At one time it was believed that this similarity indicated that the mammals were direct descendants of the amphibians. This, however, proves not to be the case, for we now know that early amphibians had only one condyle; the frog and mammal have in this regard merely evolved in parallel fashion.

Whereas the human jaw consists of but a single bone, that of the frog

contains three bony elements of the ten or so primitively present in land
vertebrates. In the throat region between the jaws lies a series of cartilages
known collectively as the hyoid apparatus; they are the remains of the
bars which in fish stiffened the gill arches.

Girdle and limb skeleton.—In the bones of the appendages and the
girdles which support them, the frog exhibits most of the elements found
in all typical land vertebrates but in a rather specialized condition correlated with leaping habits.

The pectoral, or shoulder, girdle, supporting the front legs, includes as a
main dorsal element the scapula, corresponding to the human shoulder

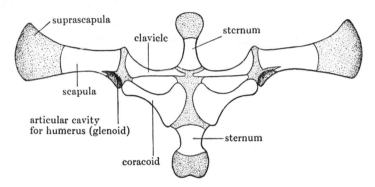

Shoulder girdle of the frog seen spread out in ventral view. Except for the clavicle (collarbone), the
elements are preformed in cartilage, and considerable cartilage may remain even in the adult. The
major elements of the girdle are scapula and coracoid. In the center are elements of the sternum,
which pertains to the axial skeleton rather than the shoulder girdle.

blade; above is a feebly ossified extension, the suprascapula. Below the
arm socket a bone extending downward and backward is the coracoid (the
human bone of this name is a mere nubbin). Directly down below the front
of the scapula is the clavicle, the equivalent of the human collarbone. In
the ventral midline is a fore-and-aft series of bones and cartilages which
forms the sternum.

The pelvic girdle includes a more or less circular structure into the
middle of which fits the head of the thigh bone, and a long process which
projects upward and forward to articulate with the sacral vertebra. The
upper bone of the girdle is the ilium; a bone at the posterior end is the
ischium. Besides these two elements the human pelvic girdle includes a
third, the pubis; in the frog (as in many amphibians) this bone tends to
remain unossified.

In the front leg the first segment contains in normal fashion a single element, the humerus. Beyond the elbow a typical vertebrate exhibits two bones, radius and ulna; in the frog these are fused into one. The wrist, or carpus, contains half-a-dozen small elements; beyond are the metacarpals and phalanges making up the "fingers," of which four are typically developed, although short; an inner spur, sometimes regarded as the missing "thumb," is probably not of this nature (it may be much enlarged in the breeding male, serving as a clasper).

The long hind leg begins proximally with a characteristic femur, or thigh bone. But the second segment (as in the front leg) consists of a single bone, regarded as a fusion of the tibia and fibula normally present. Beyond lies the tarsus, or ankle bones. Here is still another peculiarity of the frog.

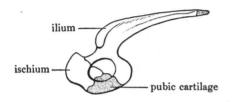

ilium

ischium

pubic cartilage

The right pelvic girdle of the frog. The pubis remains unossified.

Typically, the tarsal bones (like those of the wrist) are tiny structures. In the frog, however, two proximal elements (astragalus, calcaneum) are much elongated. This development has the result of adding a third segment to the hind leg and greatly increasing its effectiveness in hopping. Five toes are present in the hind foot.

LIFE-HISTORY

The development of the common species of frog takes place in a water environment, the streams or ponds in which the eggs are deposited in the spring breeding season. The egg contains a modest supply of nutrient yolk and furnishes a food supply for the embryo during the stages in which the major body structures and the tadpole body shape are acquired. Hatching soon occurs, however, and the larva, hardly larger than a pin, must seek its own living. The tadpole is, of course, markedly different from the adult both in general appearance and in many structural features. There is a powerful tail, used as a swimming organ; limbs, on the other hand, are absent until a late period. Lungs are not developed, and breathing is accomplished by means of gills. The adult frog is an eater of animal food, but the tadpole, on the contrary, feeds on vegetable material. In relation to this we find many differences in the mouth and jaws between tadpole and adult, and even the intestine is built on a different plan.

The period of tadpole existence is of variable length in frogs. In some species the entire development takes place in a few weeks; on the other hand, the bullfrog spends several years in the tadpole stage. Eventually,

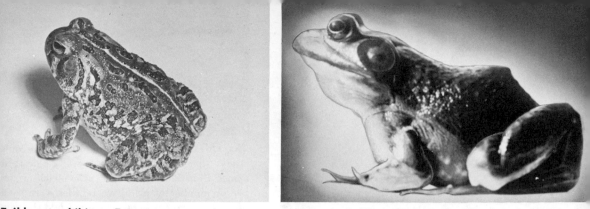

Tail-less amphibians. *Left,* the common American toad; *right,* the bullfrog. Both are familiar forms; the large bullfrog is frequently used in laboratories. (Toad photograph by Isabelle H. Conant; bullfrog copyright General Biological Supply House, Chicago.)

Small but poisonous. A South American toad (*Dendrobates*) shown at the left, natural size on a penny. Liquid from its skin glands is used by natives to poison arrow tips (Photograph courtesy New York Zoölogical Society.)

The hairy frog (*Astylosternus*) of Africa (*below*). In the male, at breeding season, hair-like projections grow out of the sides of the trunk and legs. These are not hairs, but accessory breathing organs, containing blood vessels. (Photograph copyright General Biological Supply House, Chicago.)

The Surinam toad (*Pipa*). In this large and ungainly South American toad the eggs are inclosed in the skin of the mother's back, each in a little pocket covered by a flap. When development is completed, small but maturely formed little toads emerge—the water stage is eliminated. (From a model, courtesy American Museum of Natural History, New York.)

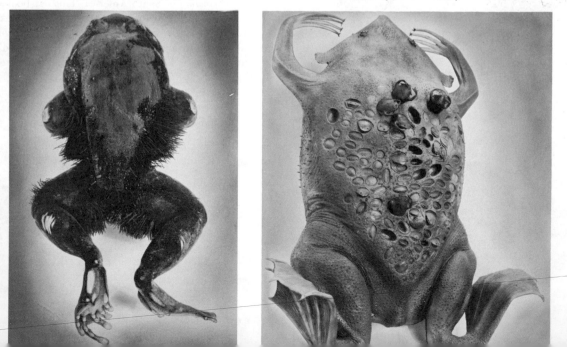

The oldest egg. As proved by microscopic study, this fossil egg is the world's oldest; it comes from Texas deposits 225,000,000 years old. A number of primitive reptiles were then present; we do not know which one laid the egg. (Specimen at Harvard University.)

An archaic reptile (*left*). *Seymouria*, named from the Texas town near which its remains were discovered, is a fossil form that retains numerous amphibian features. (Painting by F. L. Jaques, under the direction of W. K. Gregory; photograph courtesy American Museum of Natural History, New York.)

Familiar turtles. North America has many pond and marsh turtles. *Left*, common musk turtles (*Sternotherus*), showing both upper and lower shields (termed carapace and plastron, respectively). *Lower left*, a pond turtle (*Clemmys*). (*Sternotherus* courtesy American Museum of Natural History; *Clemmys* photograph by Isabelle Hunt Conant.)

Below, perhaps the most ugly turtle, the Matamata (*Chelus*) of South America. This flat-bodied form has head and limbs covered with warty excrescences and nostrils inclosed in projecting tubes; it belongs to the tropical "side-necked" turtle group. (New York Zoölogical Society photograph.)

in any case, comes the period of metamorphosis, the change of bodily shape which turns tadpole into frog. This process involves a marked reorganization of almost every part of the animal's body. The tail is resorbed; the limbs grow out (the hind legs appearing first); the gills disappear, and lungs develop; the long, spirally-coiled tadpole intestine changes into the relatively short adult structure; many head structures are radically reorganized.

These changes accomplished, the frog becomes a potential land dweller. With further development in size and proportions and the growth of the sex organs, the frog, by the time of the breeding season of the next spring, has arrived at maturity.

CHAPTER 6

The Origin of Reptiles

In contrast with the conservative mode of development of typical amphibians is that of the reptiles and the higher tetrapods derived from them. The reptile has evolved a type of egg which can be laid on land. No aquatic life-stage is necessary; emancipation from the water may be complete.

THE LAND EGG

This reptilian egg has a complicated structure to which we must devote some attention, for many of the architectural features first attained here are still persistent in the wrappings of the human embryo.

The developing amphibian obtains its oxygen and most of its food from the water and excretes its waste matter into the water again; the water also protects it from drying and against mechanical injury. In the reptile, if the water-dwelling stage is to be omitted, substitutes must be provided for these advantages which will carry the youngster through to a stage where it can make its own way on land.

As a food supply for the embryo, the egg of a reptile or bird contains a large amount of nourishing yolk which is contained in a yolk sac, connected with the digestive tract of the growing embryo. A second, larger sac, the amnion, develops about the body of the embryo. This liquid-filled sac affords protection against injury and desiccation; it is a substitute for the amphibian's natal pond. Out from the back end of the embryo's body there grows a tube and a third sac, the allantois, in which the waste matter of the body is deposited. The whole egg structure is stiffened and protected by a firm shell on the exterior. The shell, however, is porous; beneath a portion of it lies a membrane pertaining to the allantois, richly supplied with blood vessels. This acts as a lung, taking in oxygen and giving off carbon dioxide; and it is an easily confirmed fact that a reptile or bird egg, if submerged in water, will drown as surely as an adult; the contained embryo ceases to grow, and dies.

By these structures—yolk sac, amnion, allantois, and shell—all the needs of the developing reptile can be met. The water-dwelling stage has not been eliminated, but it is now spent within the protecting shell of an egg. We have here for the first time an animal which can assume land existence immediately upon hatching, the first possibility of a purely land animal.

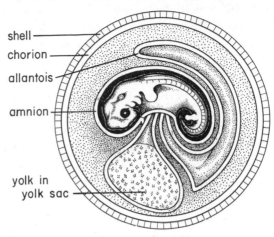

How, when, at what stage, did this crucial reproductive improvement appear? The story once seemed clear to me in the form I told it to many a student audience: "Well before the close of the Carboniferous period, the fossil record shows us, there had appeared advanced amphibian types with well-developed limbs and other features indicating that, as adults, they could be, and were, mainly terrestrial forms rather than water dwellers. A sole obstacle lay in the path of their conquest of the land—their mode of development, through which they were chained to the water (a lovely and dramatic phrase!). At long last there came the final stage in their release—the development of the terrestrial amniote egg. Their bonds were broken, and, as true terrestrial forms, the early reptiles swept on to a conquest of the earth!"

Generalized diagram of the embryonic membranes of the higher land vertebrates (amniotes). The developing embryo is surrounded, except ventrally, by a liquid-filled cavity which is inclosed by the amnion, a membrane continuous with the skin of the embryo. Developing from this and continuous with it at an early stage is a second membrane, the chorion, which lies beneath the shell. Two membranes grow out from the embryonic digestive tract: a yolk sac, directly below the embryo, is filled in reptiles and birds with a large mass of yolk; from the back end of the gut grows out the allantois, which can function as an embryonic bladder. Blood vessels surrounding it may carry to the embryo oxygen which has passed in through the porous shell of the egg in reptiles and birds, so that the allantois may function as an embryonic lung. In typical mammals the chorion comes into close contact with the surrounding maternal tissues to form a placenta whereby the embryo obtains nutriment from the mother; in such cases the blood vessels of the allantois carry the nutritive materials from placenta to embryo.

This is a fine story. However, I now suspect that it is far from the truth. It assumes that the *adult* first became a land dweller and that terrestrial reproduction was a later development. It now seems more probable that the reverse was the case—that the *egg* came ashore first and that the adult tardily followed.

This skepticism arises from a study of the oldest, adequately known, reptile faunas. Later in this chapter we shall talk of the cotylosaurs, the "stem reptiles" from which all later land vertebrates were derived. If we still assume that the land egg developed only after the pre-reptiles had become fully land types rather than leading an amphibious existence, then all early fossil reptiles should be good terrestrial types. But while the stem reptiles all possessed good limbs and were capable of walking on land, many of them, as far as we can tell from the skeletal evidence, spent much of their lives in the water or in swampy environments. Can this have been due to a reversion? This is very doubtful. The cotylosaurs are such early and primitive reptiles that it is much more probable that their ancestors had never fully abandoned an aquatic life.

Still stronger skepticism is induced by a study of Permian pelycosaurs (described later, in chap. 9). This group consists of early forms which are not merely reptiles but reptiles that are already separated from other major lines and are on the way to becoming the ancestors of mammals. Here, in this progressive group, one would think that we would be dealing with purely terrestrial amniotes. Some pelycosaurs are reptiles of this nature. But the more primitive pelycosaurs were not, to any degree, terrestrial. They were aquatic fish-eaters; they possessed limbs which would enable them to climb the banks; but their home, like that of their amphibian and fish ancestors, lay in the Permian streams and ponds.

Can this be a secondary reversion to the water? Again, this is highly improbable. This type of pelycosaur is known well back into the Carboniferous period, and the only obvious conclusion from the facts is that despite the phylogenetic position of these pelycosaurs—well advanced up one major branch of the reptilian family tree—they had never left the water.

The fossil evidence, then, strongly suggests that, although the terrestrial egg-laying habit evolved at the beginning of reptilian evolution, adult reptiles at that stage were still essentially aquatic forms, and many remained aquatic or amphibious long after the amniote egg opened up to them the full potentialities of terrestrial existence.

If we accept this as a reasonable conclusion from the paleontological evidence, we are, nevertheless, faced with a major puzzle. In the light of the earlier point of view, one could readily account for the success of the amniote type of development as being strongly favored by selection in animals which were otherwise terrestrial in habits. But what strong advantage could there be in terrestrial embryonic development in the case of forms which were still aquatic or, at the most, amphibious in adult life?

The answer can be found by considering, as we did a bit earlier, the types of reproduction found in modern amphibians. "The typical" mode of amphibian reproduction involves laying the eggs in the water and the presence of a water-dwelling tadpole stage in the development of the young. But, as we have seen, a very large proportion of modern amphibians fail to follow this pattern and, in fact, go to any extreme to avoid laying their eggs in the water. In none of these varied adaptations does there develop a shelled egg; but in one or another of these modern forms we find, singly or together, the other principal features seen in the amniote type—a large yolk, embryonic breathing organs, protective membranes analogous to an amnion.

These amphibians, then, are traveling today along a path of embryonic adaptations in a manner parallel to that followed many millions of years ago by the ancestral amniotes. But they are not obviously influenced by any "urge" toward a purely terrestrial existence, for the amphibians which show these trends toward direct development are as varied in adult habits as are amphibians as a whole.

There appear to be two major advantages. (1) Eggs and young in a pond form a tempting food supply, open to attack by a variety of hungry animals, including insects; furthermore, the larvae are in heavy competition for food with other small water dwellers. If eggs are laid in less obvious places, the chance of survival is greatly increased; if guarded or carried by a parent, they are under protection. (2) In some regions there are annual dry seasons, when the ponds and pools in which "normal" amphibians would lay their eggs tend to dry up. Reduction or elimination of the water stage increases the chances of survival of the young, which might be destroyed if they were living as tadpoles in a drying pond.

May not the amniote type of development have been similarly evolved to gain some immediate advantage rather than as any sort of "preadaptation" for land life? For modern amphibians, protection of the eggs from enemies is by far the more important of the two major advantages that are gained by changes in reproductive methods (although, in certain instances, adaptations which shorten larval life appear to be related to protection against potential drought conditions). For the Paleozoic reptile ancestors, the reverse was probably the case. Potential egg devourers were then presumably less abundant, but danger of desiccation was far greater.

Today, there are only limited regions of the tropics in which the annual weather cycle includes seasons of heavy rains alternating with droughts. But as Barrell first pointed out, large areas of the earth in late Paleozoic days appear to have been subject to marked seasonal drought (the presence of numerous red-bed deposits in the Upper Paleozoic appears to be correlated in great measure with drought phenomena). Under such conditions, the life of the amphibious vertebrates of the day was a hazardous one. Particularly hazardous was the developmental process. The old-fashioned methods meant that the young must, perforce, spend a long period of time as gill-breathing larvae, in grave danger of being overtaken by the oncoming of the rainless season and of being killed in their drying natal ponds. Any reproductive improvement which would reduce or eliminate this danger had a strong survival value. It is probable that various essays in this direction were made. The one truly successful one was that which led to the development of the amniote egg and the resultant origin of the reptiles, which from that time on became increasingly successful over their less progressive amphibian relatives. Today, a variety of amphibians are struggling (so to speak) to attain some type of development comparable to that which the reptile ancestors achieved eons ago, but their efforts are too little and too late.

Deductions from the study of climatic history are thus consonant with the facts of the fossil record. The fine story of the reptile ancestor as an animal which had become fully terrestrial in adult life and needed only, as a final step, to improve its reproductive habits in order to conquer the earth is, apparently, pure myth.

We may picture the ancestral reptile type as merely one among a variety of amphibious dwellers in the streams of late Paleozoic days. All were basically water dwellers. All alike found their living in the water with fishes and invertebrates as the food supply, for there was little animal life on land to tempt them. In most respects the early reptile had no advantage over its amphibian contemporaries. Only in its new type of development was the reptile better off. This advantage, however, did not at first imply the necessity of any trend toward increased adult life on land. It was only slowly, toward the close of the Paleozoic era, that many (but not all) of the reptiles took advantage of the new opportunities which amniote development offered them and became terrestrial types, initiating the major reptilian radiation in the Mesozoic—the Age of Reptiles. This potentiality of conquest of the earth by the reptiles was not the result of "design." Rather, it was the result of a happy accident—the further utilization of potentialities that had been attained as an adaptation of immediate value to their amphibious ancestors.

Teleology and tetrapod evolution.—It is sometimes said that the evolution of terrestrial vertebrates from fish ancestors cannot be explained by purely natural processes. How, it is asked, could there have been effected in the ancestral fish the whole series of structural and functional changes which were necessary for a successful existence on land but which appear to have been of no immediate value to the animal in its piscine existence? How could these changes have occurred unless there were some supernatural directive force behind the process, some mysterious "urge" that made for "pre-adaptation"?

In this case, as in other cases where the evolutionary history seems so complex that natural explanations appear improbable, the story can, I think, be explained on quite natural grounds with involvement of only the simplest of recognized evolutionary principles—selection for characters that are immediately useful to the animal in its actual environment without reference to their possible use in some future mode of life.

The complete transition from water to land involves a long series of structural and functional changes. For present purposes, however, we may select three of the most striking and outstanding changes that are necessary to enable a fish ancestor to become a successful terrestrial animal: (1) the development of lungs for air-breathing; (2) the development from fish fins of limbs capable of supporting it on land; and (3) the development (as discussed earlier) of a type of egg which will free the reproductive processes from the water.

How could these changes be of use to a water dweller? The clue, I believe, lies largely in the environmental factor already noted—widespread seasonal drought in late Paleozoic times. Let us discuss these three factors in turn.

Lungs(?) We think of lungs as essentially characteristic of land animals. What need does a fish, breathing with gills, have for such structures? We have already considered this matter in an earlier chapter. Lungs, we have seen, are useful to a fish under conditions of seasonal drought and were hence highly useful to a fish, as a fish, in Paleozoic days, without the necessity of dragging any mysterious "pre-determination" toward a future land life into the picture.

Limbs(?) This, too, we have already considered in our discussion of amphibian evolution. Limbs, in an early amphibian, were quite surely not a mysterious pre-adaptation for a life on land; such an existence in the first stages of limb development was (so to speak) the last thing it thought of or desired. Limbs were an immediately useful adaptation for life in the water; only gradually, as a terrestrial food supply developed, would the amphibians take advantage of the potentialities of becoming land dwellers.

The land egg(?) Our discussion above has, I hope, been sufficient to show that we do not have to account for the origin of the amniote egg by assuming any sort of mysterious "urge" toward a more completely terrestrial existence. For an amphibious animal, laying eggs safely ashore would be immediately advantageous.

We can thus see that some of the most prominent characteristics of land vertebrates can be accounted for as a series of adaptations that were of practical advantage as soon as they were acquired, while the animal was still partially or even entirely aquatic in its mode of life. The entire major evolutionary progression from fish, through amphibian, to terrestrial reptile—seemingly mysterious—can be interpreted in simple, natural terms. And it is probable that many another evolutionary development which appears difficult to understand without the introduction of teleology will likewise prove, when sufficiently investigated and studied, to be interpretable in the accepted framework of current evolutionary theory.

REPTILE ORGANIZATION

Before beginning a consideration of the various reptile types and their evolution, let us consider the general nature of reptiles as a whole. While some of the early forms, as we have just seen, lived an amphibious mode of life and others returned to the water, the group as a whole constitutes the basic stock of land animals and while varying a great deal from one form to another have many common attributes. They have advanced over the the amphibian condition and been modified from it in many regards other than the mode of development. No modern reptile is a truly generalized form, but in many respects some of the less-specialized lizards give a fair picture of a reptilian "norm."

Reptiles are suited to life on land only under warm conditions. Unlike their avian and mammalian descendants, they lack the power of maintaining a constant internal temperature, and their behavior is in great measure determined by this fact. Lizards are absent from the far north and are rare even in much of the temperate zones. Snakes are nearly as restricted in distribution and are able to withstand the winters of temperate zones only by hibernating—spending the cold season in a state of partially suspended animation in holes or burrows away from the cold air, frequently massed in rocky dens. Turtles are similarly restricted, and the crocodilians hardly venture beyond the tropics.

Heat as well as cold, however, can be fatal; a crocodile or lizard exposed too long to a hot sun will suffer or die. The ancient amphibians, we have noted, were covered like their fish ancestors with an armor of bony scales; modern amphibians have almost completely lost this covering and have a moist, glandular skin. Not so the reptiles. With a resultant reduction of water loss and potential desiccation in the open air and sunlight, the reptile skin is hard and dry and deficient in glands. The body is covered with horny scales or scutes, and although the original armor has been lost, bony scales or plates have redeveloped in many lizards, crocodilians, and turtles and are seen in various extinct types. In many reptiles the skin is strongly colored, often in patterns, and in lizards color changes can often be effected (the chameleons are especially notable in this regard).

With the exception of the snakes and some lizards, modern reptiles have limbs which are sturdy and well developed. In modern amphibians we have seen that the toes have but two or three joints, and there are never more than four toes on the forefoot. In primitive reptiles, however, and in many of their descendants there are five toes both fore and aft and a high count of toe joints—2, 3, 4, 5, 3 (or 4), counting from the thumb side out. In the trunk there is, in contrast with amphibians, a well-developed neck region and generally a well-developed tail.

There is, of course, no "breathing" through the hard dry skin, and the lungs, although still rather simple in structure, are better developed than in amphibians; except in turtles, the lungs are expanded by movements of ribs. We have noted the problems in circulation of the blood caused by the shift from gill to lungs in amphibian breathing. Reptiles have not completely solved the problem, but the heart is partly divided into two halves, so that a blood stream containing oxygen from the lungs is kept partially separated from the "stale" blood from the body, destined to pass to the lungs. Since the shell is added to the egg before it leaves the female reptile's body, internal fertilization by apposition of the cloacas of the two parents is necessary, and in most living reptiles the male has developed an intromittent organ of one type or another, somewhat comparable to the mammalian penis.

The reptilian brain is somewhat advanced over that of amphibians. The cerebral hemispheres are larger, showing the beginning of the modifications of that area of the brain which were to become very pronounced in birds and mammals. The behavior of reptiles, however, is of a relatively simple type.

Stem reptiles.—Living reptiles include only a few groups, mainly turtles, crocodiles, lizards, and snakes; they are vastly overshadowed in importance today by the birds and mammals descended from them. In the

Mesozoic, the Age of Reptiles, however, the reptiles were exceedingly numerous and varied and included many groups now extinct. These forms will be described in succeeding chapters; here we shall merely consider the earliest reptiles, first types to attain a land type of egg.

These extinct progenitors of the reptile dynasties are technically known as cotylosaurs but popularly and quite appropriately have been termed "stem reptiles." The first traces of them are found in the Carboniferous, in the deposits formed in the great coal swamps when the amphibians were at the peak of their development; stem reptiles were still flourishing in the Permian, last of the Paleozoic periods.

A typical stem reptile is *Seymouria*, shown in our illustrations. (The creature takes its name from the Texas town near which its remains were discovered.) In appearance it might be compared with some of the larger and more sluggish lizards; but structurally it was much more primitive

Limnoscelis, a cotylosaur, which was surely on the reptilian level reproductively but was still an amphibious, fish-eating animal.

than any sort of lizard. The limbs were still of the short, stubby type found in the oldest amphibians and sprawled out sideways from the body. Indeed, so close is *Seymouria* in many skeletal features to the early amphibians that, in default of actual fossil knowledge of eggs of this form, the academic question has been raised as to whether *Seymouria* was an amphibian which was almost a reptile or, on the other hand, a reptile which had just ceased to be an amphibian. This debate, however, merely emphasizes the point that it is the change in development which is the important feature in the history of reptiles; without this, the great radiation of the group and the many structural advances and changes could never have been accomplished or effectively utilized.

During the Permian the stem reptiles branched out into a number of varied lines. But their fate has been that of the founders of any great line of evolutionary development; from them arose many more progressive groups, and they themselves soon vanished from the scene as the true Age of Reptiles began.

Reptile classification.—Although living reptiles include members of but four orders, the extinct forms were highly varied. How to classify and arrange these groups is a difficult problem, for although many of the extinct types are well known, there are still many missing links in the evolutionary chain.

It was long ago pointed out that one readily observable clue to reptile relationships lay in the structure of the temple region of the skull. In fishes, early amphibians, and even stem reptiles, the walls of the skull in the temple and cheek region behind the eye were solidly built. Beneath the cheek wall lie the powerful muscles—the temporal muscles that close the jaws. When the jaws shut, these muscles contract and necessarily bulge

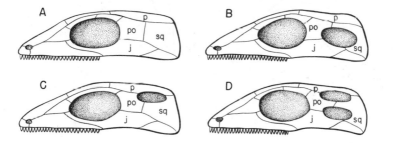

Diagrammatic side views of reptile skulls, to show diagnostic types of openings in the temporal region. *A*, the primitive anapsid type, without openings; *B*, a lateral opening is present; the synapsid type, found in the ancestors of mammals; *C*, an upper temporal opening, present in varied form in plesiosaurs and ichthyosaurs. *D*, the diapsid type, with two openings (and two arches), found in *Sphenodon* and in crocodilians and other archosaurs. The lizard and snake structure is derived by reduction from this condition. The bones of the temple region are labeled: *j*, jugal; *p*, parietal; *po*, postorbital; *sq*, squamosal.

outward in the process. Space is limited in this region. It would be a considerable improvement if the walls of the skull developed openings through which the muscles could bulge outward when contracted; and in most reptiles such openings are present. There are several ways in which reptile temporal openings have developed; these are shown in an accompanying figure. A single opening may develop low down on the side of the cheek, as in the case in the reptilian ancestors of mammals. One may develop high up in the temple region, as in the varied aquatic reptile groups including the ichthyosaurs and plesiosaurs. Third, openings may develop in both positions; this is the situation found today in a peculiar New Zealand reptile *Sphenodon* and crocodilians and, in the past, in the dinosaurs and their kin; further, lizards and snakes, although having a much modified temple region, have descended from forms with these two openings. Most students divide the reptiles into a series of subclasses defined, in part at least, on the nature of the temporal region. In giving names to types of temporal openings and, in some cases, to the subclasses of reptiles possessing them,

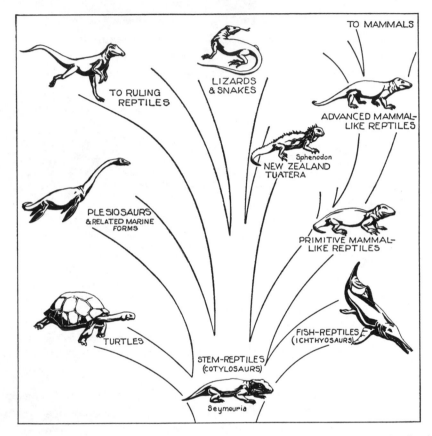

The family tree of the reptiles (except the archosaurs—"ruling reptiles"—shown elsewhere)

the attention of early workers centered on the arches of bones lying below the openings rather than on the holes themselves. "Apse" is a Greek root meaning arch. In consequence reptiles with a solid temple region—and hence lacking an arch—have been termed anapsids; mammal-like reptiles with a single arch, synapsids; reptiles with two arches, diapsids.

CHAPTER 7

Some Water Reptiles

In this and the two chapters that follow we shall discuss various types, living and extinct, into which the reptiles radiated following their appearance in late Paleozoic days. In the present chapter we shall consider the turtles, a group in which water life plays a large part, and the two most spectacular orders of marine reptiles, both extinct—the plesiosaurs and ichthyosaurs.

Reptiles are in general purely land animals; one tends in consequence to assume that the aquatic reptiles are descended from ancestors that had become fully terrestrial and that in their evolution they had done a complete about-face to re-enter the water. It is probable, however, that this assumption should be somewhat qualified. In discussing the origin of the amniote type of egg, it was pointed out that the animals which first walked ashore to deposit this type of egg and hence, by definition, were to be called reptiles were at that time still more or less amphibious in habits and had not completely abandoned life in pools and streams and swamps. It is not at all improbable that the evolutionary lines leading to such animals as the plesiosaurs and ichthyosaurs departed from the main course of reptile evolution at this stage and that their ancestors had never completely abandoned the water. Again, although a few turtles are pure land types, a large proportion are primarily aquatic, and it is reasonable to believe that the turtles' early reptile ancestors had never progressed farther toward land life than the swamp type of habitat in which many chelonians find their homes to this day.

TURTLES

The turtles are the most bizarre of reptilian groups. Because they are still living, turtles are commonplace objects to us; were they extinct, their shells, the most remarkable armor ever assumed by a land animal, would be a cause for wonder.

Turtle structure and habits.—The armor plate of a typical modern turtle is composed of two materials—horny scutes representing the ordinary reptilian scales and bony plates underneath. The outlines of these two sets of armor materials do not in general coincide; there is an alternation of joints, which gives greater strength to the combined structure.

The shell is divided into upper and lower portions (carapace and plastron, respectively), connected by a bridge at the sides. In the live animal the bone is concealed, and only the horny scutes are seen; on the upper surface, three lengthwise rows of a few scutes each cover most of the carapace, and there is a marginal ring of smaller scutes. In the underlying bony carapace we see, running down the middle of the top, a row of almost square bony plates, most of which are fused tightly to the joints of the backbone beneath them; on either side is a row of longer plates fused to the corresponding ribs; a ring of small plates around the margin completes the top shield. The undershield is composed of a much smaller number of large scutes and bony plates. The shell is widely opened at the front for the withdrawal of the head and front legs, behind for the hind legs and stubby tail.

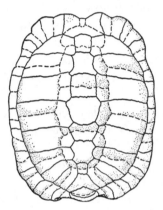

Diagram of the upper shell, or carapace, of a turtle. The major structure consists of plates of bone, outlined in continuous line. These consist of a middle row of plates, one over each joint of the backbone; paired rows of rib plates on each side; and a marginal row encircling the whole. These bones, however, are invisible in the living turtle, for they are covered by horny scutes, the outlines of which are in broken lines in the diagram. Note that in most cases the joints of the two types of covering alternate, thus strengthening the structure.

The turtle gait is a very awkward and lumbering one, with the limbs sprawled outward at the side. With the broad ventral shield no other type of walking is, of course, possible. But it is probable that this style of walking antedates the construction of the turtle's shell. This sort of locomotion was, we have noted, characteristic of primitive land dwellers in general. Apparently the turtles evolved their shells before styles in walking had improved; once the armor was in place, no improvement was possible. A unique feature of limb construction in turtles lies in the fact that, whereas in other land animals the shoulder girdle lies outside the ribs, here it is tucked away inside shell and ribs.

In turtles (as in modern birds) the teeth have been entirely lost and replaced by a stout and sharp horny bill bounding the margins of the jaws.

There is no true temporal opening in the skull, and partly for this reason the chelonians are usually bracketed with the stem reptiles in a subclass termed, with reference to this condition, the Anapsida.

In primitive land animals the eardrum was placed in a notch at the back of the skull. In most reptiles such a notch is small; in turtles it is very conspicuously developed, with a great curved ridge of bone nearly surrounding the drum. Despite this prominent development, turtle hearing appears to be very poor. (But this is a relatively small loss, for turtles make few noises for their colleagues to hear; the biblical statement that "the voice of the turtle is heard in the land" appears to refer to quite another creature—the turtle dove.)

Most reptiles have a long backbone; many primitive forms, for example, have about twenty-seven vertebral segments in neck and trunk combined, and snakes have vastly higher counts. Turtles run to the other extreme; there is a rather normal number of segments in the neck but only ten in the whole trunk, eight of which, together with their attached ribs, are tightly fused to the overlying plates of the carapace. The tail appears to be of little use and is typically short.

In other reptiles breathing is accomplished by movements of the ribs which fill and empty the lungs. This is, of course, out of the question in the turtle shell. Here the same result is obtained by movement of belly muscles inside the shell, which cause alternate contraction and expansion of the body cavities in which the lungs lie. Most turtles can remain under water for long periods. In great measure this is associated with the fact that a motionless reptile of this sort uses up very little oxygen; it "lives" at a very slow tempo. In many cases, however, the animal is able to get oxygen from the water in which it is submerged, despite the fact that gills disappeared far back in pre-reptilian days. A gill is essentially a large area of soft moist membrane through which oxygen can be absorbed. Various turtles have evolved fairly effective substitutes for gills, being able to absorb oxygen from the water through membranous areas in both throat and cloaca.

Some chelonians specialize to some degree in animal or vegetable food materials, but in general members of this order are omnivorous and will readily take in any sort of food that comes to hand. Egg-laying usually takes place annually. The eggs are deposited in shallow holes scooped out in the earth and covered.

The horny scutes of the shells of many turtles appear to grow by adding annual rings around the margin. To a limited degree, study of such rings enables us to tell the age of a turtle and thus aids in a study of growth. We may insert here a note regarding reptile growth in general. In the case of a mammal or bird, one can make fairly positive statements as to the size of the adult individuals of any given species. Not so with reptiles. The difference lies in the manner in which the bones of the skeleton grow. In birds and mammals, increase in bone size not merely slows down as the animal approaches maturity but stops completely early in adult life; no further increase in proportions can occur (let us exclude the delicate subject of waist measure!). Not so in reptiles. Bone growth slows down at about the time of sexual maturity but never ceases entirely; a turtle or other reptile may continue to grow as long as it lives. As a consequence it is impossible to state *the adult size* of a reptile of any sort. Since in nature, however, few animals live to a ripe old age, one can state an "average" size for mature individuals of any form; they will mostly be young adults. On the other hand, if one wants to make an impression, one can cite the considerably greater figure for the largest known specimen—which will, of course, be a patriarch of the tribe.

Turtle types.—Compared to the wealth of lizard and snake species, the chelonians play but a modest role in the modern world; there are only two to three hundred species representing the order. Despite a uniformity of basic body pattern in most chelonians, they exhibit a considerable variety in the nature of the shell, limbs, and skull and vary greatly in habits. In the United States and Canada all members of the order are commonly called turtles. Englishmen use the word "turtle" to refer to marine forms only; marsh or river forms are frequently termed "terrapins," and land types are dignified by the name "tortoise." They are speaking of foreign animals, however, for curiously enough there is not a single member of the order in the British Isles.

Like other reptiles, turtles are limited in their distribution by temperature. Sea turtles may wander far to the north and south in warm currents, but they are mainly tropical and breed only on tropical shores. Of freshwater and land types, a number are present in temperate zones, but most inhabit tropical or subtropical areas. There is an abundance of chelonians on every continent except Australia and Europe.

Living chelonians are sharply divided into two groups. The fossil record indicates that the older turtles were unable to pull their necks into their shells for protection. All modern forms can do this (except some large aquatic types, which are in little danger of attack and appear to have secondarily lost this ability). There are two different ways of doing the

Hitch-hikers (*above*). A pair of remoras stealing a ride with a marine turtle. (Photograph courtesy New York Zoölogical Society.)

The leathery turtle (*Dermochelys*) (*upper right*). A giant marine form, not closely related to those shown on the next page. Of the bony shell only small nubbins of bone remain, and the horny shield is replaced by a tough leathery hide. This is the largest living turtle, with a record weight of nearly a ton and a length of 8 feet. (Photograph courtesy American Museum of Natural History, New York.)

The snapping turtle (*Chelydra*) (*right*). The snappers are vicious water dwellers in North American streams, notable for large heads and large tails—unusual in turtles. (Photograph courtesy New York Zoölogical Society.)

The Australian snake-necked turtle (*Chelodina*) (*right*). This form is representative of a primitive group, the "side-necked" turtles (pleurodires), found only in southern continents. Instead of pulling the neck straight into the shell, it is tucked in sideways along the shoulder. (Photograph courtesy American Museum of Natural History, New York.)

A giant tortoise. The tortoises of the genus *Testudo*, in contrast to most turtles, are purely land dwellers, characterized by a high, domelike carapace. Several large species are present in the Galápagos and others on islands in the Indian Ocean. The form figured is from the Aldabra Islands. Record weights are over 500 pounds and lengths run to 6 feet. How these land dwellers reached such oceanic islands is a mystery. (Photograph courtesy New York Zoölogical Society.)

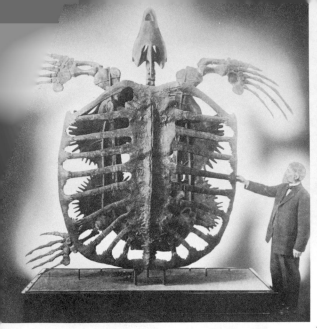

A giant Cretaceous marine turtle (*Archelon*). A restoration of this form is shown in a later plate. Although related to modern sea turtles, it is much larger than living genera. (This skeleton is in the Peabody Museum, Yale University.)

The soft-shelled turtle (*Trionyx*). These turtles are pond dwellers of modest size in which the horny covering is lost and the bony armor somewhat reduced not so much, however, as in the big leathery turtle (Photograph courtesy New York Zoölogical Society

Living marine turtles. The green (soup) and hawksbill (tortoise-shell) species. Somewhat smaller than the leathery turtle, also found in the sea, the green and hawksbill turtles retain a horny shell. (Photographs courtesy New York Zoölogical Society.)

trick. The great majority of turtles, including all those with which the reader will ordinarily be familiar, pull the head straight back, bending the neck (in which the vertebrae have complicated joints) into an S-shaped curve. These forms are termed the Cryptodira. In contrast, a few turtles of the southern continents swing the head sideways and tuck it in alongside the shoulder. They are reasonably termed Pleurodira ("side-necked" turtles).

We may begin with a review of the numerous cryptodires. Most familiar and on the whole probably the most typical and primitive of living chelonians are the numerous marsh and pond turtles, placid little animals widespread in the northern continents. Omnivorous in food habits but far from voracious, they are not uncommonly seen ashore but spend most of their time in ponds or swamps or sunning themselves on the banks. Turtles of this sort are absent from Australia, rare in Africa and South America. In North and Central America, however, there are nearly a score of genera; Asia—particularly southern Asia—where we find an even larger series of genera, is another stronghold of cryptodires.

From a primitive existence in a swampy environment, chelonians have spread in two directions as regards their mode of life. By far the greater number have tended to an aquatic existence, but we may first note the tortoises proper, which have, on the contrary, abandoned the water and become pure land dwellers, even ranging into dry desert conditions. Most of these forms are of a very uniform build and are ranged in a single genus *Testudo*. The tortoises tend to grow to rather large size; the shell is highly rounded, or domed; the legs are sturdy, massive, with stumpy feet. The diet is a purely herbivorous one. Land tortoises are particularly abundant today in Africa but can be found in all other major areas except Australia and much of South America. A tortoise from the southern United States, called the "gopher," is remarkable in that it digs long burrows (mainly for hibernation) which may run to as much as forty feet. Its name is an excellent example of variability in the use of popular names. In most of the United States, "gopher" means a ground squirrel; another type of burrowing rodent is the "pocket gopher." But in the deep South "gopher" refers to this tortoise—sometimes emended to "hard shell gopher"—and the pocket gopher is called a "salamander"!

Tortoises are of further interest for having reached a number of islands far from continental shores, such as the Galapagos and Mauritius, where they flourished greatly until the coming of man. Galapagos tortoises were for decades the main fresh-meat supply for American whalers in the

Pacific. The land tortoises generally are of substantial size, and some of the island forms reach a yard or more in shell length. The giant of the tortoise group, however, was an extinct species from India with a shell length of over 6 feet.

The island forms are of considerable interest to the student of animal history. How have these purely land forms reached islands in the Indian Ocean or the Pacific, which appear never to have been connected with the mainland? In the Galapagos the tortoises have, since their arrival there, split up into a number of species on the different islands. The study of these species by Darwin, during his journey around the globe in H.M.S. "Beagle," was one of the factors which led to his work on the problems of evolution.

But although the tortoises migrated out of the marshes onto dry land, most other chelonians have moved in the other direction, toward a purely aquatic existence. Typical of fresh-water aquatic forms, which are nevertheless relatively primitive in structure, are the snapping turtles, now confined to North America but found in fossil form in Europe as well. In most structural features they are fairly similar to the various marsh dwellers. However, they have large heads, powerful beaks, and, as the name indicates, are aggressive stream-dwelling carnivores. The ordinary snapper is of good size; the larger species—the alligator snapper—is perhaps the largest of fresh-water turtles, its weight reaching 200 pounds.

From an aquatic life in fresh waters, it is not a long step to the development of marine life. The typical marine turtles of modern days form a compact group, including the green turtles (the source of turtle soup), the hawksbill (tortoise-shell), and the loggerhead (not of much economic importance). Buoyed up by salt water, the marine turtles can—and do— grow to considerable size. Typical sea turtles have retained their horny shell covering, but the underlying bony armor is somewhat reduced. The limbs are transformed into flippers with which, despite the clumsy body shape, the sea turtles can swim through the seas at good speed and with a graceful, winglike motion. They come ashore only to lay their eggs on tropical beaches. Here, out of their native element, they are an easy prey to enemies—particularly man; once turned on their backs they are helpless.

Sea turtles evolved far back in the history of the order, and a variety of large forms of this sort (one of them illustrated here) had evolved by the end of the Cretaceous.

For the most part, turtles have retained the horny shell covering. But this is not universally the case. In one widespread group of fresh-water turtles, the horny shell has gone, to be replaced by a leathery skin. They are usually termed soft-shelled turtles; this is, however, somewhat of a misnomer, for a well-developed bony shell is present beneath the skin. The body is considerably flattened, and one writer has compared them to an animated pancake. The head ends in a peculiar projecting snout. The neck is long, and these turtles, which are active predators, strike out aggressively at their prey. The soft-shells are particularly abundant in the warmer regions of southern Asia and the West Indies; there are outliers in Africa and two species in North America, but they are absent in all other areas.

The peak of shell reduction is seen in a marine turtle, the leatherback. While it resembles other marine turtles in its general form and mode of life, this large animal, which may gain a body length of seven feet, has lost not only its horny shell but most of the bony carapace. Instead of a connected shield of plates fused to ribs and backbone, there is merely a mosaic of bony nuggets, buried in the skin; the larger elements are embedded in a series of ridges, running lengthwise of the back. Some workers have advocated the view that this rudimentary condition of the shell is a primitive one. It is, however, much more probable that there has been a secondary reduction of armor; there is little need for protection from enemies in the case of such a large marine form and the less weight the better.

The side-necked turtles, few in number, are poorly known and confined to the tropics of the southern continents. They are aquatic, fresh-water forms, some of grotesque appearance, combining some primitive features with specialization. Mainly eaters of animal food, they tend to be active predators, somewhat after the fashion of the soft-shelled turtles. The necks are long—sometimes longer than the length of the shell.

As said above, the side-necks are confined today to the southern continents. There are half a dozen genera in South America and, at the other end of the world, four in Australia; the side-necks are the only turtles to have reached the latter region. In between, two genera inhabit Africa, and one is found in Madagascar. This odd distribution is, as in many other cases mentioned in this book, explicable in terms of previous history, for the pleurodires are an ancient group, going back to the days of the dinosaurs in the Cretaceous; during Tertiary times side-necked turtles were widespread in Eurasia and North America.

Early turtle history.—Turtles are such common fossils in some formations, such as the badlands of South Dakota, that their remains are a source of irritation rather than of pleasure to the fossil collector. Many of the living turtle groups can be traced back through the Tertiary and even into the Cretaceous. Mesozoic turtles, however, were mostly of a somewhat different and more primitive sort, of which the most obvious feature is that they had not yet acquired the power of retracting the neck. But the typical turtle build persists throughout. The oldest known turtle, *Proganochelys* of the Triassic (a contemporary of the earliest dinosaurs), had already acquired all the essential features of the order, although retaining a few primitive characters such as vestiges of teeth.

Before this we have no certain knowledge of the chelonian pedigree. We are fairly sure that they arose from the cotylosaurs, some of which were rather clumsily built marsh dwellers with a tendency for developing

A restoration of the oldest known turtle, *Proganochelys* of the Triassic. Apart from the presence of extra bony spines protecting the neck and limbs, all the essential features of modern chelonians were already present. (After Jaekel.)

armor on the back, but there are no connecting links. The only possible clue is a little animal from the Permian of South Africa with the long name of *Eunotosaurus*. This has eight pairs of expanded ribs, suggesting the eight ribs which support the turtle shell, and the beginning of armor in the presence of a mosaic of small thin plates on the back. Possibly it is allied to turtle ancestry; but the students of modern turtles do not look with favor on the suggestion of the paleontologists that this odd creature be included in the chelonian domain.

The turtles, once within the shelter of their armor, became the conservatives of the reptilian world. The oldest forms were contemporaries of the earliest dinosaurs. The ruling reptiles grew to dominate the reptilian scene, but the turtles persisted unchanged. The dinosaurs passed away, and the mammals took their place, but the turtles went calmly on their placid way; they learned to pull in their heads, but otherwise remained much the same. Now man dominates the scene, but the turtles are still with us. And if, in the far distant future, man in turn disappears from the earth, very likely there will still be found the turtle, plodding stolidly on down the corridor of time.

PLESIOSAURS

Marine side branches have developed in several groups of reptiles which have also terrestrial or, at least, amphibious representatives—turtles (as we have just seen) and lizards, snakes, and crocodilians (as we shall see in later chapters). We shall conclude this chapter with a consideration of some extinct reptilian groups which were exclusively marine in nature.

The oldest remains of one of these groups, the plesiosaurs, came to light in Europe over a century ago. The scientific name applied to them means "near reptiles" and refers to the fact that it was once thought that they were forms coming up toward the reptilian level from a former water-

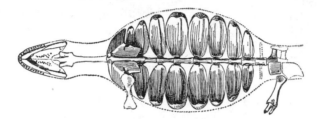

The skeleton (incompletely known) of a small fossil reptile (*Eunotosaurus*) from the Permian, which may be related to the ancestry of the turtles. Eight pairs of the ribs are expanded into broad bony plates; this may represent the beginnings of the turtle carapace. (From Watson.)

dwelling life. We now know, however, that the reverse was really the case and that they are descendants of four-footed reptiles which had become terrestrial (or nearly so, at any rate), then reverted to an aquatic and, eventually, a purely marine mode of life.

The plesiosaurs were common inhabitants of the Jurassic oceans and survived until near the close of the Cretaceous, where they are found as contemporaries of other marine reptiles—marine lizards and early sea turtles. Many plesiosaurs were of fairly modest size, with lengths from a dozen to twenty feet or so. The giant of the group was an Australian Cretaceous form (*Kronosaurus*) whose skeleton, mounted in the Harvard Museum, stretches a good forty feet. The plesiosaur build was a curious one. An old writer described a plesiosaur as "a snake threaded through the shell of a turtle," and in some instances the comparison is not inapt. The trunk was very broad, flat, and inflexible and was well plated with bones, so that there was a superficial resemblance to a turtle body. These bones, however, were not armor but areas for attachment of the powerful muscles of the paddles. Since the trunk was a rigid structure and the tail little developed, the plesiosaurs, like the similarly constructed turtles, had to "row" themselves along by powerfully developed limbs. These were highly specialized, for the number of joints in the toes was greatly increased. In no normal land reptile does the number of joints in a toe exceed five; in plesiosaurs there were sometimes over a dozen joints per toe.

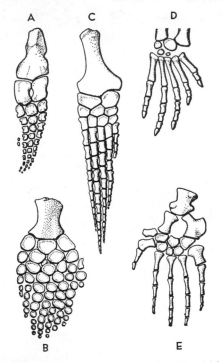

An unwieldy body of this sort was obviously a disadvantageous structure for animals which pursued elusive fishes; sharp turns and delicate steering were as impossible as they are in a rowboat. In compensation we find that many plesiosaurs had much-elongated and flexible necks (one form had seventy-two neck vertebrae!) which could be readily turned to dart the head at the prey. Such forms usually had short heads; another type of plesiosaur had a shorter neck but a much-elongated beak which appeared to serve just as well.

The limbs of various aquatic reptiles. *D*, a fairly primitive type of reptile, little adapted to marine life. Of this form only the foot is drawn; in the other figures the entire limb is shown, the long joints shortening up to make a compact paddle. *A* and *B* are ichthyosaurs (*Merriamia* and *Oph-thalmosaurus*) in which the number of toe joints has been considerably increased, while in *B* extra toes have been added. In the plesiosaurs, as in *C* (*Elasmosaurus*), there are always five toes, but the number of joints may be much increased. In the marine lizards, as in *E* (*Clidastes*), the toes are fairly normal but were spread out and were presumably webbed. (Mainly after Williston.)

We feel sure that the plesiosaurs have descended from more typical four-footed reptiles, and part of their ancestral story is clear. Plesiosaurs, we noted, do not appear until the Jurassic, the second of the three periods of the Age of Reptiles. In the period preceding this, the Triassic, we find—particularly in Central Europe—abundant remains of much smaller aquatic animals—the nothosaurs. These have technical characters which ally them with the later plesiosaurs. They were, however, much less specialized than the plesiosaurs. The body was relatively longer and less broad, and the tail was better developed than in plesiosaurs. And in the

A Jurassic sea. European deposits of this age have revealed an abundant fauna of marine animals of all types. Particularly interesting are the two types of reptiles illustrated here which left the land and returned to the water to become highly specialized aquatic types. On the left is a group of plesiosaurs. These forms had a broad flat body and swam by strokes of their limbs, which had been transformed into powerful paddles. Frequently, as in the form illustrated, the neck became greatly elongated. A second and more highly specialized type is that of the ichthyosaurs, two of which are shown at the right. These were good reptiles but had reassumed a typical fish shape and had reacquired fishlike fins. They seem to have been similar in form and habits to modern porpoises, although of course they are not related. (From a mural by Charles R. Knight; photograph courtesy Chicago Natural History Museum.)

A plesiosaur skeleton, viewed from above. This specimen shows well the small head, slender neck, broad flat body, and powerful swimming limbs of typical plesiosaurs. (Photograph courtesy American Museum of Natural History, New York.)

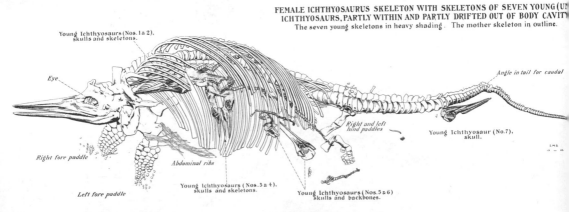

FEMALE ICHTHYOSAURUS SKELETON WITH SKELETONS OF SEVEN YOUNG (U)
ICHTHYOSAURS, PARTLY WITHIN AND PARTLY DRIFTED OUT OF BODY CAVIT
The seven young skeletons in heavy shading.   The mother skeleton in outline.

Young Ichthyosaurs (Nos. 1 & 2),
skulls and skeletons.

Eye

Angle in tail for caudal

Right and left
hind paddles

Young Ichthyosaur (No.7),
skull.

Right fore paddle

Abdominal ribs

Left fore paddle

Young Ichthyosaurs (Nos. 3 & 4),
skulls and skeletons.

Young Ichthyosaurs (Nos. 5 & 6)
skulls and backbones.

Ichthyosaur with unborn young. (Courtesy American Museum of Natural History, New York.)

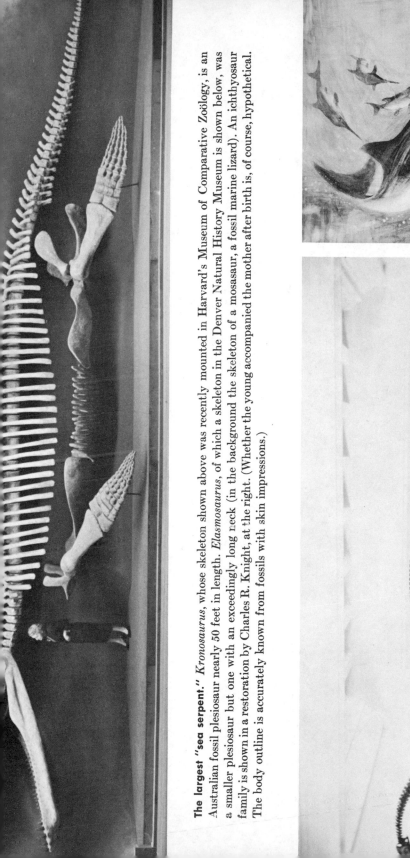

The largest "sea serpent." *Kronosaurus*, whose skeleton shown above was recently mounted in Harvard's Museum of Comparative Zoölogy, is an Australian fossil plesiosaur nearly 50 feet in length. *Elasmosaurus*, of which a skeleton in the Denver Natural History Museum is shown below, was a smaller plesiosaur but one with an exceedingly long neck (in the background the skeleton of a mosasaur, a fossil marine lizard). An ichthyosaur family is shown in a restoration by Charles R. Knight, at the right. (Whether the young accompanied the mother after birth is, of course, hypothetical. The body outline is accurately known from fossils with skin impressions.)

F.M.N.H.

Kansas in the Cretaceous. Near the end of the Age of Reptiles much of America was covered by shallow seas filled with an abundant vertebrate life. Chalk rocks of this age bearing numerous fossils are well represented in western Kansas. Some of the characteristic forms are shown here. *Center,* a mosasaur (*Clidastes*), a great marine lizard distantly related to the terrestrial monitor lizards of the Old World. *Right,* a giant marine turtle (*Archelon*), an early representative of a group that has survived to grace the soup tureen of today. Soaring above the waters are a number of giant flying reptiles of the genus *Pteranodon*—tail-less forms characterized by a long crest at the back of the head. Other common members of the fauna not illustrated are plesiosaurs, toothed birds, and a great variety of fishes. (From a mural by Charles R. Knight; photograph courtesy Chicago Natural History Museum.)

limbs the pattern was still essentially that of a normal reptile rather than a plesiosaur paddle, although the toes were somewhat elongated (and probably webbed in life). Most nothosaurs appear to be a bit off the direct line to the plesiosaurs, but as a group they quite surely include the plesiosaur ancestors.

Back of this, the trail is faint. As we have said, we look in reptiles to the temporal region of the skull to give clues as to relationships. Nothosaur and plesiosaur skulls have a characteristic temporal opening high up on the skull, of a sort not repeated in other well-known groups. (In the appendix on classification at the end of the book you will find that they are, in consequence, placed in a distinct Subclass Euryapsida.) In the Permian there have been found some small and rare reptiles with a similar temporal region. But there are no indications in their skeleton of an aquatic trend, and we cannot be at all sure that they are plesiosaur ancestors.

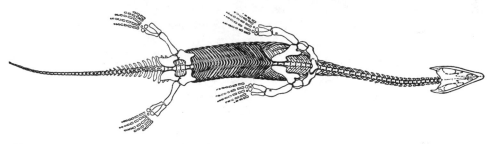

The skeleton of a Triassic nothosaur, seen in ventral view; from this sort of animal, only partly adapted to a marine life, arose the plesiosaurs. (After Peyer.)

ICHTHYOSAURS

Of all reptiles, the group most highly adapted to an aquatic existence was that of the extinct ichthyosaurs, which well deserve their name of "fish reptiles." The life-span of these forms covered most of the Mesozoic, or Age of Reptiles; they seem to have occupied the place in nature now taken by the dolphins and porpoises.

The superficial appearance was very fishlike; the body was short and rather deep and compressed laterally; the neck was very short, with the head set closely on the shoulders, so that there was a reappearance of the torpedo-like fish type in body form. In black shales in southern Germany a number of specimens have been found in which the skin outline has been preserved. These specimens show that there was a large fishlike dorsal fin. Even more interesting is the tail construction. Early finds all showed an apparently broken tail with the end of the backbone sagging downward.

The older restorations "corrected" this and showed a normal straight-tailed animal. But specimens with the body outlines preserved prove that the seeming break is a normal condition; there was a sharklike tail with, however, the backbone extending down into the lower lobe, whereas in fishes it tilts upward into the upper half of the fin.

The ichthyosaurs undoubtedly swam in fishlike fashion by lateral undulations of body and tail, and the limbs, used only for steering and balancing, are comparatively small. They are very highly modified; the individual bones assumed circular or polygonal contours and were packed very closely together so that, although the limb as a whole was flexible, there were no free movements between the joints. As in the plesiosaurs, there was a considerable increase in the number of toe joints. But while the plesiosaurs retained the original number of toes, the fish reptiles varied widely. Some had but three toes in a fin; in others, new toes were budded off, giving as many as eight digits; this is the only group of four-footed animals which has exceeded the orthodox number of toes. It seems certain that the ichthyosaurs were agile swimmers, as are the modern porpoises, which they parallel closely in body build. Some of the Jurassic water reptiles made much of their living by eating ancestral squids, as is shown by stomach contents preserved in some of the Jurassic slab specimens.

These creatures were so extreme in their marine adaptations and their limbs so obviously unsuited for use on land that the problem of ichthyosaur reproduction was early raised. A reptile's egg will drown in the water as surely as an adult. In some snakes and lizards the eggs are retained in the mother's body until they hatch, and it was suggested that the reproduction of ichthyosaurs must have been of a similar nature. In agreement with this idea that the fish reptile's young were born alive are a number of specimens which actually show skeletons of young ichthyosaurs inside the body of an adult individual. It has been argued that these may have been youngsters which had been eaten by mistake. But in several instances the young have been observed partially emergent from what would have been the outer opening of the reproductive tract in life. The mother here apparently died during childbirth or (there are human parallels) labor may have taken place after the death of the mother.

Ichthyosaurs became rare in the Cretaceous and appear to have died out well before the close of that period. It is difficult to understand why such seemingly highly adapted forms lacked success in the Cretaceous, when other marine reptiles, such as the plesiosaurs, continued to be numerous to the end of that period.

The pedigree of the ichthyosaurs presents even a worse problem than that of the plesiosaurs. The typical ichthyosaurs are those of the Jurassic. Back of this, there have been found in the middle and late Triassic (particularly in Nevada) remains of ichthyosaurs of a somewhat more primitive build, with nearly straight tails and with limbs somewhat more primitive in nature. But further back than this we draw a complete blank. There is no evidence to connect the ichthyosaurs with any other reptile group whatever; they occupy a completely isolated position. Even recourse to that old favorite, the nature of the temporal opening, is no help. As in plesiosaurs, there is a single opening high up on the head, but the pattern of the bones here is different from that of plesiosaurs and probably

developed independently within the group. Very likely they are descended from some advanced cotylosaur type. But there is no trace of intermediates. Some day, I hope, we may find the answer.

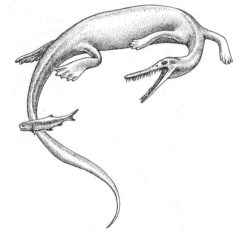

MESOSAURUS AND GONDWANALAND

We shall end this chapter by some notes on a little fossil aquatic reptile, *Mesosaurus*, which deserves little attention on its own account, but is of interest as introducing a major point of debate about the geography of continents in the distant past.

Mesosaurus, a small Paleozoic aquatic reptile of South Africa and South America. (After McGregor.)

Mesosaurus was a small reptile, two or three feet in length, whose remains are known only from certain beds in southeastern Brazil and, directly across the Atlantic, in western South Africa. The date is a bit uncertain but apparently about the beginning of Permian days. This reptile was reasonably well adapted for life in the water. The tail was very well developed and could have been a good swimming organ, and the feet on the long hind legs were large with spreading, and probably webbed, toes. (This is curious, incidentally; marine reptiles typically have *either* a good tail, as do the ichthyosaurs, *or* good propulsive limbs, as in plesiosaurs; here we have both, a situation comparable to a man wearing both belt and suspenders.) The jaws were exceedingly long and armed with a battery of very long, sharp, and slender teeth. The animal has no identifiable relatives and presumably represents an early but sterile little side branch of the basic reptile stock. And there is no trace of it in any fossil deposits except in two beds facing each other across the South Atlantic. How are we to account for this curious distribution?

Now for Gondwanaland. Far back in the last century it was discovered that Permian and Triassic beds in peninsular India contained a series of plants quite unlike any found in northern continents but very similar to contemporary floras in Australia, Africa, and even South America. It was further found that where today we find a series of mountain chains, including the Alps and Himalayas, running west to east across southern Europe and southern Asia, there was in those days exactly the opposite condition—a broad seaway separating most of Eurasia from the southern continental regions (plus India). To account primarily for the plant distribution, it was assumed that these areas formed a united southern land mass. Since many of the plants which furnished the evidence were found in an Indian formation termed the "Gondwana beds," this hypothetical continent was called "Gondwanaland."

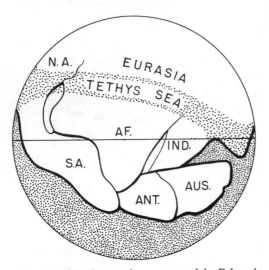

The world as it may have appeared in Paleozoic times, according to the hypothesis of sliding continents. The land masses of the world may at that time have formed a single unit, from which the Americas have since pulled away to the westward and Antarctica and Australia to the southward. While the continents were still united, a shallow sea running east and west along the line of the present Alps and Himalayas may have separated the southern land mass, Gondwanaland, from the northern regions.

The Gondwanaland question has been long and hotly debated, pro and con. Its supposed existence fits in beautifully with the Wegener theory of sliding continents, mentioned early in this volume; if one looks at a globe, one sees that South America, Africa, India, and Australia can be neatly tucked together into a single land mass, and much of the geology of eastern South America, for example, agrees excellently with the geological history of South Africa. But most geologists like to have their continents nailed down tightly. If the continents had been in their present positions in the Permian and Triassic, however, the floral similarities between the southern continents, and India as well, could be explained only by building a most unlikely series of land bridges.

Vertebrate evidence has been brought into the discussion to some extent. There are too few fossils of this age from Australia and India to make them useful as regards the eastern part of Gondwanaland. South Africa has been notable for nearly a century for the presence of Permian and Triassic beds with a rich fauna of mammal-like reptiles (of which until recently there were few traces in northern continents) and eastern Brazil has a fauna which fits well into the South African series. It was assumed that these mammal-like forms had evolved in isolation in Gondwanaland. But forms comparable to many of the South African types have now been found in—of all places—northeastern Russia, well toward the opposite pole, and the theory of complete southern isolation must be abandoned.

Mesosaurus, however, still remains one of the stumbling blocks to those who like their continents pure and undefiled and hence would do away with Gondwanaland. *Mesosaurus* was aquatic and may have ventured into estuarine and coastal waters but does not have the high degree of specialization which we normally expect—and find—in truly marine vertebrates. It is impossible to conceive of its having migrated from South Africa to Brazil (or vice versa) over an orthodox land route via North America and Eurasia without leaving the slightest trace of its presence in these continental areas. And it is equally difficult to imagine this little creature, as an ancient miniature Columbus, breasting the two thousand miles of ocean waves that separate South Africa and the Brazilian shores. *Mesosaurus* gives one vote, at any rate, for Gondwanaland.

CHAPTER 8

Lizards and Snakes

Most of the present chapter will be devoted to lizards and snakes, the most abundant and, in the eyes of most readers, the most typical of modern reptiles. Before discussing them, however, we shall mention briefly some related forms, living and extinct, which shed light on their origin.

THE TUATARA

Living today only on a few isolated islets off the coast of New Zealand is the tuatara, scientifically known as *Sphenodon*, a small reptile which appears superficially much like the lizards and resembles that group also in many structural features. There are, however, some characters more

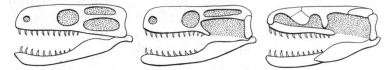

Diagrams to show the evolution of the skull in lizards and snakes. *Left*, a side view of a primitive diapsid ("two-arched") reptile. *Center*, in the lizards the lower arch of the temporal region is lost, allowing free movement of the bone connecting with the lower jaw (the quadrate). *Right*, a snake skull in which there is much greater freedom of jaw movement and flexibility for swallowing. The upper temporal arch is lost, still further freeing the quadrate bone and also a bone connecting the quadrate with the braincase; as indicated here and in the next figure, there are further areas of movement or flexibility in both skull and lower jaws.

primitive than those of any lizard; most notably, in the temple region, there are two perforations in the bones roofing the skull and giving play to the jaw muscles. The tuatara is thus a diapsid—a term which we defined at the close of chapter 6—as are the crocodilians today and as were the dinosaurs and other extinct reptiles to be considered in the next chapter—the ruling reptiles. It is, however, much more primitive and generalized than those forms, and while it has a few peculiarities of its own, it seems to be a reptilian "missing link," a survivor of an ancient group of primitive two-arched reptiles from which the ruling reptiles may have sprung. Further, while lizard and snake skulls are somewhat differently constructed in the temple region, they appear, as we shall see presently, to have been derived from a two-arched (diapsid) type, so that the tuatara may well be close to the ancestry of the lizard-snake group as well. In agreement with these conclusions is the fact that here and there in the fossil record forms similar to *Sphenodon* can be traced back to the very beginning of the Age of Reptiles, before the appearance of dinosaurs and lizards.

But there are two minor peculiarities of the tuatara and its fossil relatives which show that, primitive as they are, they are slightly off the main evolutionary lines to ruling reptiles or lizards. For one thing, these forms all have a bit of a beaklike projection of the upper jaw (not very pronounced in the tuatara itself), and this is not a primitive character. Further, the teeth in most other reptiles are either set in sockets in the jaw bones or, in most lizards and snakes, are fastened to the inner side of the jaws. In *Sphenodon* and its relatives they are, in contrast, fused tightly to the rims of the jaws. For these reasons the tuatara and its allies are set slightly to one side on the family tree as constituting a distinct order of beak-headed reptiles or, to put this in proper learned form, the Order Rhynchocephalia.

The tuatara was, at the coming of man, widely distributed over New Zealand, but it is now confined to a few almost inaccessible islands in the Bay of Plenty, where it is preserved by law. Averaging, at most, two feet in length and a dull olive or yellowish-brown in color, it is, as was said above, much like a lizard in superficial appearance and in most anatomical characters. One interesting feature which it shares with certain lizards is the retention of a median eye atop the head. We have noted that eyes of this sort were common in the most ancient vertebrates. Later they appear to have gone out of fashion, so to speak, and in most land animals today they have disappeared, unless preserved as glandlike structures inside the skull. Only *Sphenodon* and some lizards retain them as very small organs situated above a small opening in the skull roof. But even though present, they are vestigial affairs, which at the most can merely distinguish between light and darkness.

The tuatara is a dull and indolent creature, which, if it moves at all, usually progresses in a sluggish crawl. However it can, and does, dig effectively, making burrows several feet in length, terminating in a living chamber. Curiously, on the islands where the tuatara now survives, the burrow is generally shared (apparently with no great love lost) with a small petrel and its single egg or nestling. Large insects, particularly locusts and beetles, are the tuatara's main source of food.

Why has this "living fossil" survived in this one locality when its relatives have otherwise perished? This is an extreme case of survival due to isolation. Archaic members of any group usually persist only when comparatively free from competition with more progressive types. New Zealand has probably been completely separated from other bodies of land since some time in the Age of Reptiles. The development of mammals in other continents has probably been a factor in the destruction of the tuatara's relatives over most of the world. But not a single mammal (except the bat) appears to have reached New Zealand until man arrived, and the lack of competition has presumably been the reason for the preservation of this archaic reptile.

Sphenodon and its fossil relatives, we have seen, are close to being proper ancestors for other two-arched reptiles and lizards but are a bit off the main line. Can we find, in the fossil record, still more primitive forms which do fill the bill? We can. Far back in the reptile story—as far as the late Permian, before the Age of Reptiles proper had begun—there appeared some small reptiles of a rather lizard-like appearance, presumably descended from the still older cotylosaurs, with two temporal openings in the skull but without the two specializations of the tuatara and its fossil relatives. These little fellows were long ago named the Eosuchia, meaning "dawn crocodiles." This name is rather unfortunate, for although they may be remote forebears of the crocodiles, the eosuchians, as truly primitive two-arched reptiles, may equally well be regarded as ancestors of all the ruling reptiles, the rhynchocephalians, and the lizards and snakes as well.

LIZARD CHARACTERS

Although lizards and snakes have many distinctive characters, it is universally agreed that the snakes are derived from some lizard type which lost its legs (among other changes); hence most scientific workers unite the two types in a single large group, the Order Squamata ("the scaled reptiles"), the lizards forming the basic section of the order.

We alluded earlier to the fact that the most important technical feature distinguishing lizards and snakes from other reptiles lies in the structure of the temporal region of the skull; this is illustrated in an accompanying figure. They are derived, quite surely, from reptiles with two temporal openings and hence two arches of bone running back across the cheek. In lizards the lower arch, forming the original lower margin of the cheek region, has vanished, leaving the region of the jaw joint connected with the

Some American lizards. The common, large iguana of Central America (*above*) is the typical member of a family of primitive lizards abundant in the Americas. The Anoles (genus *Anolis*) are common forms, present in the southern United States as well as the tropics; they show color change and hence are frequently (but incorrectly) termed chameleons. The form illustrated (*lower left*) is a large Cuban species. The little basilisk (*lower right*) has a crested back somewhat like that of the large extinct reptile *Dimetrodon* seen in an earlier plate. Still other iguanids are the "mountain boomer" and horned "toad" shown on later plates. (Iguana photograph by Isabelle Hunt Conant; *Anolis*, *Basiliscus*, New York Zoölogical Society photographs.)

The tuatara of New Zealand (*Sphenodon*), an archaic reptile, resembles the lizards in many ways, but the two-arched construction of its skull shows that it is the only surviving member of a separate order (Rhynchocephalia) which has existed since the Triassic. This condition was typical of dinosaurs and ancestral to modern lizards and snakes. The tuatara's survival was due perhaps to its geographic isolation in New Zealand. *Left*, the skeleton of an almost identical form from the Jurassic (*Homoeosaurus*). (*Sphenodon* courtesy American Museum of Natural History, New York; *Homoeosaurus* from a specimen in Harvard University.)

Geckos. Many of these small tropical lizards are agile climbers; the name imitates the cry of the common form, shown at the left, which has become a nocturnal house dweller in search of insects to eat. Their climbing ability is due to the development of a series of cross-ridges on the soles of the feet (as illustrated below); the ridges are an aid in adhering to trees, walls, or ceilings, which they can traverse upside-down, thanks to these friction pads. (Gecko a New York Zoölogical Society photograph; Gecko foot photograph by D. Dwight Davis, Chicago Natural History Museum.)

rest of the skull surface only by a vertical prop. And in snakes, as discussed later, the skull suffers further reduction in this region.

Although most readers of this book are probably inhabitants of temperate zones, where lizards are scarce to absent, the general appearance of typical lizards is familiar to everyone. They are the most "normal" looking of modern reptiles with a well-proportioned head, neck, trunk, and long slender tail. A peculiar feature of the tail of many lizards (and of the tuatara as well) is that there is a fissure across the middle of the body of a good part of the vertebrae. A bit of pressure, or even a contraction of muscles, may cause the tail to break off. This appears to cause little harm to the lizard, which usually grows a new tail (although never as good a one as the original). This peculiar trait is useful when the lizard is attacked; a predator which has seized the tail is left with this squirming appendage to occupy his attention, while its former owner hastily departs. As in primitive reptiles, the limbs of lizards are turned well out laterally, but they are in general much more slender than in early reptiles. A peculiarity of the typical lizard pose is that the shin is slanted backward and the hind foot turned outward, not forward—hardly a primitive situation. Within the lizard group, however, we find numerous departures from this generalized body pattern. Especially notable is a trend, seen in family after family, for the development of a snakelike build with a long slender body and the limbs reduced or absent. This trend is generally associated with burrowing habits.

As the ordinal name indicates, lizards are covered with horny scales, frequently overlapping. In many lizards the horny scales are reinforced by bone growing beneath them in the skin and the development of bony plates beneath the head scales, over the normal skull. Many lizards are omnivorous; a few are vegetarians; but many favor animal food, although the smaller ones' needs must be satisfied with insects and the like. Most lizards lay typical shelled eggs, and while a few forms guard the nest, there is no parental care of the young. A percentage of them hatch the eggs internally—a feature which, as in the case of sharks and skates, may come about as a result of internal fertilization. The lizards and their ophidian relatives are the most "modern" of reptiles. Most reptile groups grew to prominence in the Mesozoic or even earlier, and they had become extinct or at least had passed their heyday before the close of that era. But while there were a few lizard-like forms in the Triassic, it was not until the Cretaceous—and the late Cretaceous at that—that lizards appeared in any numbers, and they are today in a flourishing condition. It is estimated that there are more than 2,000 lizard species.

WHO'S WHO AMONG THE LIZARDS

In a short space we cannot even make a roll call of all the lizard families (about two dozen), to say nothing of the highly numerous genera and species. We will limit ourselves to the more prominent groups and to typical or striking representatives of them.

For a primitive position among lizards, the iguanid family, mainly American forms, probably can make the best claim, with their Old World cousins, the agamid group, as close seconds. In almost every case the iguanas retain the typical lizard proportions and appearance; none of them tends toward the snakelike type of body, and none reinforces its scales with bone. The typical common iguana of the American tropics is a large lizard, running up to six or seven feet in length. A smaller form, common in the North American Southwest, is the brightly colored collared lizard, or "mountain boomer." Still smaller are the anoles, common little lizards in the southern United States and on southward into the tropics. The anoles are among the most apt of lizards at changing color, and hence they are often sold at fairs or pet shops as "chameleons"—which they are not. One iguana inhabiting the Galapagos Islands is almost unique in spending much of its time in the sea; its food consists of seaweeds found in the shallows along the island shores. Another iguanid sets a lizard record for tolerance of cold weather, being found up to a 15,000-foot elevation in the Andes and ranging south to inhospitable Tierra del Fuego. Most striking in appearance of iguanids is the male "basilisk" of Central America, with a helmeted head, and crests extending up from the back and tail, somewhat in the fashion of the ancient reptile *Dimetrodon*. Most aberrant in body proportions is the incorrectly named "horned toad" of the North American dry and desert areas, with a short flat body and a series of horny spines on head and back to give protection to this small, slow, and harmless little lizard.

The agamids parallel the iguanas closely in many regards. Very probably they have come from the iguanids, which they have superseded in the Old World where iguanids have survived only in relative isolation in Madagascar and the far Fijis. The technical difference between the two families lies in the fact that the iguanas have the teeth bound to the inner surface of the jaws, as in lizards and snakes generally, whereas the agamids have them fused to the rim of the jaws (as in the tuataras). Many of the agamids have a normal body shape, but, as in the iguanids there are spectacular variants. For example, the frilled lizard of Australia has a structure of skin, supported by unusual growth of slender throat bones, which,

The horned "toad." This little animal, so abundant in the drier regions of the Southwest, is not a toad but a lizard (*Phrynosoma*) with a peculiarly broad and flattened body. Considerable protection is afforded it by its horny spines. Were it of larger size its odd appearance would be considered as grotesque as that, for example, of the horned dinosaurs. (Photograph by Raymond L. Ditmars, courtesy New York Zoölogical Society.)

The flying dragon. This lizard (*Draco volans*), a native of the East Indies, is able to expand membranes on either side of the body like a parachute. The membranes are supported by specialized movable ribs. (Photograph by Raymond L. Ditmars, courtesy New York Zoölogical Society.)

Limbless lizards. In many lizard families limbs have been lost and a snakelike appearance results. These forms can be distinguished from the true snakes by the structure of the skull and various other characters. *Above*, a species of the "glass-snake" (*Ophisaurus*), a limbless lizard found in both North America and Eurasia. Still more specialized is the "blind worm" (*Anguis*) of Europe (*below*). This tiny creature is, as the comparison in the photograph indicates, no larger than an earthworm. It is, however, not "blind," for the eyes, though small, are well developed. (*Ophisaurus* courtesy New York Zoölogical Society; *Anguis* courtesy American Museum of Natural History, New York.)

The chameleon. The true chameleons, of the Old World tropics, principally Africa, are specialized in numerous features such as prehensile tail, cleft feet for grasping limbs, and long tongue for seizing insect prey. (Photograph courtesy New York Zoölogical Society.)

A common American lizard, the "mountain boomer," or collared lizard (*Crotaphytus*), abundant in the semiarid regions of the Southwest. This is a form of fairly good size, averaging a foot or so in length, and related to the large iguana lizards of the American tropics. (Photograph copyright General Biological Supply House, Chicago.)

A short-legged lizard (*Neoseps*). In various lizard groups there has been a tendency for reduction of the limbs. The little Florida reptile shown here belongs to a family known as the skinks, most of which, however, have normal limbs. Overpage are shown lizards in which with further reduction the limbs have been entirely lost. (Photograph courtesy American Museum of Natural History, New York.)

The giant lizard of Komodo. The monitor lizards (genus *Varanus*), so called for no known reason, inhabit parts of tropical Africa and the Orient. Though usually land forms, like most lizards, one variety takes to the sea of its mososaur ancestors. The monitors are generally forms of large size; largest of all is the species found on the little island of Komodo in the Dutch East Indies. Early reports of these "dragons" said that they were 30 feet or so in length. However, like most monsters, they shrank upon investigation. The largest is but 10 feet in length. Even so, they are very large reptiles. A pig is acceptable bait although most monitors subsist on smaller prey, such as insects and rodents. (Photograph courtesy American Museum of Natural History, New York.)

when the animal is faced with danger, can be expanded to a great—and brightly colored—frill surrounding the head and presumably giving the enemy the impression of an unexpectedly large and dangerous creature. In this family is found the only lizard which has made a step toward aerial navigation, the "flying lizard" of the East Indies, which does not fly but can glide from branch to branch by means of membranes that can be extended out on either side of the body. But whereas every other vertebrate type that has taken up gliding or true flight has adapted its limbs for this purpose, this lizard, uniquely, has developed gliding membranes supported by a series of extensible ribs.

Despite their grotesque appearance, the chameleons, essentially an arboreal African group, are quite surely an extreme development of the agamid stock. They are, of course, universally cited for their ability to change color. The head is variably horned and crested; the eyes project from the head, almost as if on stalks, and may rotate wildly in search of the insects which form their diet. The tail is prehensile, and a further aid in climbing lies in the unusual structure of the chameleon's feet. Most vertebrates use claws in climbing, but a monkey is aided by being able to grasp a limb between thumb and fingers, or big toe and the other toes. The chameleons have a somewhat comparable specialization. The feet are split, with two toes turned in one direction and three in another, the two sets opposed. The chameleon has, in effect, a double thumb.

Most remarkable is the tongue which, armed with a sticky bulb at its tip, can be shot out with almost lightning rapidity to a distance as great as half a dozen times the length of the animal's head. A chameleon may be seen sitting quietly when a fly alights at a seemingly impossible distance from its head. A flash—the fly has disappeared, and the chameleon is swallowing contentedly.

A group which is sometimes claimed to be of fairly primitive position but which is obviously highly specialized in some ways is that of the geckos. These are little lizards with short, stout but flattened, bodies and small legs, found throughout the tropics. They are extremely abundant; geckos, iguanids, and the skinks, described farther on, each include almost a third of all described lizard species, leaving only a relatively few representatives of all the other families. A notable adaptation of many geckos is that the tips of the toes are expanded and supplied with friction pads which enable them to run up vertical surfaces or even traverse a ceiling upside-down. Certain of the geckos have become habitués of human habitations in the Orient, where they perform a useful function as eaters of insects. They are noctural and quite vocal (the word "gecko" is an imitation of one of their cries), and their nightly activities in a dwelling tend to make an eerie impression on the mind of one not accustomed to them.

The skinks are representative of a series of more advanced lizard families, characterized mainly by technical characters of the skeleton such as, for example, the usual presence of bone ossicles in the skin beneath the scales and the reinforcement of the skull roof by the addition of superficial bony plates. The skinks are one of the largest lizard groups, with a particularly high concentration in the Oriental region and in Australia. Despite their abundance they are little known to the non-scientist, for they are generally small, without striking specializations, and unobtrusive, tending to hide from sight and in many cases becoming more or less burrowing by nature. The body, covered generally by smooth and shiny scales, tends to be long and slender, and the limbs at the most are relatively small. The trend toward a limbless condition, seen in a variety of lizard types, is well marked in the skinks; in many the limbs are much reduced and may be lost entirely. This tendency is associated with burrowing habits.

A number of other but smaller family groups on the general evolutionary level of the skinks are to be found in both Old and New Worlds. Of these we will mention only the lizards in the narrow sense of the term. Our word lizard is derived from the Latin "lacerta," in Roman days the popular name—and retained today as the technical name—for a series of agile little forms, which are the common lizards of Europe. As their distribution over most of that continent suggests, the lacertids are better able to withstand cold climates than most lizards; one species even extends northward beyond the Arctic Circle in Scandinavia, as a northern rival in hardiness to a South American iguanid mentioned earlier.

A final upward step in lizard evolutionary progression leads to a group often termed the diploglossans, or "double-tongued lizards." We are familiar with the forked tongue of snakes. The tongue is of normal construction in such primitive lizards as the iguanids. In the diploglossans the end of the tongue is cleft and in some is seen to be very snakelike in its slender, deeply forked appearance and rapid movement in and out of the mouth. Among the more primitive double-tongued lizards some have a normal body build, but here again there is a trend toward limblessness, and the best known forms of this stage are the slow worm, an Old World burrower, and the glass "snake" of Eurasia and America. This last lizard gets its name from the fact that it gives the impression of being brittle because of the ease and frequency with which it breaks off its tail—at the drop of a hat or with even less incentive. Somewhat more advanced is the Gila monster of the deserts of the southwestern United States and Mexico. This stoutly built and sluggish lizard has a scale structure which gives it the appearance of being covered with a pattern of colored beads. Its fame is due to the fact that it alone of all lizards has poison glands.

In many regards the most advanced of all lizards is a group of oriental and African forms called (for no known reason) the "monitor lizards." The monitors are actively predaceous; small forms and young individuals may content themselves with insects and the like, but the larger ones may attack vertebrates of considerable size. Many monitors are large forms, for example, an East Indian monitor which (most exceptionally for lizards) takes kindly to water life and has even been seen well out at sea, reaches eight feet in length. The giant of the group, however, is the "dragon" found on Komodo and other small islands in the Sunda region of the East Indies. Early reports were that the animal reached a length of twenty-eight feet. As usual, the length contracted when actual measurements were made, and ten feet appears to be the maximum. Even so, this is a large animal, weighing several hundred pounds and able to attack deer and wild pigs in its search for food. Pleistocene monitors were even bigger, an Australian fossil having an estimated length of twenty feet, thus reaching the proportions of many dinosaurs.

Related to the monitors is the one group of lizards to achieve prominence in the fossil record, the mosasaurs, huge marine lizards of the Cretaceous. These great extinct forms are found in marine deposits in almost every region of the earth; the best skeletons have been obtained from the chalk rocks of the seas which once covered western Kansas. The name, however, derives from the fact that the first specimen came from the banks of the Meuse River (Latin *Mosa*) in Holland, whence it was taken to Paris in 1795 as one of the major spoils of a French revolutionary army campaign.

The large mosasaurs reached a maximum length of thirty to forty feet; about half the length was included in the long flattened tail, which was obviously the main swimming organ. The limbs were very short, with broad spreading toes which presumably were webbed in life and served a steering function. The numerous pointed teeth of the typical mosasaurs lead us to believe that they subsisted on fish, plentiful remains of which are present in the same deposits.

Burrowing specializations in lizards reach their peak in a remarkable group of tropical forms which are technically called "amphisbaenids" but may be termed "worm lizards." These secretive little animals seldom appear on the surface of the ground and, like the caecilians among the amphibians, have attained a remarkably wormlike appearance combined with wormlike habits. The head is a compact structure, its bones solidly fused to make it an efficient burrowing organ. With one exception, legs have disappeared completely. The tail proper is very short, but the body is extremely elongate, with up to 150 or more segments in the backbone; adding to the wormlike appearance is the fact that the segments of the body appear on the surface as a series of conspicuous rings. The worm lizards surely belong to the lizard group in a broad sense, but they have specialized so far in their subterranean adaptations that there is no evidence of special relationship to any of the more normal lizard types.

WHAT IS A SNAKE?

An even more modern reptilian group than the lizards is that of their evolutionary offspring, the snakes. There are a few Cretaceous forms which are possibly primitive snakes but may be advanced lizards; there are no unquestionable ophidians before the Eocene, and the more evolved snake types do not appear until fairly late in the Tertiary. The snakes, thus, are latecomers and appear to be today at or near the peak of their evolutionary development.

The most prominent character of snakes is their elongate limbless body and its sinuous movement. But we have just seen that many lizards have taken the same path, and a number of characteristic features of the snakes' sensory organs are also shared with a fraction of the lizards. The really diagnostic features of a snake are to be found in specializations of its skull and jaws which enable these predatory animals to distend the mouth to an enormous degree and swallow a prey of large size. We have noted that even in lizards the skull has been "loosened up" to some extent, so that the bone (the quadrate) which articulates with the lower jaw can move somewhat freely, thus widening the gape. In addition, in snakes (and in some advanced lizards as well) the two halves of the lower jaw, usually tightly united, are connected only by a ligament, so that the lower jaws can be spread sideways, and there is, still further, a flexible joint part way back along the lower jaw.

But the snakes have gone far beyond this loosening of the lower jaws by (so to speak) breaking up the skull into flexibly jointed pieces so that possible distension is greatly increased. In the back part of the skull, the braincase is a solid structure surrounding the brain and protecting it from

Still more lizards! *Above*, a chameleon in action. This striking photograph has caught a chameleon midway in the lightning-fast extension and withdrawal of the tongue to capture an insect with its sticky tip. *Left*, an amphisbaenid (*Rhineura*). These are tiny, burrowing, limbless reptiles, still more remarkable in many regards than the limbless types on an earlier plate. There may be as many as 150 body segments, represented by surface rings. *Below*, a striped skink of a genus (*Eumeces*), common in the United States; the skinks in general are small and slender lizards, many of which tend toward a reduction in size of limbs. (Chameleon, San Diego Zoo, photograph by R. Van Nostrand; amphisbaenid and skink courtesy Charles M. Bogert.)

The Gila monster (*Heloderma*), the only poisonous lizard. (Courtesy American Museum of Natural History; New York.)

The King cobra of southeastern Asia, giant of poisonous snakes, with a maximum length of 18 feet. (Photograph courtesy New York Zoölogical Society.)

The snake skull. This is a nonpoisonous form, the royal python. Teeth are numerous and directed backward, every movement forcing the prey down the throat. The whole jaw apparatus is loosely constructed, rendering easy distention of the throat. (Photograph courtesy American Museum of Natural History, New York.)

A giant snake. The anaconda of South America is a boa somewhat amphibious in habits. It has been reported to reach a length of 46 feet; this is doubtful, however, and 30 feet may be a maximum. (Photograph courtesy New York Zoölogical Society.)

harm during the swallowing process. But everything else in the skull can move. At the back, not only is the quadrate bone freely movable, but with the loss of both original cheek arches, the bone that connects quadrate and braincase is itself only loosely attached. The front part of the skull can generally bend flexibly on the braincase; the bones of the palate can be moved; and the main bones of the upper jaw, still further, are rather loosely joined to the rest of the skull. As a result of this remarkable flexibility of the whole head skeleton, a snake can distend its mouth to swallow an object much bigger than the diameter of its head—or even of its trunk.

The swallowing process is aided by the tooth battery. In typical snakes the teeth are curved backward and are arranged in rows on the lower jaws and, above, in similar rows on the palatal bones and (primitively) on the jaw margins as well. As a result of the backward curve of the teeth, the prey, once seized, can never slip forward and, with alternating backward movements of upper and lower jaw structures, is gradually worked backward through the mouth and throat—to await a leisurely digestion in the expanded abdomen of its comatose "host." Incidentally, with this feeding habit, snake mealtimes tend to be highly irregular. If a large animal

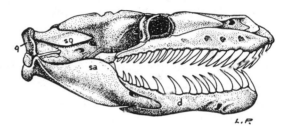

The skull of a python to show some of the mechanisms for swallowing large objects. The two jaw halves are loosely connected; the front bone of the lower jaw (d) can be bent on the back part of the jaw (sa). In addition to the ordinary movement of the jaw joint the two bones above this region (q, sq) can also be moved on the skull; the front part of the skull is also flexible. The teeth are directed backward; any motion tends to push a seized object down the throat, and it cannot slip outward.

has been caught, the food supply so obtained may last the snake for weeks or months, and there is a record of a captive python which went for nearly three years between successive meals.

Many lizards belonging to a great variety of families have tended toward the development of a snakelike body form, but few have gone as far as the snakes themselves. The body is generally very long and slender, usually with two hundred to four hundred segments in the backbone; there are complex joints between the segments, facilitating sinuous movement, and there are stout curved ribs for attachment of the necessarily strong trunk muscles. Some primitive snakes have vestiges of pelvic girdles, and in boas and pythons there is a pair of small claws where the hind legs should be, but otherwise all trace of limbs is lost. Aiding in the "slenderizing" process, only one lung is developed in most snakes. All snakes are covered with stout overlapping horny scales, and there is an especially broad and stout paired row of belly scales to give good contact with the ground (or other substrate). Fishes swim, we have seen, by body undulations, moving forward by pressing the body curves back against the water. Some trace of the body undulation survived in early land vertebrates. The snakes and other forms which have tended to lose their limbs and take on body elongation are, in a sense, returning to the original locomotor fashion. They are essentially swimming on dry land, with the belly scales pushing against resistant objects and preventing any backward slip of the body. Typically, progress is an undulatory motion, with successive curves traveling backward in the line of the long axis of the body. There are, however, variants. The "sidewinder," an American desert rattlesnake, for example, navigates through sand with its body undulating at right angles to the line of march. On occasion some heavy-bodied snakes move straight forward without undulating; this is accomplished by a rather caterpillar-like motion of the ventral scales, which have special muscles attached to them, each scale, in wave succession from front to back, moving slightly forward and then pulling the body slightly to adjust to its new position.

The major sense organs of snakes depart widely from those found in more generalized reptiles, although here, too, some of the lizards may show parallels. The eyes depart from the pattern seen in most vertebrates. The snake gazes at you with a fixed, unblinking stare. It could not blink if it wished, for it has no movable eyelids. Instead, the eyeballs are completely sheathed and protected externally by "spectacles," stiff transparent structures formed from fused upper and lower lids. The deeper parts of the visual apparatus appear to function well in typical snakes, but its structure is so unusual that there are strong reasons to believe that the eyes of modern snakes represent a secondary redevelopment following an ancestral reduction to a rudimentary state. Snakes have lost their eardrums and, while able to detect ground vibrations, appear to have little normal hearing; "deaf as an adder" is a common English proverb. We have noted that some lizards have a forked tongue. In snakes the tongue is very long and slender and deeply forked and, when the animal is faced by foe or potential prey, darts rapidly in and out of the mouth. If, on meeting a snake, you observe this action, realize that this is not a gesture of derision or defiance; he does it to smell you the better. In many land verte-brates (not, however, in man and higher primates) the nose has accessory pockets which appear to have as a main function smelling the mouth con-tents. In snakes these pockets open up from the roof of the front part of the mouth. The tongue, when extended, picks up from the air tiny smell-producing particles; with the tongue pulled back, the forked tips are in-serted in these pockets, and produce smell sensations reinforcing those gained through normal olfactory means.

From what lizard group and in what fashion did the snakes arise? In default of satisfactory fossil evidence, these questions cannot be surely settled. In a number of anatomical regards the members of the lizard group termed the "diploglossans"—particularly the monitor lizards—ap-proach the snakes most closely and appear to be the best candidates for relationship; but monitors and advanced diploglossans show no evidence of a trend toward limblessness.

Various suggestions have been made as to the "reasons" for the snake reversion to a limbless condition. For example, limbs are ineffective for a small animal trying to navigate in a deep-grass pampas type of country, whereas the snake type of progression is effective; the snake "swims" over and through the grass. The monitor lizards and their relatives—such as the mosasaurs—show a strong trend toward the water, and snakes might have evolved from this group as swimming eel-like forms (there are many water snakes today). However, the majority opinion—a well-grounded one—is that the ancestral snakes were burrowers. Most of the lizards which have assumed a snakelike form are burrowers, and, as noted below, many of the more primitive snakes living today are of a secretive and often burrowing nature. Such an origin agrees well with the nature of the snake ear and eye. The loss of the eardrum in snakes is paralleled by a similar loss in many burrowing lizards (there is little to hear in a burrow except ground vibrations), and the curious structure of modern snake eyes is readily explicable on the assumption that ancestral snakes had become burrowers, with little need for eyes, and upon return to the surface, in the case of most snakes, they had redeveloped the "remains."

As in other reptile groups, we find, of course, the greater concentration of snakes—some two thousand species strong—in the tropics, with di-minishing numbers in temperate zones. They appear, however, to be able to withstand cold weather rather better than lizards so that, for example, in the northern United States, where lizards are rare to absent, there is a fairly varied snake fauna. Temperate-zone snakes frequently hibernate by gathering, en masse for mutual warmth, in a "den" below the frost line. The eggs are generally well covered, and in some cases they are said to be brooded or guarded during incubation; there is no known instance of care of the young. Here, as in lizards, a small minority of the forms bear their young alive.

Although most snakes adhere closely, on the whole, to a uniform pattern, there is considerable variation in habits, coloration, and dentition. We can sort them out to some degree into more primitive, typical, and advanced, poisonous types.

Primitive snakes.—Best known of snake forms with a relatively primitive structure are the boas and pythons. In them, for example, the various skull parts are not as freely movable as in more advanced snakes. The term "python" is generally applied to Old World forms, "boa" to those of the New World, but apart from the trifling question of the presence or absence of a supernumerary bone lying above the orbit, there is little if any difference between the two. Most boas and pythons are constrictors, winding their bodies around their prey. (This habit, however, is also present in some of the more advanced snake forms.) The pythons and boas include the largest of all snakes. The reticulated python of the Orient and the anaconda of South America run to a bit over thirty feet in length in the largest measured specimens; the African rock python is not far behind with a twenty-five foot record; an Eocene fossil boid, however, far exceeds any of these with an estimated length of sixty feet. But not all of the group are giants; many are of modest size, and some small boas and pythons never grow to lengths of more than two or three feet. Many boas and pythons are arboreal to terrestrial; some of the smaller forms are burrowing types.

Allied to the boas and pythons are a limited number of small snakes which differ in that the adaptations for wide opening of the mouth are somewhat less developed—possibly a primitive feature. These forms, like some of the true pythons, are burrowers, a fact fitting in well with the hypothesis that the primitive snakes led a subterranean life at one stage of their evolution. Still more significant evidence is that furnished by a number of small tropical forms for which there are no popular names but which may be called the "blind snakes," since the tiny eyes are concealed beneath the head scales. The best-known genus is *Typhlops* ("blind face"). Here we have a third group of exceedingly wormlike insect-eating burrowers, quite comparable in habits and general appearance to the wormlike amphibians and lizards already described but readily identifiable on

The rock python of A... (upper figure), member of ... Old World group of sna... related to the New Wor... boas. Except for an Asiatic python and the anaconda, this is the largest of known snakes, reaching a length of 25 feet.

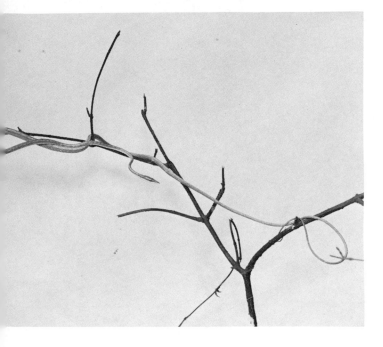

The most slender snake (left). The gray vine snake ranges from Arizona to the Amazon region.

Major man-killer. The common Indian cobra (below, left), responsible for a large fraction of all deaths from poisonous snakes.

A blind snake (Leptotyphlops) (below). A small worm-like burrowing reptile, with snakelike structural features. Possibly the ancestral snakes were similar burrowers. (Python photograph by Isabelle Hunt Conant; other photographs by New York Zoölogical Society.)

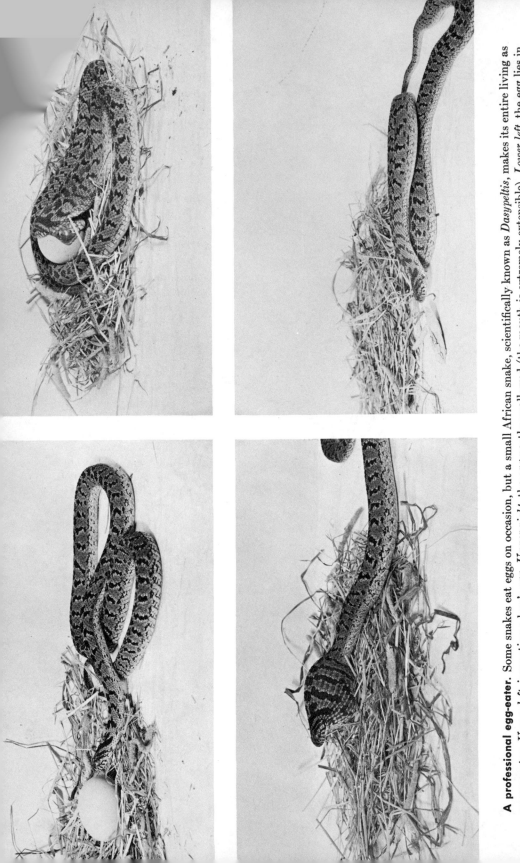

A professional egg-eater. Some snakes eat eggs on occasion, but a small African snake, scientifically known as *Dasypeltis*, makes its entire living as an egg-eater. *Upper left*, inspecting a hen's egg. *Upper right*, the egg partly swallowed (the mouth is extremely extensible). *Lower left*, the egg lies in the throat; here there are peculiar downward-projecting prongs from the neck vertebrae which can crush the shell. *Lower right*, the egg has been crushed, the contents swallowed, and the shell regurgitated. (New York Zoölogical Society photographs.)

close inspection through their body covering of typical, if small, over-lapping snakelike scales. These little snakes range from about six inches to two feet long; the jaws are short and the teeth reduced; the short, blunt skull, as in the worm lizards, is solidly built as a burrowing organ, and the swallowing adaptations are little developed. Some of their characters are specialized, but others may well be primitive. Indeed, so many of the advanced snake features are absent that some would deny them a place among the snakes and consider them only as somewhat snakelike lizards.

Typical snakes.—Most common of serpents, advanced in skull motility over the boas and pythons but lacking the specializations (such as tooth reduction) seen in advanced, poisonous types, are a great host of snakes for which as a whole no popular name exists but which are grouped by scientists in the family Colubridae. These include all the harmless snakes with which a reader in the North Temperate Zone is familiar, such as the grass snakes, black snakes, racers, and so on, most of which in rural America are lumped in popular terminology as "garter snakes." Most are of modest size, generally lacking the constricting habit and, in a majority of forms, lacking poison glands. Frogs are favorite food for many species, with small reptiles, birds, mice and other small mammals, and fish in some cases making up the diet; small snakes may of necessity favor insects and other invertebrates. There are many food variants. Various forms are fish-eaters; such types are most abundant in the oriental region. One small subgroup specializes in snails. Two forms—one in Africa, one in Asia—are specialized in structure for egg-eating. The eggs are swallowed whole; special spines projecting downward from the neck vertebrae crush the eggs as they pass down the throat. As to habitats, some are arboreal, many terrestrial, a variety of forms are aquatic, a few are burrowers. The family includes the great majority of all snake species; they are abundant in both Old World tropics, not uncommon in temperate zones; curiously, however, there are almost none in Australia.

Most colubrids are typically modest, non-aggressive snakes, not harmful to man and generally, in fact, useful as destroyers of "vermin." However, it turns out that a large percentage of tropical types are poison-bearers. Most members of the family have a considerable row of teeth along each upper jaw margin. In the poisonous forms a mouth gland has shifted from the normal production of simple mucus to the secretion of a poisonous liquid. The gland opens out at the base of several of the back teeth; these are grooved, so that poison may trickle into the wound when the prey is bitten. However, with the exception of a South African tree snake ("boomslang" in Africaans), which runs to five feet in length, most of these "back-fanged" snakes are too small and non-offensive to be dangerous to man.

Advanced poisonous snakes.—Most specialized in many regards are three families of snakes, all of which are highly poisonous and have developed specialized fangs for poison injection. One such group, almost entirely confined to the tropics of Africa and Asia has the cobras as its best-known member. These snakes are slenderly built, but in a number of cases the head can be expanded into a "hood," which appears to function as a threat or warning. We noted above that in the "back-fanged" snakes the poison may be injected along grooves in several teeth of normal size at the back end of the tooth row. In the cobra group the situation is a more specialized one. The upper jaw bone, the maxilla, bears a small number of teeth; the most anterior one, with which the poison gland is associated, is a fairly good-sized fang, fixed in position. It bears along its length a deep groove which is closed for most of its length to form a tiny injection duct. Many members of the cobra group are deadly poisonous, but their danger to man is a variable matter, for some are small, secretive in habits, or non-offensive. Further, in even the most dangerous forms the "strike" is much less effective than that of the vipers and their kin; generally the head and neck are simply raised vertically off the ground and then swung down and forward at the prey, without the swift forward strike of the rattlesnake and its relatives.

The cobras and forms closely related include a considerable number of species in Africa and from India eastward to the Malay region. Best known is the Indian cobra, responsible for a large annual death toll. The giant among poisonous snakes is the king cobra of Siam and adjacent regions, largest of all poisonous snakes, reaching eighteen feet in length. Several species of cobras, including the "ringhals" of southern Africa, have the terrifying habit of "spitting" venom at their adversaries; the head is raised, the fangs pointed at the enemy, and a fine spray of poison sent out which can travel several feet; if it enters the eyes, it can cause temporary or permanent blindness.

The Australian snake fauna is a peculiar one. We have noted that typical harmless colubrids are rare in that continent; on the other hand, cobra relatives are numerous and varied and make up the greater part of the native snake fauna. Many of the Australian forms are small and inoffensive. Not all, however; for example, the taipan of the northeastern part of the continent may reach ten feet or more in length and is said to attack on sight, and the tiger snake, while not as large (up to six feet) is common and widespread and has an extremely potent venom.

Differing to some degree from the typical members of the cobra family are the mambas of Africa. These are highly poisonous arboreal forms; the larger species of mambas, particularly the large black mamba which may reach seventeen feet in length, are among the snakes most dangerous to man because of their agility, activity, and rather aggressive habits. Quite unlike the cobras in appearance, although definitely to be included in the same family on technical grounds are the little coral snakes, the only American members of the group. There are a few species in the oriental region, but they are mainly found in the American tropics, reaching north to the southern United States. The coral snakes are relatively small with an attractive coloration of red, black, and yellow stripes. The venom is extremely potent, but the coral snakes cause few fatalities because of their non-offensive nature and their short fangs, which cannot penetrate shoe leather or heavy clothing.

Generally placed in a separate family but closely related to the cobra group are the sea snakes (Hydrophiidae). The tail and posterior part of the body are flattened from side to side as an effective swimming organ. They are purely marine forms; in most cases the young are born alive, so that it is not even necessary for them to come ashore for reproduction. These snakes are not responsible for sea-serpent stories, for none grows to a greater length than six feet. They are extremely poisonous but of little danger to man, for they appear to be of sunny temperament, and their diet is confined to fish. The sea snakes have their center of distribution in the shallow waters of the East Indies but extend into the Indian and Pacific Oceans; their extreme range carries them west and south to the tip of Africa, and to the east one species has been reported (rarely) on the shores of tropical America.

A final family of snakes is that of the vipers, which include the major poisonous snakes of northern temperate regions as well as a variety of tropical forms. The true vipers include the most familiar of Old World types; the pit vipers, a separate subgroup, are mainly American in distribution and include rattlesnakes, copperheads, and water moccasins. In this family the venom apparatus is highly developed. The main bone of the upper jaw, the maxilla, bears only the very large fang (plus developing replacements for it), which is so formed as to be a sharp-pointed hypodermic needle, the venom passing down its length to the tooth tip in a closed tube. Further, the fang (and the bone to which it is attached) is movable. A complicated series of bone articulations form a mechanism by which the fangs are tucked back inside the mouth when the jaws are closed but erected automatically when the mouth is opened. In contrast to the cobra group, the vipers typically coil the body when preparing to strike and launch the body straight forward. The vipers have a typically wedge-shaped head, broader behind, separated from the stout body by a narrower neck region; this is in contrast with most other snakes (but some harmless forms have similar contours).

True vipers are widespread in Asia, Africa, and Europe but are absent from Australia and the Americas. They are, on the whole, much less dan-gerous to man than the cobra group. This is particularly true of the com-mon European viper, the only poisonous snake in most of that continent; this form ranges from Europe across central and eastern Asia and extends north to the Arctic Circle. It is essentially a timid reptile, although it will strike when frightened.

The pit vipers are characterized by the presence, between eye and nose on either side of the snout, of a pit containing a tissue filled with blood vessels and nerve endings. This is a very specialized sense organ which re-acts to temperature. Experiment has shown, for example, that a rattler, by means of this organ alone, can detect the difference between a warm and a cold electric light bulb swinging a foot or more from its nose. It will strike at the warm bulb and remain indifferent to the cold one, even though the difference in temperature at the pit organ can be but a fraction of a degree. This is a very useful adaptation for detection of warm-blooded prey.

A number of pit vipers are present in Asia, but their center lies in the Americas. In the temperate regions of North America they include the only poisonous snakes (except for the coral snake in the South). These are the water moccasin, or cottonmouth, of the southeastern states and the lower half of the Mississippi drainage (not to be confused with harmless water snakes); the copperhead, a relatively small and inoffensive inhabit-ant of wooded areas of the eastern half of the continent; and the numerous species of ubiquitous rattlesnakes, rightly feared as the most dangerous members of the North American fauna. In the Northeast occurs the tim-ber rattler, small in size and now extinct over large areas; farther west, the somewhat larger prairie rattler is the common form. The giants of the group are two southern species termed "diamond backs." These seldom reach a length as great as six feet, although appearing of gigantic size to the person who comes upon one unexpectedly, but, with their thick bodies, they are exceedingly powerful and dangerous forms. In the American trop-

Poison! *Top*, a rattlesnake strike—a high-speed photograph, taken just after its fangs have penetrated a latex rubber bulb. A model of the rattlesnake head is shown, overpage, with the mouth wide open to expose the poison glands. Here we see an actual strike which explains why the snake is able to swallow prey of greater diameter than its own head. Not only does the lower jaw drop down but the upper jaw bends upward. In fact, all the bones of the skull, except for the small braincase area, are freely movable. This photo also shows the straight forward thrust of the strike.

Center, the fer-de-lance (*Bothrops atrox*), a large tropical American relative of the rattlesnakes, which grows to a length of 8 or 9 feet.

Below, the Australian tiger snake (*Notechis*), a 6-foot form, whose poison glands may carry at one time enough venom to kill 400 men. (Rattlesnake photograph by Walker Van Riper, Denver Museum of Natural History; fer-de-lance by Charles M. Bogert, American Museum of Natural History; *Notechis* by Isabelle H. Conant.)

North American poisonous snakes. The forms shown on this page illustrate the four types of poisonous snakes found in the United States. All other native snakes are harmless and generally useful citizens. *Upper left*, the rattlesnake (illustrated by the southern diamond back), most common and dangerous of native poisonous forms. *Directly above*, a model of a dissected rattler's head to show the poison apparatus—large glands in the cheek region and a pair of efficient fangs acting as hypodermic needles. *Left*, in order, are the copperhead, water moccasin, and coral snake. The copperhead is found in districts with good vegetation east of the Rockies. The water moccasin is a closely related aquatic form of the southern states.

The three upper snakes are pit vipers, essentially an American group. The coral snake is more closely related to various Old World forms, such as the cobras. It is a small, gaily colored, and highly poisonous form from the Gulf states. Although its venom rapidly paralyzes the nerves and brings almost certain death, it rarely kills humans, since its mouth gape is tiny and it avoids civilization. Its black, red, and yellow stripes are "imitated" by one of the harmless snakes of the same region. (Moccasin and rattlesnake courtesy New York Zoölogical Society; other figures courtesy American Museum of Natural History, New York.)

ics are other large and deadly pit vipers, such as the bushmaster and fer-de-lance. The rattlesnake, of course, is characterized by the presence of the rattle at the tip of the tail, which vibrates with a distinctive buzzing sound when the snake is aroused. An element of the rattle is formed each time the snake sheds its skin; however, the number of "buttons" is not (contrary to popular belief) a sure indication of age in years, for more than one button may be formed per year, and terminal buttons may be lost.

Snake venoms are complex protein substances of considerably varied nature. Two main types can be recognized: neurotoxins, which attack the nervous system and tend to cause death by inhibition of breathing or stoppage of the heart, and haemotoxins, which clot blood cells and attack the linings of smaller blood vessels, with damage and destruction to tissues. In general the cobra tribe and sea snakes tend to secrete neurotoxins, the vipers, haemotoxins, but there is great variation even from species to species. Danger from bites varies greatly according to the amount of toxic material received, the potency of the poison, the part of the body bitten, and whether or not it gains entry promptly or slowly into the main circulatory vessels. Cutting and suction at the wound are generally helpful measures; for many forms antivenins are available for injection. In the tropics, particularly, snakebite is a serious problem. One estimate is that as many as 40,000 people die annually from snakebite—a large proportion due to the cobra in India, where perhaps as many as 200,000 are bitten annually by one type of snake or another. In temperate regions, of course, the number of deaths is small; they are negligible in Europe, and even in the United States the much-feared rattler probably causes no more than a dozen or two fatalities annually.

CHAPTER 9

Ruling Reptiles

Once the primitive reptiles had been freed from the restrictions of their ancestors' aquatic existence, there began a great radiation of the saurian types which were to dominate the earth during the hundred million years or more of history known as the Age of Reptiles, or Mesozoic era.

REPTILIAN LOCOMOTION

Primitive walkers.—In no way is the great diversity of the numerous reptilian groups better shown than in their locomotor adaptations. The gait of the primitive land forms was an exceedingly clumsy one. The limbs were widely sprawled out at the sides of the body, the trackway broad, the step short, the walk a slow and waddling one. Such a type of walking is exhausting, for much muscular effort is needed merely to keep the body off the ground; the attainment of speed or the development of large size is impossible.

The turtles in their armor, oblivious to the world, have retained this old-fashioned mode of locomotion to the present day. The lizards have become somewhat slimmer-legged, four-footed types but have not improved their limbs much and have never played any prominent role in the world fauna. The snakes, their cousins, have abandoned limbs altogether and evolved their own peculiar style of progression. Several reptilian groups dodged the issue by returning to the water and reshaping legs into finlike structures.

Some reptiles, the forerunners of the mammals, remained four-limbed land dwellers but improved their method of walking, with important evolutionary results; in a later chapter we shall discuss these ancestors of ours which tucked their legs under their bodies and became efficient four-footed runners.

The evolution of bipeds.—Here we shall tell the story of a second reptilian group in which better running powers were attained but attained in quite a different way—the story of the archosaurs, or ruling reptiles. They constitute, as the subclass Archosauria, a major division of the reptiles. As a technical feature we may note that the skull has two temporal openings, as in the ancestral scaled reptiles, and hence they are presumably remotely related to the lizards and the tuatara.

Today these forms are represented only by the crocodile, a rather unprogressive and degenerate order, but the extinct dinosaurs and flying reptiles

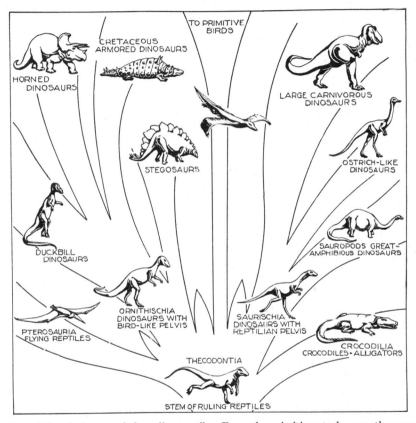

An outline of the phylogeny of the ruling reptiles. From the primitive stock came the crocodiles, flying reptiles, and the two great orders of dinosaurs, while the birds have also descended from this group.

were also archosaurs, and the birds are descendants of this group. The ruling reptiles were the dominant forms on the land and in the sky during nearly the whole Mesozoic, and the evolutionary story of the Age of Reptiles is in great measure one which tells of the rise and fall of the archosaurian dynasties.

Most of the features of the group are characters related to the development of rapid running on their hind legs—the bipedal gait. As we shall see, various members of the group failed to become pure bipeds, and many others slumped back to a four-footed pose. All, however, bear in their body architecture indelible traces of the early trend toward bipedalism.

How this process might have started may be seen among the lizards of today, such as the mountain boomer or collared lizard of the Southwest. When great speed is necessary, the front end of the body is lifted from the ground and balanced by the long tail. The inefficient front feet are relieved from duty, and the better-developed hind legs carry the whole weight of the body swiftly forward until the burst of speed is over, and the four-

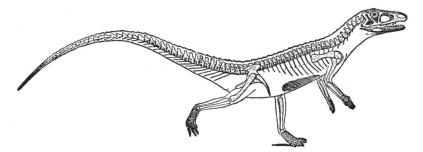

Skeleton of the primitive ruling reptile *Ornithosuchus*. This Triassic fossil reptile was about 3 feet long (tail included). (From Heilmann, *The Origin of Birds*, by permission of D. Appleton–Century Co., publishers.)

footed pose is again assumed. None of the lizards has ever become a true biped; the ruling reptiles belong to a quite different stock but had a parallel early history.

PRIMITIVE RULING REPTILES

The ancestral archosaurs come into view in the Triassic, first of the Mesozoic periods. These early forms constitute the Order Thecodontia—not a very distinctive term, since it merely means that, as in ruling reptiles generally, the teeth were set in sockets, in contrast to the condition in lizards and snakes. Many of these progenitors of the dinosaurs and birds were comparatively small creatures. The more generalized types had an average over-all length of a yard or so, a great portion of this being included in the long tail. In appearance such an animal as little *Ornithosuchus* would probably have resembled a fairly large modern lizard. Structurally, however, the differences were great. The front legs were short, the hind legs very long and much modified. Obviously we see in thecodonts the beginnings of archosaur bipedalism. These ancestral ruling reptiles, like many of their descendants, were sharp-toothed predators, and speed was useful to them in catching their prey.

It is from such ancestors that there came the crocodiles, dinosaurs, flying reptiles, and birds; with the development of these little bipeds the ruling reptiles had begun their career.

The history of the group has not been a simple one. From this stock came not only the prominent groups mentioned above but many short side-line "experiments" which were quickly abandoned. Many of the thecodonts became, so to speak, discouraged in the attempt to become

Restoration of *Ornithosuchus*, a primitive ruling reptile. Although differing greatly in internal structure, such a dinosaur ancestor would have looked rather like a lizard. Probably when unhurried, this little reptile was still a four-footed walker. (From Heilmann, *The Origin of Birds*, by permission of D. Appleton–Century Co., publishers.)

bipeds and (as did many dinosaurs later) went back to a four-footed condition (not, however, without showing clear evidences of the erstwhile attempt at bipedalism in their skeletons). Such primitive archosaurs as little *Ornithosuchus* showed a trend toward armor development in the presence of a paired row of small and thin bony plates down the middle of the back; many of the larger thecodonts developed a heavy armor covering. Further, one set of large armored Triassic forms, the phytosaurs,[1] whose remains

[1] The names given to fossil forms by scientists are sometimes of little significance and occasionally prove to be quite misleading. "Phytosaur" belongs in the latter category. It implies that they were

are common in North America and Europe, took to an amphibious, water-dwelling existence. These forms were quite similar to modern crocodiles in size and appearance and probable habits, presumably lying well submerged in the water while waiting for their prey. Breathing while more or less submerged is a problem. Here, as in some of the extinct aquatic reptiles mentioned earlier, the nostrils were moved well back from the long snout, to give a shorter air passage to the throat. But further, they were moved up on to the top of the head, sometimes even to the peak of a little "crater" above the eyes, so that nearly the whole head, as well as the body, could lie concealed beneath the surface.

CROCODILES

The crocodiles were fully evolved by the following (Jurassic) period and replaced the phytosaurs as predaceous dwellers in inland waters. The phytosaurs were not the ancestors but the "uncles" of the crocodiles, which represent a parallel evolution from the primitive ruling-reptile stock. The modern alligators, crocodiles, and gavials are a rather retrogressive group of ruling reptiles. These sluggish creatures, like many of the older thecodonts, have wandered far from the bipedal pathway which their early ancestors had taken, although they have still, one may note, the long hind legs and short front ones that are characteristic of the group. They alone of ruling reptiles, however, have survived, secure in their specialized position in the world, while their more ambitious reptilian cousins have had their splendid day and have gone.

The one conspicuous advance in crocodiles is their development of a false palate. We have commented on the fact that the primitive opening of the nostrils into the front of the mouth is an awkward position for any land dweller that has returned to a life in the water. Other reptiles, we have seen, attempted a solution of the problem by moving the nostrils toward the back part of the head. The crocodilians evolved a much better breathing mechanism. The nostrils remain at the tip of the snout. But underneath this there has been erected a partition running the entire length of the roof of the mouth, separating off the air passages. With the aid of this bony secondary palate and a flap at the entrance to the throat,

---

plant-eating reptiles, whereas they were predaceous flesh-eaters. Their numerous teeth were sharp-pointed conical affairs with blunt bases set in jaw sockets. However, the first fragmentary remains of phytosaurs found in Germany somewhat over a century ago consisted of a set of teeth with the jaw bone eroded away and with the blunt tooth-bases projecting from a block of sandstone. The finder took these bases to be, instead, the crowns of the teeth and not unnaturally assumed that the creature was a plant-eater!

the crocodile or 'gator can breathe perfectly well under water with the mouth open, provided only that the tip of the nose is above the surface.

The modern crocodilians vary little in their general body pattern and, on the whole, do not vary greatly in their mode of life. All are amphibious, spending most of their time in the water, but are competent to walk about on land, although their terrestrial excursions are usually limited. One customarily thinks of a 'gator or crocodile as crawling about on its belly. This impression comes from observing sluggish zoo specimens; in its native haunts the crocodilian ashore can stride along with the body raised well off the ground. The tail is long and powerful; it is the main propulsive organ while swimming, the limbs being held close into the body, and it is also an effective weapon, with which a powerful blow can be struck. The body is covered with a complete series of tough horny scutes, set in a flattish pavement fashion. The back and tail are further protected by bony plates beneath the scutes, and in some living and many extinct crocodilians there is a belly armor as well.

The typical crocodilian pose is one with the body immersed in the water with only the top of the head (including the eyes) visible, while the animal waits quietly for prey to appear. In popular thought, large animals are the typical prey. But while "big game" is readily accepted by some of the

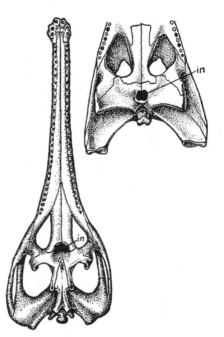

Left, a view of the underside of the skull of a Jurassic crocodile (Steneosaurus). The inner opening of the nostrils (in) lies far back toward the throat, although the external opening lay at the tip of the snout. Right, the back part of the palate of a modern type of crocodile; the internal nostril opening is almost at the back end of the skull.

larger types, even the bigger members of the group eat any sort of animal material. In cases where food habits are well known, insects appear generally to constitute a considerable fraction of the diet, with crustaceans and fish also bulking large; higher vertebrates (frogs, reptiles, birds, and mammals, usually small ones) in most cases are but a small percentage of the "catch." If not too large, the prey is bolted whole. A typical crocodilian trait is to revolve the body rapidly after a large animal has been seized. This aids in killing the prey and may also twist off a part of the animal's body for easier eating. As an aid to digestion, in the lack of any chewing of the food, the crocodiles and 'gators generally possess "stomach stones," a set of pebbles which have been swallowed and lodged in the gizzard, a muscular part of the stomach, where they perform a grinding action in reducing the flesh of the victim to proportions suitable to be passed on into the intestine. The crocodilians are, among living reptiles, the closest relatives of the birds, and it is of interest that their avian cousins show a similar adaptation.

The large eggs, on the average a few dozen in number, are laid in pockets dug in the mud of the shores or in a nest constructed of vegetation, trash, and mud. In all forms where the habits are known, the nest is guarded by the female during the incubation period, which appears to be some two to three months. In some known cases, at least, the female continues to keep a watchful eye over her defenseless young for a considerable period after their birth. Most reptiles show little interest in their progeny once the eggs are deposited; the crocodilian habits suggest that the care of the young exhibited by the birds had its beginnings while their ancestors were still archosaurian reptiles. What of the dinosaurs, who were, like the crocodilians, bird "cousins"? In most cases, naturally, we know little of their breeding habits. But in one instance, that of a small horned dinosaur from Mongolia, an extensive "nesting" site was discovered where remains of eggs, young, and parents were all found close together in a single area, suggesting that in at least this instance there was a considerable amount of dinosaur parental care.

Most crocodilians are popularly termed either alligators or crocodiles, but there are apparently half a dozen or so distinct genera. As we have said, however, the differences are generally slight. A technical difference between alligators and typical crocodiles lies in details of the position of certain of the teeth. A point of difference easier to observe in the live state (without danger to life and limb) lies in the fact that, while all crocodilians have long snouts, those of the alligators are broadly rounded, whereas those of most other forms are, to a variable degree, more slenderly shaped. Crocodilians are, as a group, the largest of living reptiles, with maximum lengths of measured specimens running as high as twenty-three feet; the giants of the tribe are some of the American crocodiles, with the Indian gavial, the salt-water crocodile of the Malay region, and the American alligator close behind in recorded lengths. Most crocodilians, however, are

Crocodilians. The alligator is the more familiar North American form. *Left, above,* a young alligator in walking pose. Contrary to the usual belief, the animal walks high off the ground rather than sprawling; the hind legs are long and powerful, as in its bipedal dinosaur relatives. *Left,* the Paraguay caiman, a small South American relative of the alligator, with a maximum length of 8 feet. *Below,* the "false gavial" (*Tomistoma*), a large slender-snouted fish-eater from Borneo, which parallels in form and habits the Indian gavial. *Bottom,* the giant saltwater crocodile of the Malay region, a dangerous man-eater. (Alligator photograph by E. H. Colbert; caiman by Frederick Medem, courtesy C. H. Pope; *Tomistoma,* New York Zoölogical Society; saltwater crocodile, Zoölogical Society of Philadelphia.)

American crocodiles. Although true crocodiles are mainly Old World forms, one crocodile species is present in the Caribbean region and reaches the southern tip of Florida. In contrast to some Old World forms, the American crocodiles are mainly fish-eaters and do not appear to be particularly dangerous to man. *Above*, a habitat group in the Chicago Natural History Museum; *left*, a live specimen. (Photograph courtesy New York Zoölogical Society.)

A crocodile predecessor. The extinct phytosaurs of the Triassic period were an offshoot of the early ruling reptiles that had habits similar to the modern crocodilians; although related, they were not ancestral to the crocodiles. Their adaptations for an amphibious flesh-eating mode of life were similar to those of crocodiles, except that the nostrils were situated at the top of the head in front of the eyes. (From a drawing by S. W. Williston.)

much smaller, and a dwarf species of crocodile from the Congo seldom exceeds three feet in length.

Crocodilians are typically tropical animals; there is scarcely a region of the tropics in which there is not a representative of the order. The alligators, exceptionally, extended northward into the warmer parts of the temperate zone. One, the American alligator, ranges from the gulf states east and north as far as North Carolina; a second, smaller alligator, is found in the marshes of the Yangtze Valley of east central China. This discontinuous distribution is readily accounted for by the fact that in earlier, Tertiary days, alligators (and other crocodilians) were present far to the north in both North America and Eurasia. Apparently (as we shall have reason to discuss again in relation to mammal distribution) northern climates were once less frigid than now, and crocodilians could live and migrate freely over much of both New and Old Worlds.

In general a given region or habitat in the tropics will contain one or possibly two species of crocodilians. Most widespread are crocodiles in the narrow sense, with a considerable number of species ranging over the Old World tropics from Australia and the Philippines to West Africa and several species in the American tropics (the common American species gets as far north as the tip of Florida). Best known of the group are the Nile crocodile (which actually ranges over nearly the whole African continent) and the Indian form, known as the mugger, of the Ganges and other south Asian rivers. Crocodilians are mainly inhabitants of fresh-water streams, lakes, and swamps, but a large crocodile of the general Malay region favors salt marshes and is known to swim far out to sea—a fact accounting for its wide distribution from the eastern coast of India through the East Indies to the Philippines and the far Fiji Islands. Only in South America—particularly the Amazon basin and the countries to the north—is there a considerable variety of crocodilians. This is due to the presence here, in addition to true crocodiles, of a number of species of caimans—alligator relatives, usually with a narrower snout and mostly of small size.

Crocodilians which specialize in fish-eating tend to have quite slender snouts, in contrast with more omnivorous relatives. Most specialized in this regard is the Indian gavial (more correctly "gharial"), one of the largest of crocodilians and so distinctive that it is usually placed in a family separate from all other living members of the order.

Prominent in popular lore regarding crocodilians is their supposed man-eating habits. There is some truth to this belief, but the matter tends to be considerably exaggerated. The majority of crocodilians, with average body (plus tail) lengths of but four to six feet, are too small to attack large animals of any sort, much less man. And even many of the bigger ones appear to be essentially harmless to us. The alligator, for example, is one of the largest members of the group which seldom attacks a man. More of a menace, however, are certain other large forms. The black caiman of South America (differing quite a bit from the usual small caimans) is dangerous, as is the Indian mugger to some extent (although charred corpses from burning ghats are the more usual form of its human fodder). It has been known from antiquity that the Nile crocodile will not infrequently attack man, and the great salt-water crocodile of Asia and the East Indies is a dangerous man-eater.

As was said, the archosaurs as a whole are much reduced from their one-time dominance, and even the surviving crocodilians are few in numbers compared to earlier days. All through the Tertiary there are records of numerous crocodilians, and back in the days of the dinosaurs crocodilians were abundant and varied.

While archosaurs were triumphant on the land and in the air, they tended little to invade the seas. Only one marine ruling-reptile group ever entered salt water—the crocodilians. We have mentioned one living crocodile which enters the sea, but he is mainly a swamp dweller. In Jurassic deposits we find highly adapted marine members of the group. These animals had turned their limbs into steering paddles and redeveloped a rather fishlike tail. They were, however, a short-lived group; successful invasion of the seas was reserved for members of other reptilian stocks.

FLYING REPTILES

With the adoption of the bipedal gait, the front legs of the archosaurs became freed for other uses. In some groups they tended to degenerate; in a few forms they functioned as grasping organs, "hands." In two cases they took on an entirely new use—that of wings, organs of flight. Flight was twice evolved in the ruling-reptile stock; once by the ancestors of the birds, as noted later; again, and for the time more successfully, by the flying reptiles.

The pterosaurs ("winged reptiles"), or pterodactyls ("wing-fingered"), were common forms during the Jurassic, middle period of the Age of Reptiles. The remains of these and many other contemporary animals have been preserved in considerable numbers in German lithographic stone deposits such as that at Solenhofen. These deposits appear to represent fine sediments which settled in the bottom of ancient coral-reef lagoons and in them are preserved many delicate structures, such as jelly-fishes and wing impressions of pterosaurs. A good example of a primitive flying reptile was *Rhampho-rhynchus* (not a hard name to spell if sprinkled liberally with *h*'s), shown in our fig-ures. This was a creature about a yard in length, with a long beak armed `with sharp teeth, a short body, and a long tail, tipped by a steering rudder. The hind legs were slim and feeble; but the front legs were very powerful wing supports.

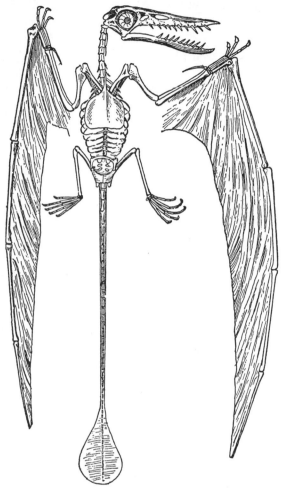

In the hand the first three fingers were short and armed with claws. The little finger had been lost. The fourth one, however, was very stout and long and was the sole support of the wing, a bat-like flap of skin which ran back to the thigh region.

This type of flying struc-ture is, of course, quite in contrast to that of birds. Nothing in the nature of feathers was present in the pterosaurs.

These aerial reptiles ap-pear to have been fish-eat-ing types, flying over the water and diving after small fishes much as do some mod-ern marine birds. What they did when ashore to rest or

The skeleton of *Rhamphorhynchus*, a long-tailed Jurassic pterosaur. About one-fourth natural size. Impressions of the soft tissues of the tail "rudder" and wing membranes are often preserved. The wings were supported by one very long and powerful finger (the fourth). The other fingers were short and, with the feeble hind legs, were probably used for clutching purposes when at rest. (From Williston.)

nest, however, has been a disputed problem. The legs were ill-suited for walking purposes but were perfectly good clutching organs, as were three fingers of the "hand." Probably, like the bats, they hung suspended from tree limbs or overhanging rocks during their times of rest.

Besides these long-tailed flying reptiles, the Jurassic lagoon deposits have also yielded remains of short-tailed relatives, some no bigger than a sparrow. Short-tailed types were destined (in contrast to the more primitive ruddered pterosaurs) to last over into the Cretaceous, the third and last of the great reptile periods. Some forms of this age grew to great size. *Pteranodon*, a form found in the chalk rocks deposited by the seas which once covered western Kansas, had a wingspread of twenty-seven feet in one specimen and was thus much larger than any bird. The beak was toothless; and a curious feature was an enormous crest projecting from the back of the head like a weather vane. *Pteranodon* and his kin were the last of flying reptiles. By that time birds were already far along in their developmental history and soon entirely superseded their reptilian cousins as aerial navigators.

DINOSAURS

The story of the dinosaurs is one of the most interesting of all the evolutionary sagas furnished by the vertebrates. Beginning in the Triassic, they increased greatly in numbers and size in the Jurassic and Cretaceous, ruled the earth without rivals for nearly the whole extent of the Mesozoic, and at its close disappeared forever from the world.

We customarily think of the dinosaurs as a group of gigantic reptiles. This, however, is not quite the case. Among the dinosaurs were the largest animals that ever walked, but other dinosaurs were no bigger than chickens. Further, while all dinosaurs were descendants of the primitive ruling-reptile stock, there were two quite separate groups of them, each containing a number of remarkable types.

The two dinosaur stocks.—These two groups are termed scientifically the Orders Saurischia and Ornithischia, meaning "reptile-like pelvis" and "birdlike pelvis." The pelvis, or hip bones, offers a fine key character for distinguishing between the two dinosaurian types, and we shall dwell for a moment on the technicalities of hip construction.

The hip bones of any land animal consist of three parts. Above the socket for the thigh bone is the ilium, fused to the backbone. Below the socket lie the pubis in front and the ischium behind. In primitive land dwellers these last two form a solid plate. In early ruling reptiles, however, the two have diverged considerably, the pubis slanting down in front, the ischium running down behind. This makes a three-pronged pelvis, each of the bones running off from the hip socket at a considerable angle to the others.

This type of construction, not so far from that of ordinary reptiles, still persisted in the saurischian, or reptile-like dinosaurs, including the flesh-eaters and the giant amphibious forms. In the other order, the Ornithischia, or birdlike dinosaurs (including among others the duck-billed dinosaurs and the armored and horned types), there has been a further complication. The pubis has swung back parallel to the ischium, just as in the birds. But, as an aid to the support of the belly, a new bony process has grown forward from the base of the pubis, giving a characteristic four-pronged type of pelvis. This structure is, as we have said, similar to that

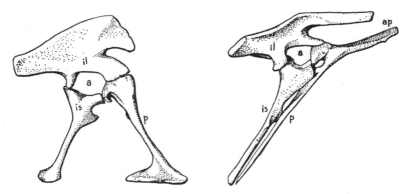

The right pelvic bones of dinosaurs, to show the contrast between the two great orders. *Left*, the triradiate pelvis of a reptile-like dinosaur (order SAURISCHIA). *Right*, the tetraradiate pelvis of a birdlike form (order ORNITHISCHIA). In each case the socket (*a*) for the thigh bone is bordered by three elements: the ilium above, the pubis below and toward the front, the ischium below and toward the back. In the birdlike form, however, the pubis is two pronged, with an anterior process (*ap*).

found in birds. But it cannot be too strongly emphasized that the birds are not descended from these dinosaurs; the two groups are essentially "cousins," descended separately from a common ancestor at the base of the ruling-reptile stock.

Footprints.—Both dinosaur groups were already in existence before the end of the Triassic, and numerous footprints have been found of dinosaurs of this age in the rocks representing old mud flats of the Connecticut Valley region. These bring out an interesting feature of dinosaur anatomy. In many dinosaurs with a bipedal gait the main reliance was placed on the three center toes; the outer one was lost, and the inner one often turned back as a rear prop. In this arrangement we have an exact parallel, even to the number of joints and comparative length of toes, to the structure of the bird foot. These footprints were discovered over a century ago. At that time dinosaurs were almost unheard of, and the tracks were (not unnaturally) long thought to be those of gigantic birds!

Small carnivorous dinosaurs.—We shall first consider the history of the Saurischia, or reptile-like dinosaurs. These were already abundant in the

A restoration of dinosaur life on the shores of the Triassic waters of the modern Connecticut Valley region showing some of the dinosaur types which may have been responsible for the numerous footprints. (The artist, incidentally, was mistaken in restoring this as a body of salt water and with seaweed; it was actually a fresh-water deposit.) (From Heilmann, *The Origin of Birds*, by permission of D. Appleton–Century Co., publishers.)

late Triassic. Some early forms were comparatively small, swift, flesh-eating bipeds, as were some of the stem archosaurs mentioned previously; little change was needed to turn one of these old forms into a primitive reptile-like dinosaur. Bipedal flesh-eaters continued to be the main stock of the saurischians throughout their history, and many of them long kept

The ostrich dinosaur (*Ornithomimus*). This was a Cretaceous bipedal flesh-eater, related to the giant carnivores but of relatively small size. The bulk was about that of an ostrich. Ostrich-like, too, were the proportions of hind legs and neck and the small head with a toothless bill. However, a characteristic dinosaur tail was present, and the three-fingered "hand" was well developed. This creature's habits are not surely known, but it may well have been an egg-stealer. (Restoration by Erwin Christman; photograph courtesy American Museum of Natural History, New York.)

The skull of *Tyrannosaurus*. The skull of this great flesh-eater of the Cretaceous measured some 4 feet in length and was armed with a powerful battery of sharp, compressed, and recurved teeth. A plate which follows shows this bipedal dinosaur in action. With a height of nearly 20 feet, *Tyrannosaurus* is believed to be the largest carnivore which ever walked the earth. (Photograph courtesy American Museum of Natural History, New York.)

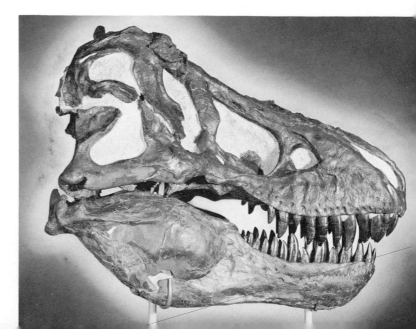

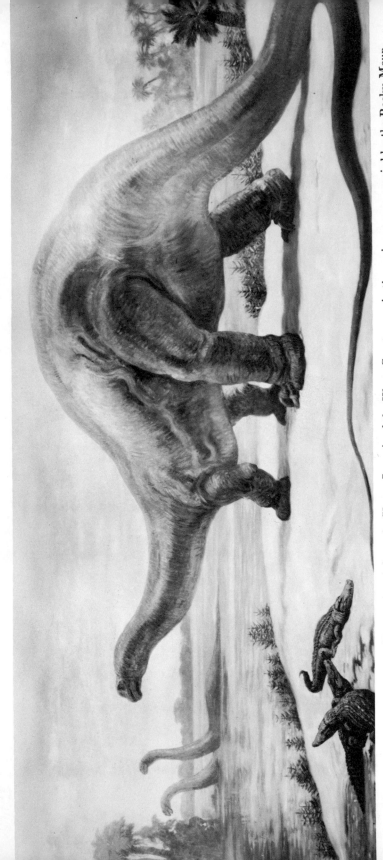

A giant amphibious dinosaur (*Brontosaurus*) from the Upper Jurassic of the West. Great areas in the regions now occupied by the Rocky Mountains and the adjacent high plains in Utah, Colorado, and Wyoming were, in the late Jurassic, lowlands with meandering streams and bayous and a rich vegetation; the region may have been rather comparable in nature to the present Mississippi Delta country. In these lagoons lived great amphibious (or sauropod) dinosaurs, such as the form shown here, and the closely related *Diplodocus* and *Brachiosaurus*. Although for pictorial purposes one of these giants is shown ashore, it is probable that they spent nearly their entire existence in the lagoons, where the buoyancy of the water would render the burden of their great weight less onerous. Primitive crocodiles were present in the same region; several of them are shown in the foreground. (From a mural by Charles R. Knight; photograph courtesy Chicago Natural History Museum.)

Upper Cretaceous dinosaurs. Two of the most spectacular types of dinosaurs are shown here in a scene that may well have occurred in western North America at the very end of the Age of Reptiles; the forms illustrated were contemporaries in Wyoming deposits. *Tyrannosaurus* (*right*) was the giant among the large carnivorous dinosaurs, with a length of about 47 feet and a height of about 19 feet. There was a large head with sharp and powerful teeth and massive hind legs; the "arms," however, were absurdly small. A pair of these creatures are shown about to attack the horned dinosaur *Triceratops*, a harmless herbivorous type. This reptile was defended by sharp horns and a bony frill over the vulnerable neck region. The rest of the body, however, was covered merely with a leather hide, and hence we may assume that a successful defense necessitated keeping face to face with the enemy. (From a mural by Charles R. Knight; photograph courtesy Chicago Natural History Museum.)

Upper Cretaceous dinosaurs (*continued*). Another scene at the close of the Age of Reptiles, but slightly earlier and from another locality—western Canada just east of the rising Rocky Mountains. This country had been previously covered by the sea and was (rather in contrast with present conditions) a lowland, well-watered, with numerous swamps and lagoons. Most of the dinosaurs shown are harmless herbivores belonging to the birdlike group (Ornithischia). In the left foreground is a heavily armored quadrupedal dinosaur (*Palaeoscincus*) covered with bony plates and with spines projecting out over the limbs; the tail, too, was a powerful, bony clublike weapon. At the right are crestless members of the bipedal duckbill dinosaur group (hadrosaurs or trachodonts). These herbivores appear to have been more or less amphibious, with webbed front feet. Many of the duckbills developed peculiar crests on top of the skull. Two crested types are seen at the left—*Corythosaurus* in the foreground, a group of individuals of *Parasaurolophus* behind. In the center background are ostrich-like dinosaurs (*Ornithomimus*). (From a mural by Charles R. Knight; photograph courtesy Chicago Natural History Museum.)

to a small size. The small form *Compsognathus*, for example, shown in one of our illustrations, was no larger than a rooster. This little type presumably preyed upon small reptiles of the lizard sort and perhaps even upon our own relatives, the tiny early mammals which were already in existence.

An interesting end development of these comparatively small and lightly built members of the carnivorous group is that exemplified by the Cretaceous "ostrich dinosaur" *Ornithomimus* and its kin. These were somewhat larger in size, with rather ostrich-like proportions, except for the development of the tail and front limbs. They were toothless, presumably

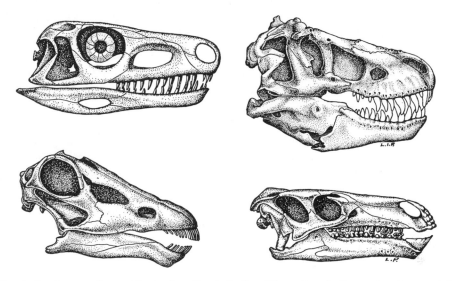

The skulls of some ruling reptiles to show the similarity of the pattern in the group and some of the modifications. *Upper left*, a small ancestral member of the group (*Euparkeria*). *Upper right*, the giant carnivorous dinosaur *Tyrannosaurus*; a large skull with huge sharp teeth suitable for a flesh-eater. *Lower left*, *Diplodocus*, an amphibious plant-eater, with a feeble dentition. *Lower right*, *Stegosaurus*, a birdlike dinosaur, a herbivore with a toothless bill and cheek teeth somewhat adapted for chewing.

having a horny, birdlike bill; the hind legs show evidences of fast-running abilities; the three-fingered hands evidently had considerable grasping power. The probable habits of this creature have been a subject of considerable debate. It was finally suggested that it made its way in the world by stealing the eggs of other dinosaurs. Teeth would not be needed in eating the egg contents, but a beak would be useful in breaking the shell. The grasping powers would be useful in egg-handling; and speed would be advantageous in avoiding enraged parents. That these thieves were, however, not always successful is shown by the fact that the crushed skull of a member of this group was found in Mongolia at the site of a nesting ground of horned dinosaurs. Very likely this reptile had been "caught in the act."

Large flesh-eaters.—The main line of carnivorous reptile-like dinosaurs tended, however, to grow to large size. By the end of the Jurassic some of these carnivores had become large enough to prey upon the gigantic amphibious dinosaurs then abundant. Still further increases in size took place in the Cretaceous. In the later phases of that period, the final chapter in dinosaur history, we find huge bipeds of the type of *Tyrannosaurus*, the "tyrant reptile." This great flesh-eater, the largest carnivore that ever walked the earth, stood some nineteen feet in height. The hind legs were very powerful; the front ones, however, had degenerated so that they were not even able to reach the mouth, much less be useful in walking, and had retained but two feeble fingers. The skull was a massive structure more than four feet in length, armed with numerous saber-like teeth that must have been highly effective biting and rending weapons.

Amphibious dinosaurs.—But in the late Triassic some of the reptile-like dinosaurs, which were already tending to quite large size, seem from their teeth to have been changing to a plant diet and (in correlation with their considerable bulk and lessened need for speed, as plant-eaters) appear to have been slumping back into a four-footed method of walking.

From such beginnings came the great amphibious dinosaurs, the sauropods, the largest four-footed creatures that ever existed. They were present throughout the Jurassic and Cretaceous but seem to have reached the peak of their development at the end of the former period. Numerous remains of amphibious dinosaurs of that age have been quarried in our western states, and one such deposit (at Jensen, Utah) has been preserved as a national monument.

Brontosaurus and *Diplodocus* are among these large creatures whose remains form a major attraction of many museums. *Diplodocus*, at eighty-seven and a half feet, holds the length record but was rather slimly built and was far from the heaviest. Twenty-five to thirty-five tons is a fair estimate of the weight of an average full-sized member of the group. In all these forms we find a massively built body, powerful limbs with a four-footed pose (but the front legs usually much shorter than the hind), a long tail, and a long neck terminating in a small head.

The head seems absurdly out of proportion to the body. The eyes were high up on the sides and the nostrils sometimes at the very top of the skull above them. This position is like that of the "blowhole" in whales and is one of the reasons for believing that these giant reptiles were amphibious in their habits, spending much of their lives in the water; the animal could breathe and see, with only the top of the head exposed above the surface.

The jaws were short and weak, the teeth feeble and few in number. It seems almost incredible that such a feeding apparatus could have gathered enough material to supply the huge body, although the fodder may have been some soft type of water vegetation which could be cropped with little effort. The brain is small in all reptiles but excessively small in these dinosaurs in proportion to their size. Very likely the brain did little except work the jaws, receive impressions from the sense organs, and pass the news along down the spinal cord to the hip region, from which came the nerves working the hind legs. Between the hips there was situated an enlargement of the cord several times the size of the brain.

This condition inspired the following flight of fancy on the part of the late Bert L. Taylor, a columnist on the staff of the *Chicago Tribune:*

> Behold the mighty dinosaur,
> Famous in prehistoric lore,
> Not only for his power and strength
> But for his intellectual length.
> You will observe by these remains
> The creature had two sets of brains—
> One in his head (the usual place),
> The other at his spinal base.
> Thus he could reason *a priori*
> As well as *a posteriori*.
> No problem bothered him a bit
> He made both head and tail of it.
> So wise was he, so wise and solemn,
> Each thought filled just a spinal column.
> If one brain found the pressure strong
> It passed a few ideas along.
> If something slipped his forward mind
> 'Twas rescued by the one behind.
> And if in error he was caught
> He had a saving afterthought.
> As he thought twice before he spoke
> He had no judgment to revoke.
> Thus he could think without congestion
> Upon both sides of every question.
> Oh, gaze upon this model beast,
> Defunct ten million years at least.

The presence of the posterior "brain" is probably associated with a more prosaic but highly useful function. We are apt to conceive, unthinkingly, of nerve impulses as traveling almost instantaneously, in the fashion of electricity. But even in mammals the speediest of nerve fibers transmits messages no faster than a few hundred feet a second, and in reptiles the rate of passage is much slower—only a few yards a second, on the average. In a small animal the brain can quite effectively send out directives for prompt responses to sensation, for the distances are short. Not so in an animal like *Diplodocus*.

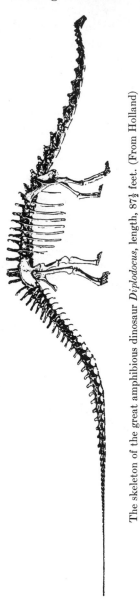

The skeleton of the great amphibious dinosaur *Diplodocus*, length, 87½ feet. (From Holland)

If, for example, some animal had the temerity to pinch this great creature's tail, a considerable part of a minute would have elapsed before the brain would receive word of this insult and be able to respond with orders to the tail to lash out at the offender; by which time the latter might be well started on his way to the next county. Even in such a form as man, with a highly integrated nervous system, many simple responses—reflexes —are brought about by nervous action in the spinal cord itself, without the necessity of referring matters to the brain. The enlargement of the cord in the hip region (the sacral region) of sauropods and other dinosaurs serves much the same sort of purpose, if to a higher degree, as a center from which much of the activity of the hind part of the trunk and of the legs and tail could be regulated without the necessity of constantly consulting the distant little brain.

A monumental construction of the backbone was necessary to carry the tons of weight of the enormous body and to transfer it to the legs. Nature had solved in competent fashion the engineering problems involved. The backbone formed an arch, supported at the top by the massive hind legs beneath the stout hips and tapered away in each direction from this pier. The dead weight of the bone of the spinal column was a considerable burden in itself; this weight had been reduced by hollowing out the sides of the vertebrae, leaving only the essential framework of these bones.

The hind legs were very massive and appear to have been quite straight weight-bearing columns like the limbs of elephants. The front legs, usually much shorter, and thus reminiscent of these animals' bipedal ancestors, bore less weight and may have been a bit crooked at the elbows. The feet seem to have been huge rounded pads with two or three large claws which would have been of aid on a slippery bottom.

Even with these stout legs it is difficult to see how these dinosaurs ever walked on land. The elements of physics show that there are natural limits set to the possible size of a four-footed land vertebrate. The weight of an animal varies in proportion to the cube of a linear dimension. But the strength of a leg, like any supporting element in engineering, is proportionate to its cross-section, which increases only by squares. If a reptile doubles in all his dimensions, his weight is about eight times as great, but his legs are only four times as strong. Hence in large animals the bulk of the legs must increase out of all proportion to the rest of the body. An elephant does not and cannot have the slim legs of a gazelle; and in the case of these great dinosaurs it seems doubtful if their limbs, stout as they were, could have effectively supported so many tons on land. It thus appears probable that the sauropods were

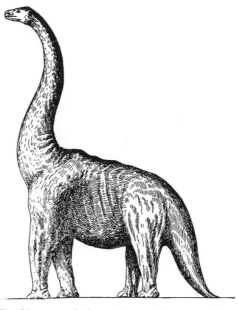

Brachiosaurus, the largest known dinosaur, with an estimated weight of about fifty tons and with a height great enough to look over the top of a three-story building. This form is known from Jurassic deposits of both Wyoming and East Africa. (From Abel.)

amphibious types which spent most of their lives in lowland swamps and lagoons where, buoyed up by the water, problems of support and locomotion were greatly simplified.

The giant of the group was *Brachiosaurus*. This form was first known from incomplete remains from the western United States, now in the Chicago Natural History Museum. There was only enough material to show that the animal was a large one, with (exceptionally) long front legs. Later a skeleton of an apparently identical animal was discovered by Germans in East Africa and has been mounted in Berlin. This form was short bodied, with a stub tail, but despite this may have reached a live weight of fifty tons. Above the long front legs there stretched up a long neck by means of which this dinosaur could easily have looked over the top of a three-story building. This giant of giants was apparently capable of living in waters of considerable depth.

The amphibious dinosaurs reached a peak at the end of the Jurassic; they played only a small part in the Cretaceous act of the reptilian drama and then vanished.

BIRDLIKE DINOSAURS

The story of the reptile-like dinosaurs was paralleled by that of the ornithischians, which resembled their avian cousins in hip structure and

A restoration of the primitive birdlike dinosaur, *Camptosaurus*. Shown here in a four-footed pose which was perhaps assumed when walking slowly; for fast travel the animal was undoubtedly a biped. (From Heilmann, *The Origin of Birds*, by permission of D. Appleton–Century Co., publishers.)

hence are called birdlike dinosaurs. It must, of course, be kept in mind that these forms were not bird ancestors and, except in this feature, did not resemble birds any more closely than did other ruling reptiles. (READER: "You said that before." AUTHOR: "I know, but you had forgotten." READER: "Nothing of the sort." AUTHOR: "Very well then; let's get on with the story.")

Primitive herbivorous bipeds.—These reptiles were rare in the Triassic but common in the two following periods. The early members of the group are well represented by *Camptosaurus*, a Jurassic type which ranged from about eight to twenty feet in length in different species. The early birdlike dinosaurs were, like the primitive saurischians, bipedal in fast locomotion, but the front legs were usually comparatively little reduced, and these animals (as the figure suggests in the case of *Camptosaurus*) may well have strolled about on all fours when not particularly hurried.

A major difference between these birdlike forms and the other dinosaurs lies in the fact that they had abandoned a carnivorous life; the ornithischians were herbivores from the beginning. In connection with this mode of life, almost all of them had lost their front teeth and replaced them, it would seem, by a stout, birdlike, horny beak. The teeth in the back of the jaws were no longer pointed but were leaflike chewing teeth with rough edges. What each tooth lacked in grinding power was often made up for by an increase in numbers; in some of the duckbills mentioned below there were somewhere between fifteen hundred and two thousand teeth in the mouth at one time.

Duckbills.—In the Cretaceous, bipedal birdlike forms were abundant. The prominent group was that of the duckbilled dinosaurs, or hadrosaurs. These were large forms, exceedingly numerous in the late Cretaceous of North America, and many skulls and skeletons are known. The popular name arises from the fact that there was a broad, ducklike beak. The limbs were massive, and probably they were far from speedy. In several cases there have been found "mummies" of these duckbills. The dinosaur corpses had dried and hardened before being buried, and the surrounding mold of rock has caused the formation within it of a natural cast of the details of the skin. These mummies show that the duckbills had webbed feet and presumably were amphibious in habits, feeding in swampy pools or about their margins.

A curious event in duckbill history was the development of crested types. In some forms the bones about the nasal opening were swollen to form a sort of "eagle beak." In another type there was a thin domelike swelling over the top of the head shaped like a rooster's comb but made of bone. In others backward growth continued, to form a sort of horn projecting up and back over the neck region. It is of interest that all these peculiar structures were formed of the bones of the nasal region; nature had, so to speak, taken the animal's nose and pulled it up over the top of its head. We have little idea what function (if any) these curious crests and horns performed. However, in one case "dissection" of a fossil skull shows that the crest contained a pair of convoluted tubes forming the air passages for the nose opening to the throat. Possibly they held a reserve store of air for use in underwater feeding.

Two armored types.—Since all the birdlike dinosaurs were herbivorous in habits, speedy locomotion was of service only in escaping from their carnivorous cousins. We find that several groups of ornithischians had (parallel to the developments in the other group) slumped back to a four-footed mode of life but with their ancestry clearly shown by the short front legs. These slow-moving types invariably were armed in some way against attack by the great flesh-eaters. Even in the stem archosaurs, we noted, there is frequently found a double row of small bony plates down the length of the back. In *Stegosaurus* of the Jurassic these protective devices had been put to good use. They had been expanded upward to form a stout double row of bony plates covering the back and two pairs of sharp spikes near the tip of the tail. These structures obviously were useful in defense against attack from above, but the flanks appear to have had little protection.

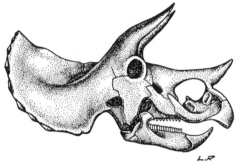

The skull of a horned dinosaur, *Triceratops*. The horns above the eye sockets were paired; the nose horn was a single structure. The back half of the skull is an enormous bony frill covering the neck. (After Hatcher, Marsh, and Lull.)

In the Cretaceous we find a different type of armored dinosaur in *Ankylosaurus* and its relatives. These forms have been not inaptly termed reptilian "tanks." The body was broad and flat and studded above with a heavy series of bony plates and nodules forming a protective layer not dissimilar to the upper shell of a giant turtle. These dinosaurs could not pull their heads and limbs into the shell as could the turtles, but these parts were also protected for there were plates of bone to reinforce the skull, and large spines projected out at the sides to fend off enemies from the limbs.

Horned dinosaurs.—A final development in four-footed birdlike types was the appearance, in the late Cretaceous, of the horned dinosaurs such as *Triceratops*. In these forms the trunk appears to have been barren of any defensive structures; everything was concentrated on head development. From the back of the skull there extended out a broad frill of bone

Stegosaurus. One of the most familiar members of the dinosaur group is this late Jurassic form, a four-footed herbivore armed with a double row of plates and spines down the back. (From a mural by Charles R. Knight; photograph courtesy Chicago Natural History Museum.)

A crested dinosaur. In many of the Cretaceous duckbilled dinosaurs peculiar outgrowths occurred on the top of the skull, formed by the bones normally surrounding the nostrils and containing air tubes which may have been of use in the partially aquatic life of these dinosaurs. These bird-like, bipedal forms are known as "duckbills" because of their broad flattened beaks. Some well-preserved fossil imprints also show webbed feet. *Corythosaurus*, whose skull is figured here, is shown in an earlier landscape plate. (Photograph courtesy American Museum of Natural History, New York.)

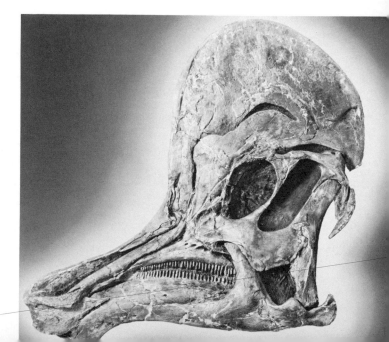

A primitive horned dinosaur (*Protoceratops*) from the Cretaceous of Mongolia. This small reptile (the largest specimen only about 6 feet long) was an ancestor of the great horned dinosaurs, but one in which the horns were still undeveloped. (From a mural by Charles R. Knight; photograph courtesy Chicago Natural History Museum.)

Dinosaur eggs, found with the remains of the dinosaur shown above. (Photograph courtesy American Museum of Natural History, New York.)

A small flying reptile (*Pterodactylus*) from the Jurassic lithographi stone of Germany. Except that this is a short-tailed form the skelet: elements can be readily compared with those shown in an accom panying text figure (p. 185).

which covered the neck region (a favorite place for attack). Then, too, there developed on the skull stout bony horns. Two of them were usually present over the eyes, much like the horns of cattle, and a third was usually present on the nose. An interesting find is that of a primitive horned dinosaur, *Protoceratops*, in Mongolia. This small animal belies its name, for there were almost no traces of horns; the neck frill, however, was already well developed.

Extinction of dinosaurs.—These horned forms were very abundant in late Cretaceous times, as were most of the major groups of dinosaurs. But at the end of the Cretaceous all of them disappeared completely; the reign of the ruling reptiles was over; the Age of Reptiles was at an end.

What caused the extinction of the dinosaurs is a question to which no single or certain answer can be given. It is obvious that the carnivores would necessarily have died out with the extinction of the herbivorous types upon which they preyed. For these latter, changes in vegetation, the gradual disappearance of the plants upon which they fed, and the replacement of these plants by others which were not suitable for dinosaur food may have been important factors leading toward extinction; and study of fossil plants shows that a major change in the world's flora was taking place in late Cretaceous times. Many dinosaurs, as we have seen, were lowland swamp and lagoon forms; and at the end of the Cretaceous land areas were rising, with a consequent reduction in such types of country. In addition, there is distinct evidence of a chilling of the climate in North America at this general time. All these factors can, more or less directly, be traced back to one major geologic event of the times—the rise of mountain systems, particularly the Rockies. This event, by raising land levels, had considerable influence on climates and physiography and secondarily on plant and animal life. Perhaps the Rocky Mountains killed the dinosaurs!

CHAPTER 10

Birds

The birds are a group of interesting and often delightful creatures which are regarded as forming a separate class of vertebrates and seem to us as unlike reptiles as any sort of animal could be. But, apart from their powers of flight and features connected with it, they are structurally similar to reptiles. Indeed, it is because they are so close to the ruling reptiles from which they are descended that we may best consider them here.

THE STRUCTURE OF BIRDS

Birds have been called by an old writer "glorified reptiles." Feathers are, in reality, almost their only distinctive feature, for almost every other character can be matched in some archosaur type. Large quills form the expanse of the wing, taking rise from the back of the forearm and from the three fused fingers in which the wing terminates. The fleshy part of the tail is reduced to a short stub; from it there arises a spreading fan of feathers, which is an effective steering organ. The rest of the body is covered with a thick, overlapping set of smaller, softer feathers which form a very effective insulation for the retention of the bird's bodily heat. The varied coloration of the feathers may be in part related to sexual display, in part related to protective coloration.

Unlike reptiles, birds are warm blooded, that is, they have a high and constant body temperature (in some cases several degrees above that found in man). This requires a large supply of oxygen, good lungs, and an efficient blood circulation. We have previously remarked upon the incomplete separation of fresh and "used" blood streams in lower land forms. The birds have evolved as efficient a circulation as our own, with a four-chambered heart and but one main vessel (or aortic arch) carrying all the pure blood to the body. But this evolution in birds has taken place independently of that in mammals, for while the main vessel in man arches over to the left of the body, the birds have put their emphasis upon a similar vessel on the right side.

Brain and sense organs are much modified in relation to flight. Birds (like men, but in contrast to most other vertebrates) depend largely upon their eyes for their information about the outside world. The sense of smell is little developed. The eyes are large; within the eyeball is a circle of bony plates. Such plates are present in many vertebrate groups (not in mammals). In part they serve to protect the eyeball from distortion by the pressure of liquids within it. In many primitive vertebrates they apparently serve additionally as a protection against water pressure, and in birds, against wind pressure in flight.

The brain is large, and, as in mammals, it is the cerebral hemispheres in the front of the brain that are mainly responsible for its large size. But there is a conspicuous difference between the two groups here. In mammals it is the "gray matter" of the surface of the hemispheres that has become enlarged, and (most conspicuously in man) the surface is often thrown into complicated foldings to increase the area of gray matter. This region is associated with intelligence (in the popular meaning of that term) or, as a minimum, with the ability to learn by experience and training which results in intelligent behavior. As may be seen from a figure in the next chapter, the surface of the bird brain is smooth, and the "gray matter" is small in amount. What has increased is a set of internal structures known technically as the corpora striata ("striate bodies"). Experimental knowledge of the function of this internal region is limited, but it appears that it is the seat of innate patterns of behavior—instincts, in familiar terms.

This brain structure is not unexpected in the light of what we know of bird behavior. Birds can be taught (pigeons in the Harvard laboratories have learned to play Ping-pong!) but in general are much more difficult to train than mammals. On the other hand, they have very complex inborn patterns of behavior. These patterns are an absolute necessity for their mode of life. Regulation of flight requires a series of complex reactions of instinctive nature, and, still more, the young must be maintained until their potentialities for flight are fully developed, leading in most cases to further complex behavior patterns in mating, nest-building, and care of the young.

In the skeleton there are many modifications in connection with flight. Lessening the specific gravity, there are not only air sacs within the body, connected with the lungs, but many of the bones of the skeleton may be hollow and air filled.

The breastbone, or sternum, is usually little developed in reptiles; in birds it is an enormous plate covering nearly the whole underside of the chest, with a strong keel in the mid-line. To it are attached the powerful chest muscles which propel the wings (these muscles form the white meat of the chicken). In the hand there are but three fingers remaining of the original five (there has been a similar reduction in some dinosaurs), and these are more or less fused, reduced, and clawless, acting as wing supports.

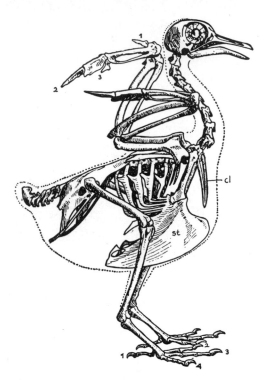

The skeleton of a modern bird. Note the huge breastbone, the short body, and tail. The wing skeleton includes the remains of the first three "fingers." The foot is four toed, with the first toe at the back, the second to fourth pointed forward, with a structure very similar to that of many dinosaurs. The bird "wishbone" is the collarbone, or clavicle (*cl*), of other animals. The enormous breastbone, or sternum (*st*), is developed to carry the powerful wing muscles (the "white meat" of the chicken). (After Heilmann, *The Origin of Birds*, by permission of D. Appleton–Century Co., publishers.)

We have noted that the pelvis is similar to the type found in some dinosaurs and have also mentioned the fact that the hind legs of birds and two-footed dinosaurs are almost identical. Modern birds are toothless, with a horny bill, a feature also found in some flying reptiles and dinosaurs.

All in all, the bird skeleton is quite similar to that of ruling reptiles. Even flight is not in itself a distinctive feature. Feathers are a unique character; but although these structures seem quite different from the horny scales which cover a reptile's body, the difference is in reality not very great. The two are identical in chemical composition; feathers may have come from horny scales in which the edges have developed into a large number of fine, interlocking subdivisions.

The oldest-known bird.—The description given above applies to the typical modern bird. Far different in many respects and much closer to its archosaur ancestors was the oldest-known bird, *Archaeopteryx* of the Jurassic. Two skeletons of this primitive bird have been found in the litho-

A Jurassic scene on a coral island in Germany. A fine-grained limestone, used for lithography, appears to have been deposited in the lagoons of coral islands in the warm seas that then covered much of Europe. The sediments were so fine that remains of many small and delicate creatures are preserved. Several of the more interesting vertebrates are shown here amid a foliage of cycads and other Mesozoic plants of tropical appearance. In the air are small flying reptiles of the genus *Rhamphorhynchus*, with long, keeled tail. Fluttering about or perched on cycads are the most primitive birds, of the *Archaeopteryx* type, known only from these deposits. The two small animals at the left are dinosaurs of the genus *Compsognathus*, no larger than a rooster but nevertheless related to the large flesh-eating dinosaurs. (From a mural by Charles R. Knight; photograph courtesy Chicago Natural History Museum.)

Dinosaur footprints. In the Connecticut Valley are great deposits of sandstones of Triassic age. In various places these have revealed footprints of dinosaurs. A major collection is that at Amherst College (including the specimen figured). The footprints shown are three-toed. They are, of course, comparable to those of birds and were long supposed to have been made by gigantic birds.

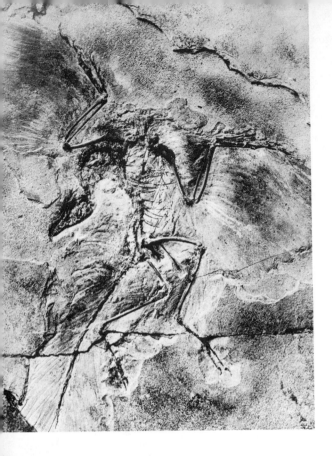

The oldest birds. The only known remains of Jurassic birds consist of these two skeletons and one isolated feather. All were found in the lithographic limestone of southern Germany, half to three-quarters of a century ago, and not another scrap has been discovered since. Although the two specimens are thought by some to be rather different from one another, they probably belong to the single genus *Archaeopteryx*. *Left*, the specimen in the Berlin Museum; a nearly complete skeleton, showing all four limbs and tail, with the head twisted back to the left; the feathers are well displayed. *Lower right*, the other skeleton, in the British Museum; this individual is less complete. *Below*, a restoration of a pair of these birds (the color scheme is, of course, imaginary in museum exhibits). The bird in life was about the size of a crow. (Restoration from Heilmann, *Fuglenes Afstamning*.)

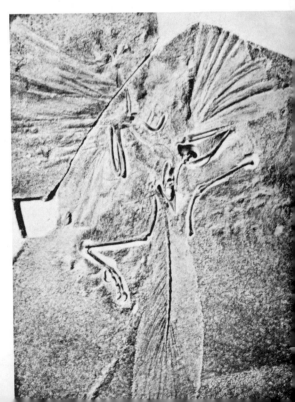

graphic stone deposits mentioned in the last chapter. It is fortunate that the delicate type of preservation characteristic of these beds has shown traces of the feather, for otherwise there would be little to distinguish this early form from some of the small dinosaurs. The tail was still of the old-fashioned reptilian type but with a double row of feathers down its length; there were wings, but these were rather feeble, and the three fingers supporting them still bore claws. The breastbone was small, indicating weak flying muscles. None of the bones were hollow, and the jaws were armed with teeth. These forms are, because of the possession of feathers, technically birds, but birds not far removed from their ruling-reptile forebears.

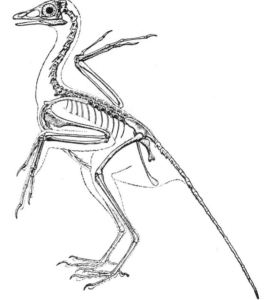

There has been much discussion as to the origin of flight. The theory most generally held is that the ancestral bird was a tree dweller and that flight started as a slight parachute effect as the Proavis jumped from branch to branch, the developing wings breaking the fall on landing; gliding would have been a subsequent stage. A second theory is that the ancestors were ground types and that the feathered arms and tail helped, by acting as planes, to increase the speed in running, lifting the Proavis somewhat off the ground.

Skeleton of one of the oldest known birds, *Archaeopteryx*. If this be compared with the recent bird shown in the last figure, the main differences noted here are the long bony tail, the clawed fingers, and the absence of the great keeled breastbone. All these are characters in which the oldest birds were still reptilian and dinosaur-like in nature. (After Heilmann, *The Origin of Birds*, by permission of D. Appleton–Century Co., publishers.)

Cretaceous birds.—A full period after the appearance of the first birds we catch a second glimpse of the developing bird life of the Age of Reptiles in the remains of some water birds from the marine rocks of Kansas. One of the types, *Hesperornis*, the "western bird," was a water-dwelling diver which had lost the power of flight; a second, *Ichthyornis* (the "fish bird"), seems to have lived a life like that of modern terns and was modernized in most respects, with powerfully developed wings. Teeth were still present in *Hesperornis*, and they have been claimed to be present in *Ichthyornis*, but the material on which this is based is dubious. In any event, teeth appear to have been lost at about the end of the Mesozoic, and during the Tertiary there took place the development of modern bird groups.

Ostrich-like birds.—Today we find in the southern continents a number of flightless birds—the rhea of South America, the ostrich of Africa, the emu and cassowary of the Australian region, and the kiwi of New Zealand.

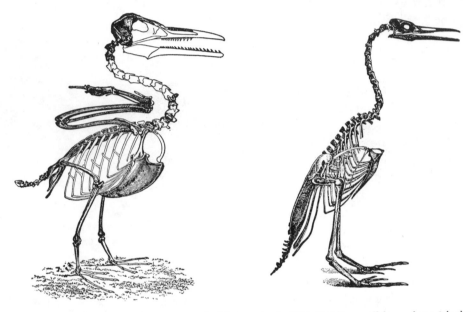

Skeletons of Cretaceous toothed birds. *Left, Ichthyornis,* the "fish bird," a small form about 8 inches in height, with good powers of flight. *Right, Hesperornis,* the "western bird," a large wingless diver. These skeletons were found in the 1870's (when the woodcuts here reproduced were made). Curiously, but little further material of these interesting birds has since been discovered. (From Marsh.)

In the Pleistocene epoch just behind us there were several other birds of the same sort, notably the "elephant birds" of Madagascar, whose enormous eggs have sometimes been found preserved entire in the swamps of that island, and the varied moas of New Zealand (one reaching twelve feet in height), which seem to have been still in existence when the Maoris reached those islands not so many centuries ago. Except for the kiwi, all these birds have much in common; they are large, with powerful hind legs and small heads, the wings have been reduced to vestiges, and the feathers are soft and fluffy; in addition, the palate is of a peculiar, "old-fashioned"

The four living great ratites. The birds illustrated on this page are all large flightless types which show many similarities in build and, with the kiwi (pictured later), form a group termed ratites. *Upper left*, the familiar ostrich, a desert type of Africa, once widespread over Asia as well. A notable peculiarity is the reduction of the toes to two rather than the three toes common in most birds. *Upper right*, the cassowary of Australia and New Guinea, with a characteristic casque of bony tissue atop the head. *Lower left*, the emu, also an Australian resident; note the very different coloration of the young. *Lower right*, the rhea of the South American pampas. (Photographs courtesy New York Zoölogical Society.)

Extinct giant birds. In past times there existed many flightless birds, some of them much larger than birds now living. *Above*, some of the moas of New Zealand which resembled the living ratites in general build but in some instances reached a height of 10½ feet. Although now extinct, they existed there so recently that "fresh" bones and even feathers have been preserved. *Lower left*, *Phororhacos*, a giant type from the Miocene of Patagonia, which stood about 4 feet in height and (unlike the ratites) had a large head and powerful beak. *Lower right*, *Diatryma* from the early Eocene of Wyoming. This bird was about 7 feet tall; the contemporary horses were but the size of a fox terrier. (Moas from a mural by Charles R. Knight; photograph courtesy Chicago Natural History Museum. *Phororhacos* and *Diatryma* courtesy American Museum of Natural History, New York.)

type, contrasting with that of nearly all other birds. The similarities are, for the most part, merely features which one would expect in any sort of large bird which has ceased to fly, but the palatal structure (although this has been questioned) seems to definitely indicate that the ostrich-like forms are a natural unit of birds of rather primitive character, technically called, because of palate and jaw structure, the Paleognathae.

Their distribution is interesting. Considerable freedom from enemies is a feature common to these types. In New Zealand there is not a single flesh-eater to disturb them. In Australia the only carnivores are comparatively harmless pouched animals related to the opossums; there are (in nature) no members of the cat or wolf tribes on that continent. The same was also true of South America until quite recent geological times, and the island of Madagascar has no flesh-eaters of any great size. Further, the rhea and ostrich dwell in open country where, once they attain any size and speed, escape from enemies is easy.

What is the significance of this group? Some have suggested that the ostrich-like birds are truly primitive birds which have never attained the power of flight. This is highly improbable, for they are in general structure far more advanced than *Archaeopteryx;* quite surely they are the descendants of birds which once flew but have abandoned this mode of life to reassume a purely terrestrial existence.

Why do birds fly at all? The search for food is a major factor; but safety from enemies is perhaps a still more important reason. If regions could be found where food were plentiful enough without flying for it and where there were no enemies of importance, birds might well abandon flying, conserve their energies, and return to the ground. Such conditions are met in the homes of most of these flightless types and may have been responsible for the development of the Pleistocene and Recent giant birds.

It would be reasonable to expect that the ancestors of the ostrich-like birds, although flyers, would have been forms in which flight was relatively poorly developed; hence they would have the more readily abandoned this mode of life. Can we find today birds of this sort which indicate the type of ancestors from which the ostrich-like forms were derived?

In South America there lives a group of birds known as the "tinamous." These are rather stoutly built birds, some of rather large size, resembling the partridge and other game birds in appearance. The resemblance to game birds extends to habits as well, for they are mainly ground dwellers. When startled, they can fly but clumsily and for only a short distance. And the palate is that of the "old-fashioned" type found otherwise only in the ostrich-like birds. Here, then, we have relatives of the modern great flightless types which are already mainly terrestrial forms, which can fly but which fly poorly and would obviously abandon this habit readily if freed from enemies. They seem rather surely to represent the group from which the ostriches and their kin have sprung.

"Modernized" birds.—Apart from the ostrich-like forms and their tinamou relatives, all living birds are usually included in a major group, the Neognathae—the name referring to the more progressive nature of the palate and other structures. They are highly abundant and varied and are arrayed in some twenty or so orders. All have basically the same structure. The differences between the orders making up the group are no greater than those found within a single order of mammals, such as the carnivores. Despite these similarities, however, we should give here some brief account of these interesting creatures. Below we shall list the orders, seriatim, with brief notes as to the content of each and some major points of interest. As a visual *aide-mémoire* an accompanying figure gives a thumbnail sketch of a member of each group. For each an ordinal name is given. There is universal agreement as to the nature and number of orders in almost every case, but there are differences as to the names to be used for them. We have here used a system of names in common use in the United States, whereby each ordinal name consists of the scientific name of a common genus plus the ending *-formes* (thus, for example, *Pelicaniformes* indicates "pelicanus-like birds"). We shall list the orders in the fashion adopted by many ornithologists, beginning with various water fowl which are thought to be somewhat primitive and ending with the advanced perching and singing birds.

Gaviiformes.—The loons or divers. Fish-eating birds, frequenting open bodies of water, mainly the seacoasts. The loons, restricted to the Temperate and Arctic Zone of the Northern Hemisphere, are excellent swimmers and divers, with webbed feet in which all four toes are pointed forward; under water both wings and feet are used for propulsion, and the birds frequently "submerge" for long distances. Although most of their time is spent in the water, once in the air, they are strong flyers. Breeding takes place in the far north, with the simple nest placed on the ground at the margin of a marsh or lake.

Podicipitiformes.—The grebes, a second small group of swimming and diving birds, of much smaller size and of contrasting structure. The feet, instead of being webbed, have separate broad flaps of skin bordering each toe, and the wings are not used under water. Even more than the loons, the

The king penguin. The Southern Hemisphere contains a great variety of these flightless, strong-winged swimmers. Mainly Temperate-Zone ocean dwellers, one form is found in the Galápagos region below the equator, while others breed on the Antarctic ice. The large king penguin nests on the Falklands, Kerguelen, and other southern islands. (New York Zoölogical Society photograph.)

The world's boldest navigators—the plovers. The tiny golden plovers migrate annually from the Arctic to the Southern Hemisphere. The voyages of the plover, which nests in northwestern Canada, include autumns in Labrador and winters in Paraguay. For an unknown reason the young plovers remain in the North an extra month, and then, with no guide or previous training, follow the exact course to join their elders! This and other mysteries of migration are discussed in the text. The Alaskan form (*right*) makes nonstop flights to the Hawaiian Islands in the course of its migration to and from the South Seas. (Photograph courtesy New York Zoölogical Society.)

A ratite relative. The large, flightless birds shown after page 202—the ratites—are believed to have descended from flying ancestors. Possibly representing such ancestors are the tinamous of South America, shown here. They are similar to the familiar "game birds" in habits but are poor fliers. Their skull structure resembles that of the ostrich-like birds. (From Lydekker.)

Water birds, ancient and modern. *Above*, a restoration of *Hesperornis*, a flightless diving bird from the chalk rocks of Kansas, similar to a modern loon but primitive in its retention of teeth. *Upper right*, a grebe; *below*, a flock of red-throated loons. These two modern types are good swimmers and divers but also fly well. They do not have teeth. *Right*, the Galápagos albatross, a powerful oceanic flyer. (*Hesperornis* after Heilmann; others, New York Zoölogical Society photographs.)

Thumbnail sketches of representatives of most of the orders of birds

grebes take to the air with reluctance. Their food, primarily animal, includes frogs, mollusks, and insects, as well as fish; unlike the loons, they are mainly fresh-water lake dwellers. The nest is commonly a floating platform of matted rushes and other plants.

Procellariiformes.—Petrels, shearwaters, albatrosses. These are birds highly adapted for oceanic life, spending the greater part of their lives on the wing over the seas; their wings are long and powerful, and their ability to journey so far and so long over the ocean is aided by highly-developed powers of soaring flight, in which there is little expenditure of energy. A curious (if minor) specialization of the order is that the external nostrils form a pair of tubes over the top of the beak. The albatrosses, mainly found in southern tropical or subtropical seas, are the largest of flying birds, with a wingspread of ten to twelve feet; the petrels and similar forms, of much smaller size, are widely distributed, and, in contrast to most members of the order, one group of Southern Hemisphere petrel-like forms may dive after their food. Except for breeding, albatrosses and petrels are almost never found ashore. The albatrosses and some of the petrel group nest on the ground on isolated oceanic islands; the petrels typically nest in burrows on rocky shores.

Sphenisciformes.—The penguins are among the most interesting of birds—flightless forms with, however, well-developed wings which are used as "flippers" in swimming through the southern oceans; the webbed feet are used merely as rudders. The penguins are often stated to be Antarctic forms, but, although the large king penguin nests on the icy shores of the Antarctic, the group is essentially limited to the South Temperate Zone. One penguin resides in the Galapagos region, near the equator. The penguins come ashore only to nest in colonies on the more inaccessible shores. They lay only one or two eggs; most make no nest, and in some cases the egg is carried on the parent's feet. All penguins are solemn-appearing birds; the emperor penguin, some four feet tall in the erect posture assumed by penguins when ashore, is especially impressive. Some fossil penguins were even larger, one reaching human proportions.

The origin of penguins has been much debated. It has been suggested that they are the descendants of birds which came ashore on Antarctica during the early Tertiary, when that region presumably had a temperate climate, but were later driven off into the sea as the ice advanced toward the end of the Tertiary. Unfortunately this attractive story is probably false; it seems more likely that they are derived from albatross-like ancestors which never reduced their wings but, instead of flying over the ocean, tended to settle upon its surface and put their wings to a new use by flying through, rather than over, the waves.

Pelicaniformes.—Pelicans, gannets, cormorants, frigate birds, tropic birds. A group of rather large, aquatic, and mainly oceanic birds which can be distinguished from other birds of somewhat similar habits in that all four toes are webbed. They are fishermen which seek their food by diving, often in spectacular fashion. Most familiar are the large pelicans, with their long bills and the great pouch below the jaws for storing their fishy "catch." The most abundant members of the order are the cormorants, rather long-necked fishing birds, with a dark glossy plumage, often congregating in flocks of great size and nesting in colonies on rocky cliffs; some are marine, but they also fish in inland waters. More compactly built, with shorter necks and typically white in color, are the gannets, which are marine forms with habits comparable to those of the cormorants. The frigate birds and tropic birds are oceanic forms which rather resemble gulls and terns in habits and general appearance.

Ciconiiformes.—Herons, ibises, storks. The members of the heron group are excellent flyers which frequent (often in colonies) the shores—generally of inland waters—where they subsist upon fish and other animal food; with their long bills, long necks, and long legs, they are adapted to a wading rather than a swimming mode of life. The egrets are members of the heron group; related are the bitterns, smaller marsh dwellers with a more compact build. The storks are a further group of marsh-feeders with long legs but a short neck; the ibises are similar but with a slender bill, curved downward at the tip. Many members of the order perform elaborate courtship ceremonies.

The flamingos are frequently included in the same order, but current opinion suggests that they should be regarded as a separate group; they resemble the ducks and geese in some regards, the heron group in others. Neck and legs are extremely long, and the very long bill is abruptly bent downward at mid-length; the color is of a remarkable pinkish or rosy hue. They are mainly invertebrate-eaters, living in shallow lagoons or marshes —generally salt-water—where they congregate in enormous colonies and build their nests atop pedestals of mud.

Anseriformes.—Ducks, geese, swans, screamers. Most members of this order form a compact group of clearly related forms. They are swimming birds which are also good flyers and may migrate long distances. The legs are short; there are three webbed toes and generally a broad, flattened bill. With few exceptions they frequent fresh waters. Their feeding habits are varied, ranging from vegetable food to mollusks and fishes. The one exception to the uniformity of the order is the inclusion here, somewhat doubtfully, of the "screamers" of South America. They gain their name from their loud voices; they are about the size of a turkey and, while resembling the duck-goose group in some features, have very large unwebbed feet and narrow bills.

Falconiformes.—The hawks, falcons, eagles, buzzards, and vultures. With this order we at length leave the water for more terrestrial and arboreal bird types. These form the major group of rapacious birds which, in contrast to the owls, seek their prey by day. They are of robust build, with powerful wings, short but stout arched bills with sharp cutting edges, generally short stout legs and sharp curved claws. Most commonly small mammals or other birds form the food supply; however, fish may be the food source—as in the osprey—or carrion, as in the Old World vultures. Similar in habits to the latter, but a very distinct series, are the New World forms including the common "turkey buzzard" of the southern United States, the condor of the Andes, and the great California vulture, which are the largest members of the order.

An anomalous form, rather doubtfully belonging to this group, is the secretary bird of Africa, some four feet high, with long legs and tail and with plumes projecting back over the neck as if he were a secretary of olden times with quill pens stuck over his ears. The secretary is mainly a ground bird, feeding on frogs and toads but with an especial fondness for snakes.

Galliformes.—The game birds: pheasants, fowls, partridge, grouse, guinea fowl, turkeys. Most of the members of this order are of fairly uniform nature, generally of moderate size and compact build, often with marked sex differences in the plumage. They are terrestrial in habits, flying but little—rapidly—but for only short distances, nesting on the ground, and feeding primarily on grain. The more typical forms can be roughly placed in four categories: (1) The pheasants, Old World forms ranging east from the Mediterranean country and especially abundant in southeastern Asia and the adjacent islands. The domestic chickens are derived from the jungle fowl of the Malay region. The peacocks of southeast Asia, with a recently discovered African species, are close to the pheasants. (2) The "upland" game birds, mainly inhabitants of the North Temperate Zone, such as partridges, quails, and grouse. The ptarmigan (with a white winter plumage) is a northern outlier; the bob-white, prairie "chicken" (similar to the extinct heath hen), and the ruffed grouse (often miscalled a

The kiwi. Of the many varieties of flightless birds which inhabited prehistoric New Zealand, all are now extinct except the kiwi (*Apteryx*) (*above right*). These nocturnal birds are somewhat larger than domestic fowl and believed related to the "ratites," such as the ostrich. The long bill is used to probe for worms, their principal food. (Photograph courtesy American Museum of Natural History, New York.)

Birds recently extinct. Within recorded history numerous birds have become extinct, the great auk, the passenger pigeon, and, shown here, the dodo (*upper left*) and the heath hen (*below*). The dodo was a flightless bird related to the pigeons and somewhat larger than a turkey. It was abundant on Mauritius when that island was first discovered, but man and his domestic animals caused its extermination before the close of the seventeenth century. Skeletons alone have been preserved; but from excellent paintings replicas such as this have been made. (Photographs courtesy American Museum of Natural History, New York.)

The heath hen was closely related to the prairie hen of the western plains and more distantly related to the grouse and other game birds. It was once found in many areas of the eastern United States, but a century ago survived only on Martha's Vineyard. Here a flock remained with slowly dwindling numbers until 1932, when the last survivor, illustrated here, died. (Photograph courtesy Dr. A. O. Gross, Bowdoin College.)

Humming birds on the wing. These tiny birds have such rapid wing movements—on the general order of 50 beats or more per second—that in flight they appear but a blur to the eye or to an ordinary flash camera. The stroboscopic attachment, however, delivers rapid-fire bright flashes. By this light these birds were photographed (by Harold E. Edgerton, of Massachusetts Institute of Technology) as if "frozen" in midair. (The "bait," inci-

"partridge") are familiar North American forms. (3) The African guinea fowls, of which there are a number of forms in addition to the common domesticated species. (4) The turkey of North America, with a second species in Central America.

There are several less characteristic types appended to the game-bird order. The brush turkeys, or mound-builders, of Australia and the islands north to the Philippines are notable as the one group of birds using artificial means of incubation. The very large eggs are laid in mounds of decaying vegetation and warmed by the heat of decomposition; one exceptional form instead buries the eggs in warm sands. Once laid, the eggs receive no further attention; the young, most unusually, are born with plumage of mature type and immediately start to make their own living. The curassows and guans of South America (and north to southern Texas) are large handsome birds, usually with a feathered crest; in contrast to most members of the order, they are arboreal in habits. A very unusual form, likewise arboreal, is the hoactzin of the American tropics, rather pheasant-like in appearance. On the first two "fingers" of the wing, the young have redeveloped claws—lost since the days of *Archaeopteryx*— which they use to clamber about the branches.

Gruiformes.—Rails, cranes, bustards. In this order and the next we reapproach the water after having been on firm ground for some time, for many of the forms to be considered are waders. The present order is a very miscellaneous assemblage, with the rails and coots as a central stock. These are small-to-medium-size wading or swimming marsh birds, omnivorous, with rather poor powers of flight, and mostly secretive in habits; the head is small, the neck rather long, the legs long and sturdy, the wings short, and the body narrow. The cranes are large birds, superficially resembling the herons or storks with their long necks, long legs, and usually long bills; in contrast to the rails, the wings are very well developed, and the northern species of the group (they are Northern Hemisphere birds) are migratory. A third major group included here are bustards of the Old World, which are (except for the ostrich-like forms) among the largest of birds, with a very bulky build. Quite in contrast to rails and cranes, they are terrestrial types, frequenting dry open plains. Because of their size they get off the ground with difficulty but, once in the air, are competent flyers. Appended to the order is a miscellaneous assortment of other birds, including the cariamas of South America, the little button "quails," and the trumpeters of the same continent; these last are the size of a hen but have the appearance of a miniature ostrich.

Thought to be related to the present order are two series of large flight-less birds which antedate in time the ostrich-like forms of modern and Pleistocene days. In the Eocene of North America is found *Diatryma*, a contemporary of the earliest horses. The horse at that time was the size of a fox terrier, but this bird (armed with a powerful beak) was seven feet tall. Later in the Tertiary a comparable group of large flightless rail-like birds flourished for a time in South America.

The giant early birds arouse speculation; their presence suggests some interesting possibilities—which never materialized. At the end of the Mes-ozoic, as we have seen, the great reptiles died off. The surface of the earth was open for conquest. As possible successors there were two groups, the mammals—our own relatives—and the birds. The former group suc-ceeded, but the presence of such forms as *Diatryma* shows that the birds were, at the beginning, their rivals. What would the earth be like today had the birds won and the mammals vanished?

Charadriiformes.—Plovers, sandpipers, gulls, terns, auks. This is a sec-ond miscellaneous group of diversified birds which have wading forms as a central stock and are separable from the last only on highly technical char-acters. A central assemblage is that of the sandpipers, plovers, curlews, snipes, and woodcocks—long-billed, generally slender in build, long-legged little birds, most of which frequent shores and marshes. A second major group is that of the gulls and terns, a cosmopolitan group of medium to large-sized gregarious birds. They are long-winged, strong-flying forms, mainly oceanic but frequenting inland bodies of water in some cases. All are good swimmers, especially the gulls, which, however, are less capable of sustained flight than the graceful terns. Fish are a main source of food, although any other animal material (including refuse) is generally wel-come. A third major group is that of the auks, including the puffins, murres, and guillemots. They are inhabitants of the colder parts of the Northern Hemisphere, taking the place occupied by the penguins in the south. (The "penguins" of Anatole France's "Penguin Island" are the auks of English-speaking peoples.) These are stockily built birds, with large heads and short wings; on land they have, in contrast even with their gull relatives, an upright pose like that of the penguins. Again as in pen-guins, the wings are used in swimming, but in living auks flight is also pos-sible. Their food consists of fish and aquatic invertebrates; most of their lives are spent on the open ocean, often at great distances from the shore, which they visit only for breeding. Nesting takes place in colonies, often of enormous size, on rocky ledges. Unfortunately for them, these breeding places are often accessible to man, and his depredations were responsible for the extinction during the last century of the great auk, or garefowl, formerly widespread over the northern reaches of the Atlantic. This was the largest of auks, with an erect height of about $2\frac{1}{2}$ feet, which had (alone of the family) lost the power of flight, thus further paralleling the penguins.

Columbiformes.—Pigeons, sand grouse. Here, in contrast with the sprawling and diversified nature of the last two orders, we are dealing with a compact and clear-cut group of birds. Most are contained in the single family of the widespread pigeon group. They are eaters of grain and fruit, with a well-developed crop for storage. Most are good flyers, but they are in the majority ground dwellers; the young are born at a very immature stage and nourished by a milky fluid secreted in the crop. Relative immunity from attack by carnivorous mammals may be responsible for the fact that a large proportion of the species are native to Australia. In North America the passenger pigeon was once an exceedingly common form. Of gregarious nature, it existed in flocks of incredible size; a reliable naturalist estimated one flock which he saw early in the last century as consisting of over two billion birds. With the advent of the blessings of civilization, the picture changed. The last passenger pigeon died some decades ago in the Cincinnati Zoölogical Garden.

Another famous extinct bird is the dodo of Mauritius in the Indian Ocean (a similar if smaller form was present on a neighboring island). A grotesque, fluffy-feathered, and stupid creature, of large size, the dodo was a member of the pigeon group which had settled down as a flightless ground dweller on this island where natural enemies were absent. With the arrival of an unnatural enemy—man—this fat and tasty bird was rapidly exterminated.

The only other forms appended to the pigeon order are the sand grouse, small birds of Asiatic and African plains and deserts, pigeon-like in technical structural details but rather like the grouse and other game birds in habits.

Psittaciformes.—The parrots. These (as was the case in the last order) form a clear-cut group of birds, almost exclusively tropical and especially abundant in South America and Australia. Their characters are familiar— the powerful hooked bill, which can cope with stout shells and husks and other vegetable food materials, the large head, the compact body, and short legs. The parrots are arboreal types, and their feet are well adapted for perching and clutching; in most birds there are three forwardly directed toes plus a variable fourth one, usually directed backward; in parrots two of the four are turned back, giving a firm clutch both fore and aft.

Cuculiformes.—Cuckoos, turacos. The cuckoos, although distinguished by various technical characters (such as the same type of foot noted in the case of the parrots) are more interesting because of the peculiar breeding habits of many members of the group. In a typical cuckoo, a single egg is laid in the nest of a bird of another species, and one of the original eggs is removed; the young cuckoo develops rapidly, hatches, throws out the other eggs, and solicits the exclusive attention of his presumably perplexed foster-parents. The turacos, or plantain-eaters, African birds generally with brilliant plumage and handsome crests, are related to the cuckoos, but they raise their own young.

Strigiformes.—The owls. Nocturnal birds of prey and mainly eaters of small mammals, they resemble the hawk-eagle group in such adaptive features as the powerful beak and claws, but the resemblances are quite surely due to convergence rather than real relationship. Notable in their appearance is the position of the large eyes, which are turned directly forward, so that the same object is clearly seen with both eyes; as a compensation for the lack of lateral vision in normal pose, the neck is so built that the head can be swiveled around in nearly a half-turn. Hearing is also acute, and the owl's wings are so fashioned that its own flight is almost soundless. The prey is bolted whole; the inedible bones and fur are regurgitated in pellet form.

Caprimulgiformes.—Goatsuckers, nightjars, frogmouths. Related to the owls, the members of this small but widespread group of birds, known by a variety of names, are seldom seen, for they generally go about only in the dusk or at night, catching insects while on the wing; however, to country dwellers in North America, the call of one member of the group, the whippoorwill, is a familiar and pleasant sound. They are medium-size birds of compact build, short legs, and soft gray to brown plumage; the mouth has a broad gape and is surrounded by a series of long tactile bristles—adaptations useful in their insectivorous mode of life.

Micropodiformes.—Swifts, hummingbirds. Thrown together here by ornithologists for technical anatomical reasons are two groups of birds which seem to have little in common except excellent flying ability. The swifts are often confused with the swallows, which have much the same form and habits. They are small birds, with long pointed wings; they capture their insect food in the air and are rapid and graceful in flight and gliding ability. The nests may be composed of bark and twigs but are always bound together by saliva, secreted by the bird, which hardens into a very firm cement; in some cases the entire nest is formed of this material, valued by the Chinese for soup-making. A North American species which once nested in hollow trees has discovered that man-made chimneys are often more favorable.

Most remarkable of all birds in many ways are the hummingbirds with their brilliant coloring and unusual powers of flight, the wings beating with almost incredible rapidity. Although they visit flowers for nectar, they generally use small insects as the mainstay of their diet. These tiniest of avian forms are exclusively American and mainly tropical, only one species (the ruby-throated hummingbird) reaching the eastern United States. Very isolated in position but possibly related to swifts and hummingbirds are the colies or mouse-birds, small long-tailed birds of the African forests.

Coraciiformes.—Kingfishers, bee-eaters, rollers, motmots, hornbills, hoopoes. This is a diversified cosmopolitan assemblage of birds, mainly tropical in habitat. Little can be said of the group as a whole, apart from technical characters, except that there is a compact body, with short legs and a short neck, and a well-developed bill of variable shape. Nearly all are Old World forms. Most familiar to those living in the Northern Temperate Zone are the kingfishers, with long stout bills, short weak legs, wings that are short but powerful, and generally bright plumage. The name is inappropriate for the family as a whole, for while the typical kingfishers live up to their name, many of their relatives seek small animal prey far from the water. The bee-eaters are an Old World group, mainly African, with long slender curved bills; their name stems from their insectivorous habits. The rollers are likewise Eastern Hemisphere birds, generally with a brilliant plumage, and also are mainly insect-eaters; the typical rollers generally capture insects in the air, and the name is derived from the peculiar revolutions they execute while on the wing. The motmots of Central and South America and the todies of the West Indies occupy much the place in the New World occupied by the rollers and bee-eaters in the Old World.

Returning again to Eastern Hemisphere forms, the hornbills are a numerous tropical group. They are striking birds of large size (some reaching four feet in length) with enormous bills and, in many cases, with a second horny structure—a casque—developed above the bill. The nesting habits are without parallel; the female, before laying her eggs, allows herself to be shut up by the male behind a mud wall in a tree cavity, where she remains until the young are fully fledged, being fed meanwhile by her mate.

A final small group within the order is that of the hoopoes (so called because of their peculiar cry), small but graceful Old World birds. The European form has a splendid feathered crest; the bill, with which they probe the ground for grubs and insects, is very long, slender, and curved downward.

The trogons, forest birds of the tropics and primarily South American, are small beautifully plumaged birds, with a short but strong bill. They are an isolated group, now frequently considered as forming a separate order. The quezal of Central America is generally considered one of the most beautiful of all birds.

Piciformes.—Woodpeckers, barbets, jacamars, puffbirds, toucans, and honey guides. Most prominent in this group are the woodpeckers, tree dwellers with (as in the order as a whole) the two-and-two arrangement of the toes (mentioned in some earlier orders) which facilitates climbing. The bill is a powerful drilling organ, capable of boring deep into trees for the grubs which, with adult insects, form most of their diet. Equally remarkable is the very long protrusible tongue; the slender tongue bones which support it can be withdrawn into a bone-inclosed sheath which curves back, up, around the head, and forward to the forehead. The numerous barbets of the tropics are also insect-eaters, with shorter but stout bills. The little African honey guides, barbet relatives, gain their name from their habit of directing attention to the nests of bees. Bills reach their maximum size here in the toucans of South and Central America; the bill is both deep and long and rivaled for size only by that of the hornbills. Further New World tropical members of the order are the jacamars, with a rather long bill, and the puffbirds, with a shorter but stronger beak.

Passeriformes.—The perching birds are last in our long list of bird groups. Last, but not least, for there are included in this order—generally considered as the most highly evolved of bird types—well over half of all living birds. Every avian form which has not been previously mentioned and with which the average reader is likely to be familiar belongs here. In general these birds are of small to medium size. The term "perching birds" is derived from the foot construction; in most there is a well-developed hind toe and three forwardly directed ones; when the "sole" of the foot is pressed against a branch, the toes, due to the tendon arrangement, automatically assume a clutching position.

This great order can be divided into a number of relatively primitive groups and a much larger and still more advanced series, the songbirds. The more primitive forms are almost exclusively tropical and are particularly abundant in South America, a continent poor in songbirds. They include, among many others, the broadbills, antbirds, ovenbirds, manakins, and cotingas; the tyrant flycatchers of America, however, have northern representatives, such as the kingbird and phoebe.

Any attempt at discussion or even enumeration of the some four thousand songbirds of the world would be far, far beyond the scope of a book of this sort. A recent review has attempted to gather the songbirds into some eleven assemblages, which we shall list: (1) the larks, which are generally considered relatively primitive members of the group; (2) the swallows; (3) the bulbuls and related Old World forms; (4) primitive insect-eaters such as the Old World flycatchers, the warblers, thrushes, mockingbirds, thrashers, hedge sparrows, wagtails, and pipits; (5) the shrikes, mainly Old World birds, an offshoot of the last group; (6) the waxwings and a few tropical relatives; (7) the creepers, nuthatches, titmice, and chickadees—forms primarily of Old World origin, which may have evolved from generalized insect-eaters; (8) Old World tropical nectar-eaters, such as the sunbirds, white-eyes, and Australian honey eaters; (9) a primarily American group, including vireos, honey creepers, wood warblers, tanagers, grosbeaks, finches, buntings, American sparrows, chaffinches, linnets, American blackbirds; (10) a somewhat comparable Old World series including the starlings, Old World orioles, and the large weaverbird family, of which the Old World sparrows and the waxbirds are members; and (11) the large and somber crows and ravens and various crowlike birds of the Australian region, among which, amazingly, the magnificent birds of paradise of New Guinea and adjacent regions are definitely placed.

BIRD MIGRATION

Most terrestrial animals spend their entire lives in the regions in which they were born. But birds can, and often do, migrate seasonally from one region to another. Although perhaps not over 20 per cent of bird species make major migratory movements, most birds do move from one area to another to some extent. For birds of warmer regions, movements may be related to wet or dry seasons, to abundance of food in different areas at different times. Even in temperate zones, movements may be simply between higher and lower altitudes. Most interesting and most studied, however, are the seasonal north-and-south movements of bird populations which frequent the colder life zones. These are mainly Northern Hemisphere forms, since the southern continents do not extend as far from the Equator as the northern ones. Little is known as to the nature of the stimulus which initiates migration in the case of forms which migrate within the warmer regions of the globe where the length of days are nearly uniform. But as regards birds which move between northern regions with short winter days and southern ones with longer hours of daylight, the stimulus is known in certain cases, at least. Migrations are associated with breeding habits and hence with the state of the reproductive organs, controlled by the major gland of internal secretion, the pituitary gland, situated at the base of the brain. And this gland is influenced by the amount of light received by the eyes. Length of daylight is thus, through a sequence of events, the primary initiating influence in migration between colder and warmer regions.

Migration may take the form of a simple north-south drift over a restricted region in the fall and a corresponding shift back to the north in the spring. Often, however, the route followed is a definite one—and a long one. Many European birds fly to and from Africa over definite "flyways," and similarly many American forms follow an Atlantic coast route to Florida, thence across the Caribbean to South America, or farther west they follow the land chain via Central America. Some of the routes followed necessitate excellent navigation. For example, the Alaskan golden plover in autumn sets its course southward over three thousand miles of open ocean to the Hawaiian Islands on its way to the South Seas. A deviation of a few degrees would cause it to miss its goal; how does it set so straight a course?

The more common American golden plover follows a more complicated route. After a June nesting season on the Arctic shores of northwestern Canada, the adult birds fly east to spend the early autumn in Labrador. Then, turning southward, they take a straight course from Newfoundland and Nova Scotia to Venezuela, never normally touching land unless driven off their course by storms. A further, more leisurely, migration carries them farther south to center in the Paraguay region by the time winter has arrived in the North.

Possibly a bird might learn such a route by following his elders. But what of the newly hatched young? They are left behind in the Arctic; there is no one to guide them. A month later they leave and follow the same complicated route east to Labrador, south over the ocean to the Orinoco, and, finally, after some six thousand miles of travel over lands and seas they had never seen, they rejoin the older birds in their wintering region. Why and how do they do this, all untaught? Does each germ cell of the golden plover carry within its tiny nucleus a marked flying chart of the Western Hemisphere?

Homing.—The acute geographical sense of birds is shown in the study of homing. It is a matter of common knowledge that many birds are able to return to their nests even if removed to a considerable distance; the homing pigeon is a familiar example of this, and all birds which have been investigated exhibit the homing instinct to a greater or lesser degree. Of course, not all birds taken far from their nests return; some may be in poor condition and unable to make the flight; others may become victims of enemies or of storms; others possibly may not strongly feel the "urge" to return even if capable of doing so. But a remarkably large percentage do return, and may return rather promptly.

How this is accomplished is, to a considerable degree, a scientific mystery. A bird may be taught a route, as is the case in training homing pigeons; even if taken over it only once, the bird might pick up enough visual landmarks. But vision can be eliminated as a necessary feature. Birds can still return if carried away blindfolded or if transported by sea, as, for example, terns taken by boat from Key West across the entire Gulf of Mexico to Galveston. Perhaps, one might suggest, the bird can, in default of vision, be able to keep track of the direction in which it is traveling, just as a man can to a degree maintain his orientation when traveling at night by keeping account of the turns of the road. But this possibility has been eliminated by carrying birds long distances (as Berlin to Frankfort on the Main) strapped to a revolving phonograph turntable. Such birds, apart from a touch of understandable dizziness, appeared to be quite undismayed and promptly set off in search of their home—to which a fair proportion returned.

A considerable fraction of the known results may be interpreted in simple fashion as the result of random search. For example, in some recent experiments birds taken far from the nest and released were followed for some time by airplane. Those which reached home did not immediately start to fly in the proper direction; wandering over a spiral or zigzag course, they appeared to eventually reach an area already known. But there are cases in which released birds almost immediately set off in the correct direction over unknown territory.

In these last cases simple explanations fail. Some have suggested that birds "feel" forces beyond the power of the normal senses. For example, the rotation of the earth would cause a small increase or decrease in a bird's weight, according to the direction of flight, and would cause a very small deviation in direction as well, thus giving clues as to position. But the effects are so slight that it seems incredible that they could be felt by a flying bird. Again, the strength, direction, and dip of the earth's magnetic force differ from point to point over the earth's surface, and it has been suggested that a bird's homing is influenced by an attempt to seek out the place where these forces are found in a combination familiar to the individual. But we know of no sensory organ of a bird's body which could be receptive to magnetic forces. Some years ago a psychologist wishing to do more work on this problem applied for financial support to a fund designed for the investigation of psychic phenomena. Such phenomena, it was decreed, were to be defined as those beyond the realm of the known senses. The investigator argued that the homing instinct was in this category as surely as anything he knew of! His request was refused, but his argument seems to us a reasonable one.

CHAPTER 11

The Origin of Mammals

We now retrace our steps from a consideration of the great dynasties of the Age of Reptiles and of the dinosaurs' avian cousins to the origin of our own closer relatives. Man is a mammal, member of a great group which includes almost all the larger animals inhabiting the surface of the earth (to say nothing of the whales and seals of the seas, the bats in the air).

MAMMAL CHARACTERS

Reproduction.—The name indicates one of the features of the group. Mammals nurse their young. Postnatal care of the young is unusual in lower vertebrates but common in birds and mammals, highest of back-boned animals. In addition, none but the most primitive of mammals lays eggs; all others bear their young alive, and in most there is a well-developed mechanism for the nourishment of the unborn young within the mother's body. These features seem to be associated with a high degree of organization of the mammal body, for the development of which there is needed a considerable period of growth.

Warm blood.—Mammals, like birds, are warm-blooded forms; a high body temperature is maintained. Hair and sweat glands are peculiar mammalian features associated with temperature regulation, and there is a highly developed breathing mechanism which is in constant use. In addition there is a very efficient circulatory system with a complete separation of aerated and impure blood streams. The heart (pp. 121, 124) is a double-barreled, four-chambered organ, as in birds (but not in reptiles), with blood on one side in transit from the body to the lungs; on the other, from the lungs to the body.

Brains.—The brain of reptiles is a relatively tiny structure tucked away in a small braincase centrally situated toward the back of the skull. The brain of mammals, even the most stupid of them, has enlarged enormously. Most parts of the brain have, however, remained fairly constant in size. It is in the cerebral hemispheres, originally small structures dedi-

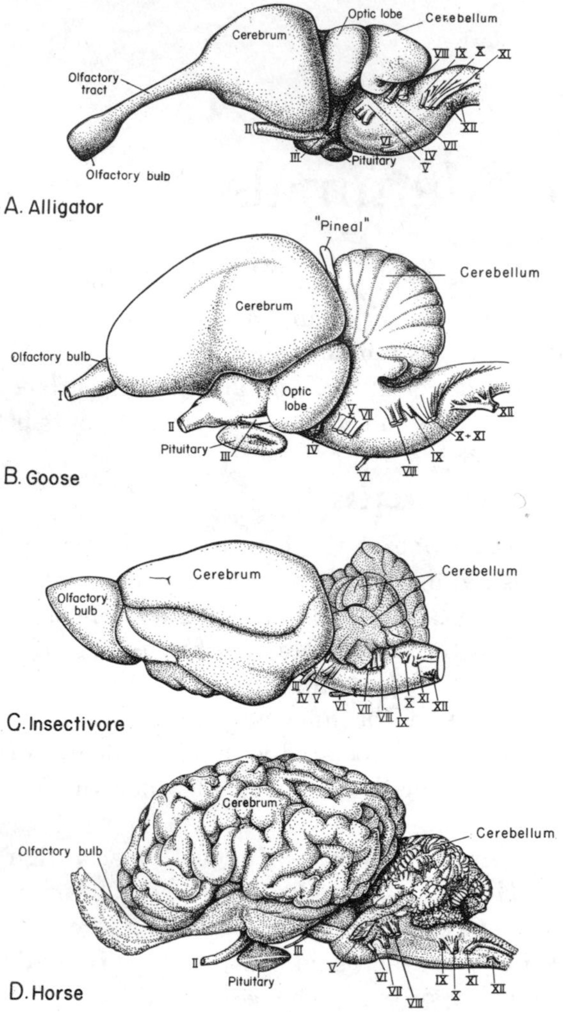

A. Alligator

B. Goose

C. Insectivore

D. Horse

cated to the sense of smell, that almost all the growth has taken place. Here there have arisen higher brain centers which have placed the mammals as a group far above any other vertebrate stock in their degree of mental development. In birds, too, we noted a great expansion of the hemispheres. There, however, the increased size was due to growth of structures lying deep within the hemispheres, the striate bodies—corpora striata—which presumably are responsible for the complex but mainly instinctive behavior patterns of birds. Mammals have these structures, but their important advance has been an increase in size and complexity of the surface areas of the hemispheres, the "gray matter," the cerebral cortex. Here, in mammals, are received all sensory impressions, which may be stored as memories,

Brains of a reptile, a bird, an archaic mammal, and a progressive mammal. In reptiles, the cerebral hemispheres are relatively unimportant, and the optic lobes of the midbrain are (as in frogs) in control of the animal's "higher" activities. In birds, the cerebrum is large, but there is little "gray matter" on the surface; the centers important for control of behavior patterns are concealed in the "striate bodies" in the interior of the cerebrum. In primitive mammals the lower part of the cerebrum—below the prominent tissue seen in the figure—is still mainly connected, as in lower vertebrates, with the sense of smell, but the upper portion has a surface covering of gray matter ("the neopallium"), which is the seat of higher faculties; the optic lobes are reduced in size and importance (and are not visible on the brain surface). In more advanced mammals the olfactory parts of the cerebral hemispheres are small and confined to the lowest part of these structures, whereas the surface, housing higher centers, is greatly increased by folds ("convolutions").

and here, on receipt of new sensory information, judgments are made (with the aid of memories of past experiences) as to what action should be taken. Here, then, is a structure in which learning and training can be accomplished as the result of "thought"; here is the realm of human consciousness.

Later on—particularly in the case of apes and primitive human types—we may cite figures of brain size in various forms, with more or less the implication that these figures are an index to intelligence. In default of anything better, such citations are permissible, but their meaning is very limited. Actually, intelligence in a mammal is not a matter of the whole bulk of the brain but of the sheet of gray matter spread over the surface of the hemispheres. In two animals with the same total brain volume, this cerebral cortex may have a much bigger area in one than in the other if (as in man) the surface is greatly folded ("convoluted"); further, in the gray matter are layers of countless millions of nerve cells, interconnected in complex fashion, and it is clear that the degree of intelligence is to some extent correlated with the degree of complexity of this marvelous structure and not merely its area. Further, brains vary more or less with the size of the animal, and bigger forms have bigger brains than small relatives. We sometimes find cited as a supposed index of intelligence the ratio of brain size to body size. But in any group of related forms we find that, while the brain is actually somewhat larger in a big animal, it does not grow proportionately to the size of the body. If this index were valid, it would seem to prove that a small monkey is smarter than a man—a conclusion with which we may disagree without conceit. All in all, then, brain-size figures are useful– but beware!

Skull.—The skull and head of mammals is a very different structure from that of reptiles. Many of the reptile bones have been lost, and the median eye no longer functions. The originally solid temporal region has been pierced for the accommodation of the jaw muscles, leaving a bar, or arch, at the edge of the cheek region; the braincase has swollen out enormously to accommodate the expanding brain. The skull of reptiles joins the backbone by a single, round, bony knob, or condyle; in mammals a pair of condyles is present. In reptiles (except the crocodiles) the nostrils generally open into the front of the mouth. In mammals there has developed a secondary palate, a bony partition which separates nasal and food passages back to the throat. This is a feature of importance in forms in which constant breathing is a vital necessity. In reptiles there are normally some seven bones in the lower jaw; we mammals have but one (the dentary), and this articulates with a different bone on the side of the skull, the squamosal. Our whole jaw joint has changed.

Teeth.—The dentition of mammals has become greatly modified. Lower vertebrates have an indefinite amount of tooth replacement; mammals have but two sets of teeth, "milk" and permanent. The teeth of primitive vertebrates were usually of about the same shape in different parts of the jaw; in mammals the various parts of the tooth row are highly differentiated. There was some early variation, but in the ancestors of the higher mammals the dentition came to be made up of three sharp nipping teeth, or incisors, at the front of each half of each jaw, a single large, stout, piercing tusk, the canine, four premolar teeth behind this in the front of the cheek region, and three grinders, the molars. This gives a total of forty-

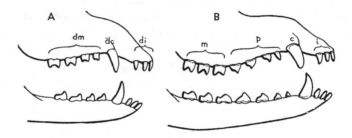

The teeth of a generalized mammal (primitive placental) seen from the right side. *A*, deciduous ("milk") teeth, consisting of incisors (*di*), canines (*dc*), and "milk molars" (*dm*); *B*, permanent teeth, including incisors (*i*), canines (*c*), and premolars (*p*) replacing the milk teeth, and, in addition, the permanent molars (*m*).

four teeth. Most mammals have lost some of this set of teeth (we have, e.g., but thirty-two); few have exceeded this number. This type of dentition is one suitable for a carnivore, and the ancestry of all the mammals lies through a long line of flesh-eating types.

Ears.—In reptiles the eardrum lies almost at the surface, near the back of the jaws; in mammals it is deeply sunk in a tube, and an external ear flap for the concentration of sound waves has made its appearance. In reptiles there is but one small ossicle, the stapes, for the transmission of sound from drum to inner ear; in mammals there are three. The two new ossicles have been derived from the old reptile jaw joint, abandoned for a new jaw articulation in mammals.

Locomotion in mammals.—We have discussed the cumbersome, waddling type of walking common to all primitive land animals and noted how the ruling reptiles solved the "problem" of fast movement by taking to a bipedal mode of life. Our ancestors also came to a successful solution but in quite another way. All four limbs were retained but swung around into a fore-and-aft position, the knees brought forward, the elbows back. In this mammalian pose, we have a much more efficient sort of apparatus than the primitive one. In an early reptile type much of the energy expended was used to keep the body from collapsing on the limbs; here all the muscles can be used for straightaway forward propulsion. This change in posture has been accompanied by many changes in the bones and muscles of the limbs, of which we will note but one feature here. In primitive reptiles the count of the toe joints was, from the inner toe out, 2-3-4-5-3; in mammals the middle fingers have shortened up, giving a count of 2-3-3-3-3.

In reptiles the inner toe in either front or hind foot is short and not remarkable in any way. In mammals, on the contrary, they were originally stout structures, diverging from the other four digits as "thumb" and "big toe." These toes are retained in this specialized form in monkeys and apes, and in man the thumb, at any rate, is very highly developed. They are useful to monkeys for gaining a grasp on a limb, and their presence in primitive mammals has suggested that the ancestral forms were arboreal in habit. In terrestrial mammals (as is readily observed by looking at the family dog or cat) these digits, relatively useless on the ground, are usually reduced or lost.

Action.—If we attempt to evaluate the meaning of all the features in the structure of mammals which distinguish them from reptiles, we may perhaps sum it up in one word—activity. The ancestors of the mammals were carnivores, leading lives in which speedy locomotion was a necessity. The limb development has given effectiveness to this potential activity. Brain growth has given it intelligent direction. The maintenance of a high body temperature and the various changes associated with this are related to the need of a continuous supply of energy in animals leading a constantly active life. Even the improvements in reproductive habits, which are a prominent feature of mammalian development, seem related to the needs for a slow maturation of the complex mechanisms (particularly the brain) upon which the successful pursuit of an alert and active life depends.

OUR REPTILIAN ANCESTORS

We might at first imagine that the evolution of such a highly developed group as the mammals would have been a late feature in reptilian history. This, however, is exactly the reverse of the case. The reptilian stem from which the mammals sprang was one of the first differentiated from the primitive reptile stock; and the first mammals themselves appeared nearly as early as the first of the dinosaurs.

Pelycosaurs.—A first stage in the differentiation of mammals from other reptile stocks is that of the pelycosaurs, well represented by fossils in the Texas redbeds. These beds date from the late Carboniferous and Lower Permian, a time when the stem reptiles still flourished and ruling reptiles were unheard of. Among the pelycosaurs were all the dominant carnivores of the day (as well as a herbivorous side branch). In many features they still had not departed far from the primitive reptile stock, and some forms (such as *Ophiacodon*, figured here) were so primitive that they appear not to have fully emerged from the ancestral waters and were still mainly aquatic fish-eaters. Pelycosaurs were more slim in build than the average stem reptile, but the limbs still sprawled out at the sides in primitive fashion. Definite indications that we have here the first approach to mammal structure are seen in the skull. In the dentition we find, instead of a uniform tooth row, a few front teeth comparable to incisors, large canines (usually two), and back of them, a series of cheek teeth corresponding to mammalian molars and premolars. Still more distinctive is the cheek region. There is a lateral temporal opening, with a bar of bone below it corresponding to the arch at the lower margin of a mammal skull.

Some of the more generalized types were rather lizard-like in appearance although not in structure. Others, such as *Dimetrodon* and *Edaphosaurus*, were remarkably specialized. These odd forms (the former, illustrated early in this book, the most common carnivore of the Permian) had long spines growing out from the back, supporting a sail-like flap of skin. Various speculations have been made as to the possible function of this curious structure. The original describer jocularly suggested that the animal used it to go sailing on the lake. This is as good a theory as certain others that were brought forth, such as that the spines simulated the rushes among which *Dimetrodon* lay in wait for its prey. (The reply to this, of course, is that the animal would be still better concealed if the spines were sawed off and he would be completely hidden.) The correct solution seems to be that the sail was a first crude attempt at a heat-regulating device. Here was a broad area of skin, richly supplied, we are confident, with blood vessels. If exposed broadside to the sunshine, it would rapidly warm the body; turning endwise to the light or moving into shade would stop the process. Cutting down the blood supply—and hence "blanching" the skin—would conserve heat, and conversely, if the animal were overheated, opening

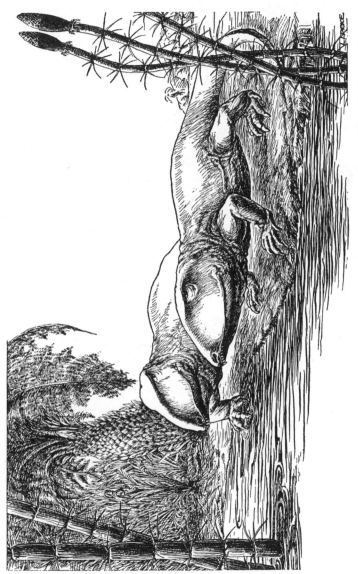

A primitive mammal-like reptile, the pelycosaur, *Ophiacodon*, from the late Carboniferous of North America. Some of the members of this group developed peculiar sails along the back and other specializations; *Ophiacodon*, still partly a water-dweller, is more generalized and represents a very early stage in the differentiation of our ancestors from the primitive reptile stock. (A drawing by L. I. Price, from Raymond, *Prehistoric Life*.)

wide the blood flow to the sail would turn it into an excellent radiator. This adaptation was successful for the time but was later abandoned in favor of less spectacular regulating devices, and the ancestors of more progressive forms were pelycosaurs with more normal body contours.

Therapsids.—In the Karroo beds of South Africa, dating from the later Permian and Triassic, we find that the most common of animals were mammal-like reptiles descended from the pelycosaurs and termed therapsids, in reference to the mammal-like build of their temporal arch. Close to the main line of evolution of the group and far along toward the mammalian condition were a number of aggressive flesh-eaters whose skull and body structures are illustrated here. In advanced forms the skull was in-

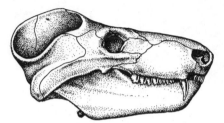

A side view of the skull and jaws of a therapsid from South Africa. In many ways this reptile skull approaches that of mammals, with a large single opening in the temple region, the lower jaw bones reduced except for the single tooth-bearing bone (dentary), exclusive in mammals, and a dentition divided into incisors, large canine teeth, and cusped cheek teeth.

termediate in type between that of a primitive reptile and a mammal; many of the bones absent in mammals were on their way toward reduction or were already lost. A small third eye was still generally present on the top of the skull, but its opening was a tiny one. The brain cavity was still small, and the brain was still presumably reptilian in type. In the jaw all the original elements were still present, but the dentary— the only jaw element retained in mammals—was far larger than the other bones. The old single condyle, joining skull with backbone, had been replaced by a double one, and in the roof of the mouth a secondary palate had developed, just as in mammals. The teeth, too, were approaching mammalian conditions. There was already a differentiation into incisors, single stout canines, and cheek teeth which were sometimes cusped instead of being primitive reptilian cones. In the ear, however, there was still but a single auditory bone for sound transmission.

Markedly changed, too, were the limbs. These had already shifted far toward their fore-and-aft mammalian position, and the bones were already much modified to meet the new conditions. In many of these forms the count of the toe joints was still that of primitive reptiles, but in some cases the joints which were destined to be lost were tiny, and in still other mammal-like forms the number had been reduced to that of mammals.

These therapsids were thus very advanced, mammal-like forms in most

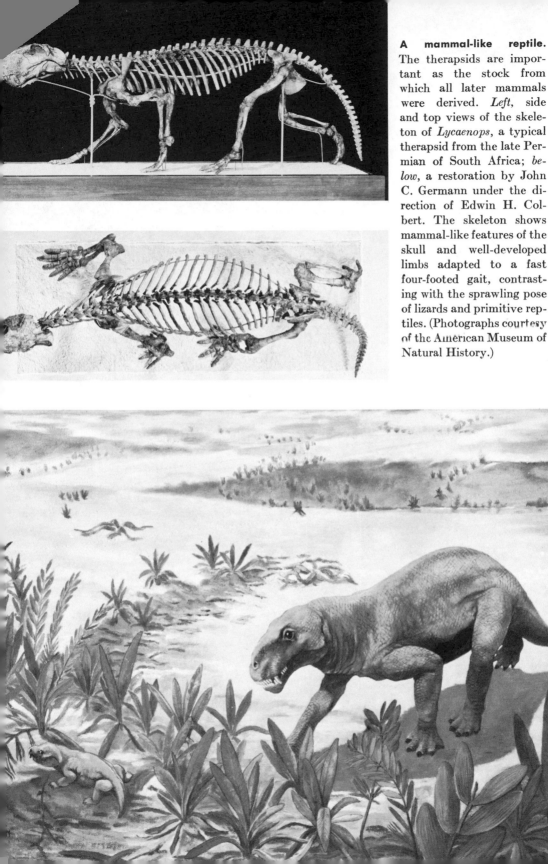

A mammal-like reptile. The therapsids are important as the stock from which all later mammals were derived. *Left,* side and top views of the skeleton of *Lycaenops,* a typical therapsid from the late Permian of South Africa; *below,* a restoration by John C. Germann under the direction of Edwin H. Colbert. The skeleton shows mammal-like features of the skull and well-developed limbs adapted to a fast four-footed gait, contrasting with the sprawling pose of lizards and primitive reptiles. (Photographs courtesy of the American Museum of Natural History.)

South Africa in Triassic days. A majority of our finds of mammal-like reptiles have been made in the rocks of the Karroo Desert of South Africa. *Above*, a restoration by Charles R. Knight (in the Chicago Natural History Museum) of a scene in South Africa in early Triassic days. A large and ponderous dicynodont, a herbivore, is being attacked by a pack of its carnivorous "cousins" related to *Lycaenops*, shown overpage.

An ancient mammal jaw. Mammals from the Age of Reptiles are known from very fragmentary materials, such as the jaw shown here, somewhat enlarged. It consists of a single bone; in contrast to the several bones making up the reptilian lower jaw. These older mammals were almost universally of tiny size, in strong contrast with the contemporary giant reptiles. (After Simpson.)

Triconodonts. Largest of Mesozoic mammals, these sometimes reached the (relatively) gigantic size of a kitten and hence were not confined to a diet of insects, worms, and grubs but could (as suggested by the restoration at the left) attack small reptilian contemporaries. (From the *Illustrated London News*.)

Such restorations are highly speculative since we have no complete skeletons and must infer non-bony traits in any case. But we do know that when the giant reptiles died out, highly evolved mammal forms suddenly came forth.

of their skeletal parts. But what of the other features of soft anatomy which are so important in mammal development? Were these animals already warm-blooded? Had they hair or scales? Did they nurse their young? We cannot, of course, give any positive answer to these questions. But the general progress shown in the skeleton suggests similar advances in other respects; the development of a secondary palate, so useful in a warm-blooded animal, appears significant. We arbitrarily group the therapsids as reptiles (we have to draw a line somewhere), but were they alive, a typical therapsid probably would seem to us an odd cross between a lizard and a dog, a transitional type between two great groups of backboned animals.

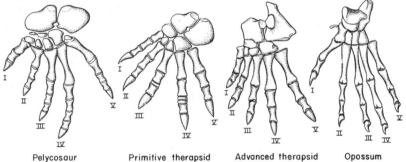

Pelycosaur   Primitive therapsid   Advanced therapsid   Opossum

The left hind feet of mammal-like reptiles (pelycosaurs, therapsids) and of a primitive mammal (the opossum) to show the change in number of toe joints. In primitive reptiles the count of joints in the successive toes, from the "big toe" out (not including the long bones in the sole of the foot), was, in order, 2–3–4–5–3 (or 4). In primitive mammals (as in ourselves) the count is 2–3–3–3–3. Within the therapsid group the change to the mammalian count occurs; the primitive form illustrated has the reptile count, but in toes III and IV the "extra" joints are reduced to nubbins.

There were many variants among the mammal-like reptiles in these Karroo beds of South Africa. Some were still not far from the pelycosaur pattern; others were even more mammal-like in some respects than the dog-jawed reptile described above. Many were far off the main line, as, for example, the dicynodonts ("two-tuskers"), lumbering plant-eating forms which flourished greatly for the time but eventually became extinct without leaving descendants.

If the reader has followed through the line of descent of mammals and man to this point and continues it onward later in this chapter toward advanced mammals, he will be struck by one disturbing theme which runs through the entire story from the jawed-fish stage onward. This is the fact that the entire line of descent is, without exception, one of predatory flesh-eaters. The fish ancestors of land vertebrates preyed upon their fellow

fish, and the early amphibians had the same diet. Our early reptile ancestors likewise ate animal food; among the pelycosaurs and therapsids were
the dominant and most bloodthirsty animals of their day. The progressive
line of mammal evolution continued through forms that were at least
would-be carnivores, although, as we shall see, many were too small to
tackle larger prey than insects. All along the series we see side branches
developing as herbivores which (like the dicynodonts) flourished greatly—
but then died out, while their flesh-eating relatives survived. Is there a
moral to this story? Let us not draw one, for if so, it would appear to
support the doctrines of such characters as Nietzsche or the late and unla-
mented Hitler.

The mammal-like forms were the commonest of reptiles during the later
Permian and the early Triassic. But during the latter period occurred the
development of the ruling reptiles. The archosaurs soon crowded the mammal-like reptiles out, and they disappeared at the end of this, the first
period of the Age of Reptiles. They were, however, destined to live on in
their mammalian descendants; for the first faint traces of mammals appear
in rocks of late Triassic age, and primitive mammals continued to be present throughout the Mesozoic as obscure contemporaries of the great
dinosaurs.

EGG-LAYING MAMMALS

A primitive stage in mammal development is that represented today by
two curious Australian types, the platypus (*Ornithorhynchus*) and the
spiny anteater (*Echidna*). These are very highly specialized creatures,
leading specialized lives. Both are toothless as adults (the young platypus
has a few tooth rudiments). The platypus is a good swimmer, being mainly
a stream dweller, but also a good digger, nesting in burrows in the banks.
The absence of teeth is compensated for by the development of a broad
horny bill. The anteater is protected by a stout spiny covering, comparable
to that of the hedgehog; in relation to its anteating habits there are powerful digging feet and a long slim snout. Both forms are certainly mammals,
having fur and nursing their young. But there are many primitive reptilian
features; the most conspicuous and important is the fact that they still lay
eggs in reptilian fashion!

These animals thus represent the most primitive, the most reptilian,
stage in the development of mammals. It is unfortunate that they are so
highly specialized in their mode of life, for the egg-laying ancestors of
higher mammals were certainly neither platypuses nor anteaters; and the
lack of teeth (the most frequently preserved parts in fossil mammals) ren-

ders it difficult to compare them with extinct forms. It is probable, however, that most of the mammals of even Mesozoic days had already abandoned egg-laying habits and bore their young alive and that these curious living types have had a separate line of ancestry since the earliest (Triassic) days of mammalian history. And it is even possible that in the future a more adequate knowledge of the fossil record will show that they diverged still lower down the family tree, in the therapsid stage.

The preservation to modern times of archaic animals of any sort is usually attributable to isolation. An animal may attain isolation geographically, or it may become isolated by taking up a mode of life in which there is little competition. Both factors have operated to save these odd mammals. Their mode of life is extraordinary, in the proper sense of the term; and in Australia they are in a region in which most of the other mammals also are of a comparatively unprogressive type.

PRIMITIVE MAMMALS OF THE AGE OF REPTILES

With the exception of these egg-laying forms, all existing mammals bear their young alive; the egg is retained inside the mother's body, giving it additional protection. Bigger, if fewer, offspring seems to have been the mammalian trend—"quality, not quantity." It is probable that this stage had been attained by the characteristic mammal groups of the Jurassic and Cretaceous. These Mesozoic mammals, contemporaries of the great dinosaurs, are poorly known. For the length of the Jurassic and Cretaceous, a period estimated at about eighty millions of years, we know of mammals not one whole skeleton, and (until near the close of the Age of Reptiles) not even a complete skull. Our knowledge of these forms is gained almost entirely from teeth and jaws, and even these are quite rare. I strongly suspect that if all the known specimens of little Mesozoic mammals were brought together, they would fit comfortably inside an old-fashioned derby hat.

Typical Mesozoic mammals were, on the average, no bigger than a rat or mouse and may have resembled these living forms in general appearance (although not in structure). Their teeth were sharp; they were seemingly flesh-eaters in their tendencies, as had been their reptilian ancestors, but most of them were too small to attack other vertebrates. Probably insects and worms were their main diet, supplemented by buds, eggs, and whatever came to hand. Their brains, as far as can be told, were still poor by modern mammalian standards but showed great improvement over those of reptiles. Presumably these forms were very inconspicuous in their habits, dwelling in wooded or bushy regions. There are suggestive features indicating a tree-dwelling life, and, as is the case with many mammals today, they may have been nocturnal. Inconspicuous and small they had to remain, for, as contemporaries of the dinosaurs, the threat of death from the great carnivorous reptiles lay constantly over them.

But this long period of "trial and tribulation" was not altogether disadvantageous. It was, it would seem, a period of training during which mammalian characters were being perfected, wits sharpened. As a result, when, at the close of the Cretaceous, the great reptiles finally died out and the world was left bare for newer types of life, higher mammals, prepared to take the leading place in the evolutionary drama, had already evolved.

THE AGE OF MAMMALS

In our study of mammals we shall have much to say of the sequence of appearance of various forms. To do this satisfactorily, it is necessary to know something of the geologic timetable of the Cenozoic era, the Age of Mammals. We may regard this era as including two periods, Tertiary and

SUBDIVISIONS OF THE CENOZOIC ERA
OR AGE OF MAMMALS

Periods	Epochs	Estimated Time since Beginning of Epoch in Millions of Years
Quaternary	Recent	1/50
	Pleistocene	1
Tertiary	Pliocene	10
	Miocene	30
	Oligocene	40
	Eocene	60
	Paleocene	70

Quaternary, the latter short and reaching to modern times. These periods are subdivided to form seven epochs, which are indicated on the accompanying table. In the older epochs the mammals were mainly of archaic kinds, gradually giving place to the ancestors of the existing types. In the later epochs of the Tertiary there is considerable evidence of a gradual cooling of the North Temperate regions. This cooling culminated in the

Pleistocene Ice Age in which portions of Europe and North America were several times covered with great glacial sheets of ice.

POUCHED MAMMALS

In the late Cretaceous beds we find that the evolution of mammals during the reign of the dinosaurs had already resulted in the development of the two great living groups of mammals—the marsupials and placentals.

Opossums.—Of these two groups the marsupials, or pouched mammals, are the more primitive and were the more abundant in the last days of the

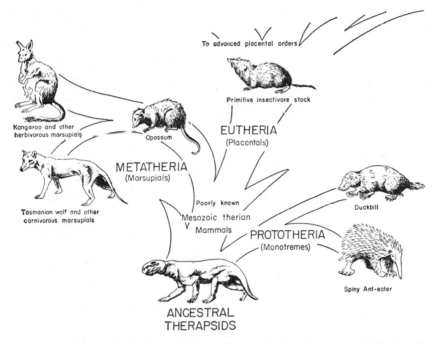

A "family tree" of the lower mammal groups. (From Romer, *The Vertebrate Body*, W. B. Saunders Company.)

dinosaurs. The living opossum is a typical marsupial, and in habits and many structures it seems to be very similar to the Mesozoic mammals. The mammal egg contains little yolk for the embryo to use in its growth, and, when mammals began to bear their young alive, there was no satisfactory mechanism by which the mother could supply nourishment to the unborn young within her. In consequence, the young were surely born at a very tiny and immature stage of development. In marsupials for the most part, this situation still exists; in the opossum, for example, the young are born only eight days after the egg is fertilized. In marsupials the mother's womb secretes a milky fluid which can be absorbed by the embryo, and there may be a connection between the embryo's empty yolk sac and the maternal tissue, through which some food materials can reach the young from the mother's blood stream. These methods of nutrition, however, are generally rather unsatisfactory.

In our own case prematurely born babies are reared in an incubator; most marsupials have evolved a substitute in the pouch, or "marsupium," from which the group takes its name. The tiny, seemingly helpless marsupial baby crawls up the mother's body and gains entrance to the pouch on the belly of the mother. Here it finds shelter and warmth; here, too, are the mother's teats from which it gains nourishment to grow to a stage where it is ready to face the world. As well as being small in size and exceedingly immature, the young marsupial has very different proportions from those of the adult. The one act that the newborn animal has to perform—or die—is to make the upward climb to the pouch, and much of the energy of the embryo, in the brief time available, is devoted to developing good front legs that will serve this function. In such a marsupial as the kangaroo, for example, the adult has small front legs, very long and powerful hind ones; the fresh-born kangaroo baby has good front limbs, while development of the hind legs has scarcely begun.

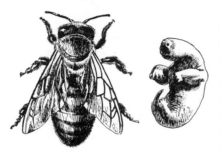

Right, a newborn opossum, compared with a honeybee, to show its tiny size. Its condition at birth is very immature, but the "arms" are well developed to enable it to climb up into the mother's pouch. (From W. J. Hamilton, Jr., courtesy McGraw-Hill.)

There were numerous pouched mammals present in the last days of dinosaur supremacy, animals much like the living opossum in build but mainly of much smaller size. At the dawn of the Age of Mammals the opossums were widespread over the earth. In most regions, however, marsupials made little progress, for they were accompanied by higher mammal types which rapidly supplanted them. In two regions, however, they had better luck.

South American marsupials.—The continent of South America is now connected with the rest of the world by only the narrow Isthmus of Panama. This connection appears to have been broken at the dawn of the Age of Mammals and did not re-form until a very late stage of Cenozoic

Egg-laying mammals. The two types which still survive in the Australian region. *Left*, the platypus (*Ornithorhynchus*), an amphibious animal with a horny bill and webbed feet. *Right*, the spiny anteater, with spinelike hair and a long beak, adapted to termite-eating. This animal is a powerful digger; the specimen shown (from New Guinea) was able to dig down vertically and bury itself in about one minute. (*Ornithorhynchus* photograph courtesy New York Zoölogical Society; *Zaglossus* photograph by S. D. Ripley.)

Pouched mammals. The American opossums are the most primitive of living marsupials, but the group survives mainly in Australia. On this page and the two next following are shown a few of the many specialized pouched forms of that continent.

The Tasmanian wolf. An animal which in its habits and almost every major structural feature parallels closely the wolves of other continents but which, nevertheless, is a true pouched mammal. (From a painting by Charles R. Knight; photograph courtesy American Museum of Natural History, New York.)

The rabbit-eared bandicoot. Bandicoots resemble rabbits in their large ears, hopping gait, and the ability to dig holes; but they have a long nose, long tail, and are insect-eaters rather than vegetarians. (Photograph courtesy New York Zoölogical Society.)

Marsupial mothers and young. *Above, left,* a wallaby with a youngster still in the pouch; *right,* a wombat and its partly grown offspring. The wallabys belong to the kangaroo family. The wombat is a herbivorous form which is in many ways the marsupial "other number" of the American woodchuck, with rather similar dentition and habits. (Photographs courtesy New York Zoölogical Society.)

Giant Pleistocene marsupials. The Ice Age seems to have been the time when mammals reached peak size. In Australia there were numerous giant marsupials in the Pleistocene. Two are shown below: a giant kangaroo, which may have stood 10 feet high in life, and *Diprotodon,* an extinct relative of the wombat but as large as a rhinoceros. (From a mural by Charles R. Knight; photograph courtesy Chicago Natural History Museum.)

Flying phalanger. The Australian phalangers are pouched mammals resembling squirrels; a "flying" form (*right*) has parachute folds of skin at the sides of the body like those of the true flying squirrel.

The koala (*below*). This Australian animal is frequently termed a native "bear." In size, however, it is comparable rather to a Teddy bear, and its habits are not bearlike, for its diet consists of eucalyptus leaves.

The kangaroo pouch (*below at right*). The pouch of a female has been opened up to show two tiny young attached to the mother's teats. (Phalanger and koala photographs courtesy New York Zoölogical Society; pouch photograph by Harry C. Raven.)

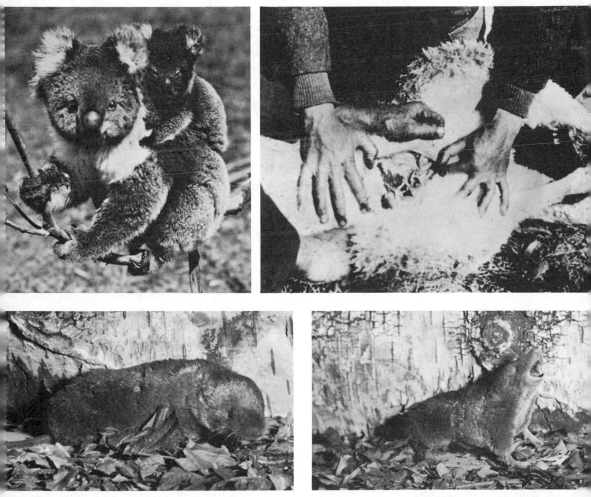

The shrew. The smallest mammal, and closest of living animals to our insectivorous ancestors, this shy nervous creature is not photogenic; the two figures above, however, give some idea of its appearance (suspicion and defiance appear to be the emotions recorded). (Photographs by Oliver P. Pearson.)

Insectivores. *Upper left*, one of the oldest known skulls of a placental mammal, in the hand of its discoverer, Dr. Walter Granger, in the Gobi Desert. Apart from the shrews, the living insectivores include a number of specialized types. *Upper right*, the hedgehog of Europe, a small, prickly animal not related to the porcupine. *Center left*, a tree shrew (*Tupaia*) of the Malay region, the only arboreal insectivore, related to the primates and often included in that order. *Center right*, a mole, adapted to live underground and eat worms and grubs. *Lower left*, *Solenodon* of the West Indies, relatively large and primitive. *Lower right*, the elephant shrew, a small African desert form with a wiggly nose. (Cretaceous skull courtesy American Museum of Natural History; others, New York Zoölogical Society photographs.)

history. A number of placentals reached that continent and in its isolation developed into many strange and curious herbivorous types considered in later chapters. None of the placental carnivore types, such as dogs and cats, entered South America before the bridge went out. But opossums are omnivorous in their food habits, and some of their descendants developed in South America into flesh-eaters paralleling the wolves and cats of other continents; one even developed into an animal very similar to the saber-toothed "tigers" which were long prominent members of the placental carnivore group. In the Pleistocene, however, North American connections became re-established, placental flesh-eaters poured in, and the carnivorous marsupials disappeared, apparently being unable to compete with them. Now, of pouched mammals, we find in South America only opossums of several sorts (some quite small and one partially aquatic) and one or two aberrant little forms. Incidentally, opossums, we have noted, were present in North America in early Tertiary days and are present to-day. It is probable, however, that they had died out in the north and that these hardy animals are re-immigrants, returned from a long sojourn in the south.

Australian marsupials.—In Australia the pouched mammals had their one great opportunity. That continent is isolated today and seems to have been isolated since the Cretaceous. Pouched mammals had entered Australia before its isolation, but not a single higher mammal of any sort, and until man arrived no land mammals had entered since (except, of course, bats and some rats, which seem to have gradually worked their way down the East Indies). There was a whole continent free for the marsupials to (literally) spread themselves in; and there they developed into a curious and interesting fauna, the members of which in many ways have paralleled groups of higher mammals which were evolved in the other continental areas.

The opossum, which seems to lie at the base of the Australian evolutionary array, is an arboreal animal with an omnivorous diet. From such a form we might have—and have had—Australian evolutionary lines leading in both carnivorous and herbivorous directions. Some small Australian marsupials, the dasyures, which are still arboreal but have started a carnivorous career, are sometimes termed native "cats"; they are catlike in habits and appearance, except for rather pointed snouts. Other forms which are in many ways equally or more primitive are the little pouched "mice," mouselike in appearance but very different in structure and habits. They resemble some of the small early fossil opossums and in habits are eaters of insects and other small animals, much like the shrews among placentals. Curiously, the mouselike parallel in external form even extends to the development of little hopping forms comparable to the kangaroo rats. Among the more specialized carnivores a prominent representative is the Tasmanian wolf, or thylacine. It is often called a "tiger" because of its striped back, but this is a very poor name, for the thylacine closely parallels the wolves and dogs of other continents in habits and many structural features. Present in prehistoric times all over Australia, it appears to have failed in competition with the dingo, a true dog introduced by man and since gone wild. The native "wolf" is now confined to Tasmania and even there is close to extermination. Also confined to that island is the Tasmanian "devil," a burly and powerful pouched carnivore rather comparable to the wolverine among placentals. Smaller animals which have diverged from the typical carnivorous marsupial stock include an anteater comparable to placental forms and—most striking of all specializations— a marsupial "mole," with powerful claws, vestigial eyes, and a conical snout protected by a horny shield. This animal is in almost every detail a duplicate of the placental golden mole of South Africa.

The Australian marsupials so far described all have dental variations adapted to a carnivorous diet, and all have a good complement of incisors. In one common type of classification they are hence called polyprotodonts ("many front teeth") in reference to this last feature.

In contrast with these is a second series of forms which branched off from the primitive omnivorous diet toward a vegetarian life. The cheek teeth develop into a good grinding mechanism. And in the front of the jaw the lower incisors, and, exceptionally, the upper ones as well, are reduced to a pair of long chisel-like teeth resembling those of placental rodents; because of this last feature, these herbivores are often termed the diprotodonts ("two front teeth"). A curious (although quite minor) character is that the second and third toes of the hind foot (both rather small) are bound together by skin to function as a unit.

Somewhat transitional in nature between the two main marsupial groups are the bandicoots, which, unlike most Australian forms, seem to have no close parallels among placentals. They have large ears and a hopping gait and, in these regards, are somewhat rabbit-like, but the resemblance goes no further. In the fusion of toes and development of "grinders" they show the diprotodont pattern; but the incisors are of normal type. A more generalized, central group of true diprotodonts are the phalangers, small but varied arboreal forms. They are often misnamed "opossums" in

Australia, but the true comparison in habits and in general appearance is with the squirrels among the placentals. Here parallelism has gone to the extreme of developing among the phalangers three different forms with membranes between front and hind limbs which are used for gliding, like those of true flying squirrels. More specialized types, in which upper as well as lower incisors are reduced to a pair of chisels, are the koala and the wombat. The former is a little short-tailed arboreal form, often called a native "bear"; but the resemblance in appearance and size is closer to a Teddy bear, and its most un-bearlike diet consists of eucalyptus leaves. The wombat, now rare, resembles the marmots or woodchucks of Eurasia and North America in many regards.

One major placental group for which the Australian marsupials have failed to create any close parallel is that of the ungulates, the hoofed mammals, such as the horse and cattle tribes. However, some of the kangaroos resemble them closely in their mode of life, although not in appearance. The essential nature of a typical ungulate is that of a large plains-dwelling grass-eater, capable of rapid locomotion. This is exactly the life of some of the larger kangaroos, although the appearance of these forms, with long hopping hind legs and a stout tail as a balancer, is far different. Among the numerous kangaroo types there are considerable differences in size and habits; one smaller form is even a tree dweller.

We know little of the fossil history of these living Australian animals except that in the Pleistocene there were a number of larger forms, now extinct, related to the living marsupials of that continent. These included giant kangaroos and giant wombats grown to the size of a rhinoceros.

This interesting Australian development was possible only because of the isolation of these rather primitive mammals; and with the breaking of this isolation by the arrival of man, the Australian fauna seems on the road to extinction. Besides the inroads made by the fur trade, the introduction by man of other animals has done immense harm. Dogs and cats have found the marsupial an easy prey; and the introduction of the rabbit has proved a calamity both to man and to the native animals.

PRIMITIVE PLACENTALS

Beyond the conditions found in the marsupials, there is one final stage in the general line of mammalian ascent—the evolution of primitive placentals.

Development of the placenta.—We have noted that a flaw in the process of bearing the young alive lay in the fact that the young, when born, were very tiny and helpless, owing to lack of sufficient means of nourishment before birth. The marsupial pouch has been fairly effective in filling this want. But the other mammals have done a better job. One of the membranes (the allantois) which surrounded the developing egg of the reptile ancestors has in higher mammals come into contact with the walls of the uterus in which the developing embryo lies. Through the walls of this fused area, the placenta, food and oxygen are carried from the mother to the embryo and permit it to grow to a far higher stage before birth than was otherwise possible. With a long period of development before birth and a long period of protection and training after birth, the placental mammal can slowly mature its complicated mechanisms in brain and body. It can, as a result, function more efficiently as an adult than lower forms in which, so to speak, the body has of necessity to be somewhat hastily thrown together.

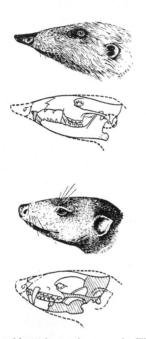

The oldest placental mammals. These figures represent two of six known skulls pertaining to the ancestors of the advanced mammalian groups. All are from the Cretaceous of Mongolia; the original skulls are but an inch or two in length and represent forms somewhat similar to the shrews of today. (From Gregory.)

The ancestors of the higher mammalian groups evolved an efficient placenta at a very early time; presumably much of their success has been due to their development of this useful reproductive mechanism. But they had no "patent" on this structure; and, while most marsupials have no placenta or a very inefficient one, a placental structure very similar to that of higher mammals may be seen in one Australian genus. However, this development in marsupials has been too slow to do the group much good; it is too late for the pouched mammals to try to compete with their true placental cousins with which we are concerned from now onward.

Insectivores.—The ancestoral placentals were seemingly, like a majority of the small early mammals, rather general in their food habits but primarily insect-eaters. Among the placental mammals there are still a few forms, mainly rather small and inconspicuous, which have retained such feeding habits to this day; they seem on the whole to have departed least from the primitive placental stock and are commonly grouped as the order Insectivora. Several tropical forms, from Madagascar, the Congo region, and the West Indies are thought by some workers to be especially primitive. Insectivore types which have some anatomical specializations but seem to represent fairly well, in mode of life and general appearance, the ancestors of all the placental mammals are the shrews—small creatures, mouselike in appearance. These are not uncommon but are unfamiliar to most of us because of their small size and shy habits—features in which

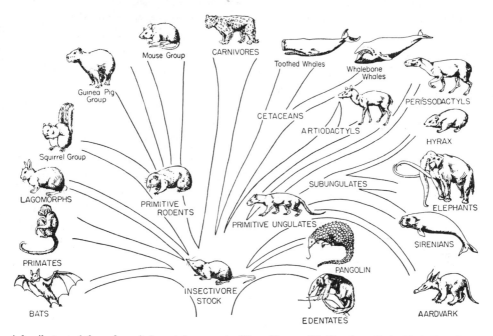

A family tree of the orders of placental mammals. (From Romer, *The Vertebrate Body*, W. B. Saunders Company.)

they presumably are similar to their ancient Mesozoic ancestors. The shrews include the smallest of mammals, one species weighing no more than a ten-cent piece. They are extremely nervous little creatures, almost incessantly active. Correlated with this and owing in great measure to the fact that their small size results in a relatively high loss of heat through the skin, shrews eat voraciously to obtain food with which to stoke the furnaces of their bodies. A typical shrew will eat its own weight in food daily. An interesting recent discovery has been the fact that one of the shrews has poison glands within its mouth, used for killing small animals upon which it preys—the only case among mammals of the development of this snakelike specialization.

There is a further modest array of living insectivore types of which the best known are the hedgehogs and the moles. The former animal (not to be confused, as is often the case in rural America, with the porcupine) is a European resident, mainly nocturnal and sleeping by day, curled up and protected by a spiny coat. It is mainly an eater of insects and slugs but varies its diet with small birds or small snakes. The moles are highly specialized for underground life, with powerful digging limbs and broad claws and with eyes and external ears reduced; the diet consists mainly of earthworms. The mole type has developed twice within the order, for in addition to a widespread series of typical moles, a South African animal, the golden mole, mentioned earlier, shows in its structure evidence that it has independently adapted itself to a subterranean career.

Of especial interest is a modest group of insectivores from the Old World tropics, the tree shrews. Although the trees may have been the ancestral home of placental mammals, most insectivores have turned toward terrestrial or even subterranean modes of life; the tree shrews alone are persistently arboreal. Apart from their retention of this possible primitive habitat, the tree shrews are important in that they appear to represent the stock from which sprang the primates. Despite the fact that their appearance shows no indication of relationship to monkeys, apes, or men, many points in tree-shrew anatomy show significant primate resemblances.

The oldest of insect-eating placentals have been discovered in Mongolia in the shape of a few tiny skulls of animals contemporary with the late Cretaceous dinosaurs. This final stage in mammal evolution had been reached toward the end of the long Mesozoic "training period." When the Age of Reptiles had closed and the dinosaurs had vanished, these efficient placentals were fully prepared to take over the world, and their spread was rapid. During the Paleocene, first of Tertiary epochs, there was a speedy differentiation from the primitive small insect-eaters into a variety of diverse evolutionary lines and a strong tendency for increase in size. By the Eocene the main lines of mammalian evolutionary history had been established. Few of the original insect-eaters have survived; but all the great array of living higher mammals, from men to whales, from horses to rats, have come from this insectivore stock.

CHAPTER 12

Flesh-eating Mammals

We may reasonably begin our discussion of the more highly evolved mammals, the placentals, by giving a brief history of the carnivores, the flesh-eaters. The early mammals were mainly insectivores, forms which presumably ate a bit of this and that but subsisted mainly upon insects and worms. This is flesh-eating of a sort; and it needed only an increase in size before some of the descendants of the insect-eating mammals were capable of preying upon their vertebrate relatives and became carnivorous successors of the flesh-eating therapsids and dinosaurs.

CARNIVOROUS ADAPTATIONS

Teeth.—The major changes which have been brought about in mammals of carnivorous habits are concerned with the teeth. The carnivore has to make its kill mainly with its teeth and has to pierce stout hide, cut tough tendons and hard bones. On the other hand, flesh is comparatively simple to digest and need not be well chewed. We find, in relation to this, that in the more strictly flesh-eating forms grinding molar teeth have been reduced almost to the vanishing-point. A cat, for example, has no chewing power whatever. Dogs and their kin, adhering less strictly to a carnivorous diet, have kept all their molars except one upper pair and have retained some grinding surface in their cheek teeth; the bears have veered sharply away from the flesh-eating habits of their ancestors and have redeveloped considerable chewing power.

The front part of the dentition is highly developed. The incisors are highly useful in biting and tearing; the canines, or "dog teeth," are long and pointed stabbing weapons in all flesh-eaters. Such cheek teeth as are left generally have sharp ridges and pointed cusps rather than flat surfaces. In all typical carnivores there has developed on either side of the jaw a very specialized pair of teeth called "carnassials," which function in an important way in cutting hard pieces of food (notice, e.g., how the house cat works a bone around to the side of the mouth to crack it). One of the upper teeth (the last premolar in living forms) and the lower tooth in back of it become very large and much elongated, with a sharp fore-and-aft ridge. The two teeth do not meet directly in a straight chopping motion but pass each other, the upper tooth to the outside, acting as a pair of shears which can crack and slice very tough materials.

Skeleton.—While the teeth are much modified in carnivores, the skeleton of the trunk and limbs is generally rather primitive in pattern. A car-

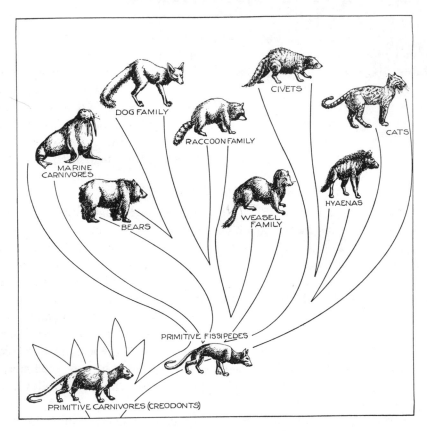

A simplified family tree of the carnivores

nivore must be speedy to catch its prey. But it cannot develop the rigid skeleton or the hoofed condition with a reduction in toes which we find in the large herbivores discussed in the following chapters. It must remain supple and retain its claws in order to attack and grapple with its prey. There are well-developed claws in all typical carnivores, and, except for a frequent loss of the "thumb" and big toe, the toes are all retained.

ARCHAIC FLESH-EATERS

Carnivores needed only, for their inception, to have some of their placental cousins develop into harmless herbivorous types upon which they might prey. The development of such herbivores, particularly the archaic ungulates, began promptly with the extinction of the dinosaurs. In the Paleocene, first of the Tertiary epochs, there began, just as promptly, the development of carnivores, to which they fell a prey.

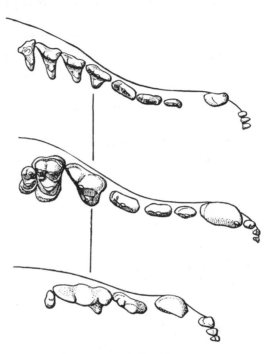

Some of the early carnivores were almost indistinguishable from the ancestral insectivores from which they were being differentiated. But quite rapidly the old carnivores spread out into a number of types with divergent adaptations. The various flesh-eating forms which were dominant in the first two epochs of the Tertiary (the Paleocene and Eocene) are often grouped as archaic carnivores, or creodonts, now entirely extinct. In almost all of them shearing carnassial teeth were developed, although the pair of cheek teeth selected for this use varied. The brain was usually small, and presumably the intelligence was nothing to boast of; but this mattered little, for the archaic herbivores which formed the main staple in their diet were equally

Crown view of the upper teeth of the right side in various carnivores. *Above*, *Sinopa*, a little-specialized Eocene creodont, with all teeth present. *Center*, a fossil dog; the last molar is lost, the other two molars have some chewing power, and the triangular tooth in front of these (the last premolar) is a specialized shearing tooth (carnassial). *Below*, a cat; shearing is highly developed, but chewing ability is gone; of the molars there remains only a single vestigial tooth. The straight line connects the last premolar, or carnassial, in the three types.

feeble in brain development. In most of these old carnivores the body was long but the limbs short and the speed consequently slight; but, again, the older hoofed animals were also comparatively slow of gait.

Some of the smaller creodonts seem to have been comparable to modern weasels in their habits; others were more wolf- or lionlike. The range in size was great. Among the large members of the old stock may be mentioned *Hyaenodon* (with stout, hyena-like teeth), which survived until the Oligocene and may have preyed upon the great titanotheres. A contemporary Mongolian carnivore had a skull a yard long; this beast was the largest flesh-eater that land mammals have ever produced.

MODERN TERRESTRIAL CARNIVORES

The end of the Eocene and the beginning of the Oligocene saw the downfall of the archaic creodonts and their replacement by more modern flesh-eaters, the fissipedes, ancestral to the dogs, cats, bears, and other living types. (The name "split feet" is used to distinguish them from the web-footed sea carnivores.)

Speed and brains.—If we seek for the cause of this revolution in the flesh-eaters, we naturally turn to an examination of the history of the herbivores forming their food supply. Here we find that at about the beginning of the Oligocene there was a similar overturn of stocks. The archaic ungulates and the more clumsy odd-toed types (such as titanotheres) were passing out of existence and being replaced by speedier hoofed mammals which would have been able to elude many of the rather slow creodonts which had previously abounded. Then, too, the older ungulates were small-brained forms; their successors seem to have tended toward larger brain development and greater intelligence. This latter feature may perhaps be the real clue to the downfall of the creodonts. It takes brains to stalk a prey; if the would-be eater is more stupid than his potential dinner, his chances are poor. Among all the creodonts there was only one comparatively inconspicuous Eocene group (miacids) in which a good brain was developing; it is from this one stock that all later flesh-eaters have arisen.

An early fissipede.—Typical of the early members of the modern carnivores which appeared at the end of the Eocene and beginning of the Oligocene was the little *Cynodictis*, about the size of a weasel or a toy fox terrier. This "dog-weasel" is usually regarded as the ancestor of the later dogs. This may well be correct; but *Cynodictis* was little specialized in any particular direction and may have been close to the starting-point of all the later carnivore lines. The body was long, the limbs rather short, as is still the case with such types as the arboreal weasels or civets. This suggests that the ancestors of modern carnivores had been tree dwellers to start with. In brain size *Cynodictis* was much better off proportionately

Civets. The civet tribe (Viverridae) are Old World tropical
forms, many of which are relatively primitive carnivores. In
general proportions the palm
civet (*left*) resembles the archaic carnivore shown above.
The binturong (*lower left*) is an
oriental shaggy-haired nocturnal variation on the same
pattern. The aardwolf (*Prote-
les, lower right*) is a long-legged,
rather foxlike civet which lives
mainly on termites. (New York
Zoölogical Society photo-
graphs.)

Various carnivores:

The otter is a member of the weasel group, which has grown to a moderately good size and taken to an aquatic life as a fish-eater. A close relative, the rare sea otter of the Pacific, has gone farther in this type of life and become a marine form, paralleling the seals.

The badger, also of the weasel group, is a powerful digger, using the long claws on its forelegs to dig and capture burrowing rodents. The many badger genera are widely distributed over the Northern Hemisphere.

The cheetah (*Acinonyx*) is an oriental member of the cat family, which is sometimes tamed as a "hunting dog." Unlike the typical cats, which depend on a sudden spring to seize their prey, the cheetah is capable of high speed over a long distance.

Hyenas. These unlovely relatives of the civet-mongoose group of Old World carnivores have become scavengers. The spotted hyena, shown below, is an African form. (Hyena photograph courtesy American Museum of Natural History, New York, other photographs courtesy New York Zoölogical Society.)

than any creodont; this increase in intelligence may have been (as we shall see much more markedly in the case of our own ancestors) developed in relation to the complexities of arboreal life.

In the Oligocene the differentiation of the modern carnivore families had already begun. Living terrestrial flesh-eaters are divided by technical characters into two main groups: one including civets, hyenas, cats, and their relatives; the other group, the weasels, dogs, raccoons, bears, and their kin.

Civets.—In the first group the cats are the most familiar forms, but the civet family, including the mongoose, genets, and many other small Old World flesh-eaters, is much closer to the basic stock. These tropical animals have never reached America. Most are still arboreal in habits, with the short limbs and small size of their early ancestors; they occupy in the Old World tropics roughly the place taken by the weasels and their kin in the North Temperate Zone. There are (as in the weasel family) many variants from the general pattern; we may note, for example, that in the East Indies one civet has, like the otters, taken to an aquatic existence and fish-eating habits. The great island of Madagascar has apparently been separated from the African mainland for a good part of the Cenozoic era, and its mammal fauna is sparse. There are no members of the dog or cat tribes, and the only carnivores found there are members of the civet family. Most notable is the foussa, an animal looking something like a good-sized cat, reaching five feet in length, with a large tail. Catlike, too, is the fact that the face is short and the teeth like those of cats, highly adapted for shearing flesh. Some students of carnivore history have suggested that the foussa is a primitive cat type; but its general structure otherwise is that of the civets, and it is probably a member of that family which, in the absence of true cats, has tended to adopt a feline mode of life.

A very aberrant offshoot of the civet family is the aardwolf of southern Africa. This creature has rather the appearance of a long-legged fox or of a small and slender sharp-nosed hyena. The teeth are tiny, a feature associated with its diet; this animal—nocturnal and a good burrower—makes its living mainly by eating white ants (and thus paralleling some of the "edentates" discussed in a later chapter).

Hyenas.—Likewise an exclusively Old World family are the hyenas. These are large and repulsive scavengers, with stout teeth capable of dealing with the leavings of their more delicate carnivorous cousins; they are also not averse to a bit of grave-robbing. They are heavily built but long-legged, ungainly creatures. They developed, at a comparatively late time, from the civets. Once widespread over Eurasia, they are now confined to southwestern Asia and Africa. In the former area, and spreading west through Egypt along the north African coast, is the striped hyena. South of the Sahara in Africa the common form is the spotted hyena, a larger and more aggressive animal, which generally hunts in packs.

Cats.—Much older in origin than the hyenas are the felids, the cats, which were already prominent in early Oligocene days. These are the most purely carnivorous of carnivores; back of their carnassials there is scarcely a trace of molar teeth. A cat is ably equipped for stabbing, biting, and slicing; for chewing, it is not equipped at all. Rather in contrast to the dogs, which often run their prey down and hunt in packs, the cat is a crafty individualist. The agile body is incapable of sustaining speedy running for long distances; stalking and a sudden jump on the prey is the cat method. The claws are highly developed as a useful aid.

Modern cats are of many kinds, but although they vary greatly in size and pelage, they are quite uniform in basic structure. The largest members of the group are the lion and tiger. (Beware, by the way, the thrilling jungle picture which shows both lions and tigers. That picture was made in Hollywood, and the animals came from the zoo, for there are no tigers in Africa, and lions are almost extinct in Asia.) There is an abundance of felid species in tropical Asia and in Africa. Next in size to lion and tiger are several species of large arboreal forms, the leopards and panthers. Smaller is the serval and then follow a number of cats of modest size. Among these the common African wildcat, termed the "kaffir cat" in the south, is of especial interest. This form resembles the ordinary gray striped cat to be found in any alley, and is probably the species that was domesticated (and venerated) in Egypt. In the more northern region of Eurasia are wildcats of modest size and one larger felid, the lynx, now rare in Europe.

In the Americas the largest member of the family is the jaguar, of leopard size, golden yellow in color with black rosettes, found from Mexico and southernmost Texas southward throughout the tropical forests. The ocelot is a small form of similar range and somewhat similar appearance. Next in size to the jaguar is the puma, also called the cougar or mountain lion, a large drab brown animal. Its far-flung range extends from southern Canada to far south in South America; it is, however, absent from the more settled parts of North America. In both Americas there are further smaller felids; most common in North America is the "bobcat," technically a lynx.

The one living felid which all agree is definitely set apart from ordinary felines is the cheetah, or hunting leopard, of Africa and southern Asia. This slender, long-legged animal is not merely an adept stalker but a swift runner, able to overtake the antelopes which are its normal prey. The cheetah is frequently captured by man and tamed for hunting purposes.

These various cats look quite different superficially; but under their skins they are very similar to one another; it is practically impossible, for example, to tell a lion's skull from that of a tiger. Our domestic cat is probably one of the latest additions to man's animal entourage.

Sabertooths.—Quite sharply marked-off from the ordinary members of the cat family were the saber-toothed cats, now entirely extinct, but for-

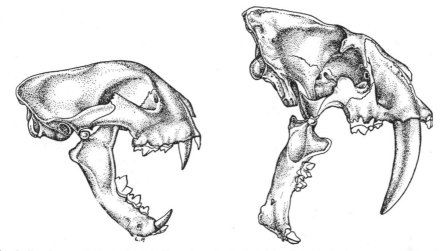

The skull and jaws of a typical cat (*left*), and a sabertooth (*right*). In both the jaws are opened to their full extent. In the sabertooths the whole skull and jaw are shaped to assist in this wide gape and to aid the downward drive of the sabers.

merly widespread. *Smilodon*, a Pleistocene form, was a common inhabitant of America until fairly recent times, geologically speaking. Its skeleton is to be found in every respectable museum, for exceedingly numerous remains have been found at the Rancho la Brea pits in California. At this place in the suburbs of Los Angeles there were, during a late stage of the Pleistocene, pools to which animals of the region came to drink. Beneath the surface were deposits of soft tar. Any unwary animal that stepped in it was trapped. The presence of struggling animals brought, in search of food, numerous flesh-eaters, notably sabertooths and an extinct wolf type and a giant vulture. These forms were very frequently trapped in their turn. In recent decades large quantities of skeletons have been excavated from the now hardened tar.

Smilodon was a powerful animal as large as a tiger. But there were major differences from the ordinary felines. The lower canine teeth were small; the upper ones, on the contrary, were exceedingly long, curved sabers, capable of inflicting a deep slashing wound. The lower jaw was so articulated that it could be dropped at a right angle to clear the sabers; and the shape of the skull indicates the possession of strong muscles to stiffen the head during the stabbing and slicing stroke of the upper canines.

It seems probable that this saber-tooth development was an adaptation which fitted these old cats for dealing with thick-skinned animals such as proboscidians. The ordinary carnivore usually tries for a quick kill—a bite into a vital structure. This, however, is almost impossible in an animal with a thick hide. But long slashes of these sabers would make profusely bleeding wounds, followed eventually by death. The sabertooth played a waiting game, it would seem.

The extinction of these animals is very possibly due to the practical extinction of the large thick-skinned animals which may have formed their prey. In the Pleistocene, for example, there were four large and common proboscidians in North America, as well as numerous large ground sloths. Today all are extinct. It may be that the mastodons—primitive elephant-like forms—were an especial favorite of these creatures, for both saber-tooth and mastodon survived to the end of the Pleistocene in this continent, but both disappeared much earlier in Europe.

Weasels.—The second great group of modern land carnivores is that of the dogs, raccoons, bears, weasels, and related types. Among these forms members of the weasel family are closest to the original fissipedes in a number of respects, such as their frequent persistence in an arboreal life and their usual small size and short-legged build. Most are inhabitants of the North Temperate Zone and, in correlation with chilly climates, have a thick pelt which makes them important in the fur trade. In general they have highly developed scent glands in the anal region—a feature most ostentatiously developed in the skunks. Weasels, polecats, stoats, ermine, and mink are small but bloodthirsty forms, with slender bodies and short but sharply clawed toes; they are widespread over the forested areas of Europe, northern and central Asia, and the colder half of North America. Rather larger but similar in habits are the martens, equally widespread; the sable is a north-Asian variant; the fisher (the name is not particularly appropriate) is a large American marten. The peak in size of this first series of weasel relatives is the wolverine, or glutton, of the more northerly forests of Eurasia and America. His appetite is indicated by his name. In disposition the wolverine is no improvement over his smaller relatives, but his build is different, with a stocky and powerful body about two feet in length and a short tail.

A second series of mustelids consists of forms of a more terrestrial, and often more or less fossorial, nature and, in many cases a trend toward less purely flesh-eating habits, with insects and even plant materials often prominent in the diet. The body is generally more stocky than in the forms so far considered (except the wolverine). In connection with digging habits the toes tend to be armed with heavy but less pointed claws and, in relation to food habits, the carnassial teeth are less powerful shearing structures. Characteristic here are the skunks, purely American in their several distinct types. Their striking black-and-white coat patterns are generally an effective warning of their potency, but natural selection has not yet bred into them a realization that the automobile is unconcerned with olfactory sensations. More broadly distributed are a number of genera of badgers, with broad flat bodies. They are awkward-appearing animals but powerful diggers; burrowing rodents are their major prey, the ground squirrel and prairie dog being favorites with the American badger.

But although the family is as a whole a northern group, a fair number of forms enter the tropics and the southern continents (but not, of course, Australia). In the Americas, one genus of skunks enters South America and there are two tropical American forms rather similar to martens. In the Old World, despite competition from somewhat parallel carnivores developed in the civet family, a number of types have ventured into the tropics. Several badger-like forms are found in the orient, and in Africa are types rather like the skunk and weasel. The ratels of Africa and India are badger-like in appearance and, although varied in diet, are said to have a fondness for honey—hence their alternative name of honey-badgers.

Besides arboreal and terrestrial forms the weasel family includes two aquatic types. The otters are widespread forms which have very much the slender weasel proportions but are, of course, of much larger size—and of far better disposition. The limbs are very short, the short toes, webbed. The otters are fish-eaters, swimming by graceful undulatory movements of the long body. Most otters are fresh-water stream dwellers; more extreme in its specialization is the sea otter of the colder North Pacific coastal waters, highly prized for its fur but close to extinction. This is a much more compactly built, short-tailed animal with large webbed hind feet; in contrast to the true otters this animal is primarily a mollusk-eater.

Dogs.—The dogs, wolves, and foxes and their relatives form a second family of this second group of modern carnivores. They are mainly plains-dwelling, pack-hunting animals, capable of rapid locomotion for long distances. They are considerably less purely carnivorous in their adaptations than the majority of their weasel relatives; they have, for example, retained two of their three upper molars and have a fair degree of chewing ability. The remains of plains animals are comparatively abundant, and hence our history of fossil dogs is relatively full, from the small Oligocene ancestral types down to those of today. There have been many side branches developed from the dog stock, such as extinct hyena-like and bearlike dog forms.

The domestic dog, wolf, coyote, and jackal are a closely related series of animals which are the typical modern representatives of the family. The dog is man's oldest companion, having been at least a camp follower since Mesolithic days. During the course of this long stretch of time selection has evolved a great series of breeds of the most varied nature. The dog is unknown in a truly wild state. The dingo of Australia, a large tawny dog, has become in part wild but was quite surely introduced by man. Very likely it is close to the original dog type, as are perhaps some dog races in India and various pariah dogs of the East, which it resembles. Possibly the dog ancestor was a strain of wolf—with a more tractable nature than the normal wolf—for dog and wolf are anatomically almost indistinguishable.

The wolf is the largest, boldest, and most widespread of canids; it ranges over the entire North Temperate and Arctic Zones through Eurasia and North America, but it has been exterminated in the more thickly populated regions. A smaller American form is the coyote, which in contrast to the wolf is not dangerous to man or adult cattle but may make depredations on the pig sty or chicken yard and can cause havoc in an unprotected herd of sheep. The coyote is, hence, hunted and killed in great numbers. But his elimination from a region is not an unmixed blessing; his usual diet consists of hares, rabbits, and small rodents, and in the absence of coyotes these "vermin" multiply greatly. The coyote is primarily a plains dweller but in recent years has spread eastward as far as the Adirondack Mountains. The jackals of the Old World are forms with habits somewhat similar to those of the coyote, but they are less aggressive and are, to a great extent, carrion-eaters. They are essentially tropical in distribution, ranging widely, in a number of species, through Africa and the Orient.

Distinct from the typical dog-wolf group is a second familiar canid type including the foxes. The red fox (the coat color varies considerably) is

Dogs. The dog family (Canidae) includes a large number of living and fossil types. *Above, left, Cynodictis*, a small, primitive form from the Oligocene, classed as a dog, but perhaps close to the stem of all later carnivores. *Above*, the dingo of Australia, a dog gone wild but introduced into that continent by early man. *Left*, an African jackal, similar in habits to the coyote of North America. *Below, left*, the Cape dog of South Africa, an aggressive form which hunts in packs. *Below*, the large-eared fox of eastern and southern Africa. (*Cynodictis* from Scott's *Land Mammals in the Western Hemisphere*, by permission of the Macmillan Co.; other photographs by New York Zoölogical Society.)

Baloo. The sloth bear is the only bear found in India (apart from the Himalayan region), and (as in Kipling's *Jungle Book*) is commonly known by the name of Baloo or Bhalu. It is a relatively small animal, with shaggy hair and very long claws; its food is mainly fruit, flowers, insects, and honey.

Raccoon relatives. The raccoon is a tree-dwelling cousin of the dogs who has tended toward a mixed or herbivorous diet. Several close relatives of the raccoon are shown here. *Right*, the coati (*Nasua*) of Central and South America, distinguishable by his long nose. *Directly below*, the true panda (*Ailurus*) of southeastern Asia, larger but showing obvious similarities to the raccoon. *Lower right*, a much larger related form (*Ailuropoda*) from the same area, commonly called a panda but properly, the giant panda. (Photographs courtesy New York Zoölogical Society.)

found, like his wolf cousin, throughout the North Temperate Zone, and farther to the north is the Arctic fox, notable in that (like several other animals of the Far North) it assumes a white pelage in the winter. The smaller gray fox is present in North America only, as is the little kit fox of the West.

So far we have confined our mention to forms familiar to the average reader who, like most of the canids mentioned, probably lives in the North Temperate Zone. But in the tropics and on through Africa and South America there are to be found numerous other members of this family not familiar to us and for the most part not worth mentioning unless we were to go into detailed technical descriptions. In South America, for example, there are half a dozen canids which are good members of the family but not particularly comparable to wolves, coyotes, or foxes; the Indian hunting dog, the dhole, is clearly set off by dental characters from the dog-wolf group; the Cape hunting dog of South and East Africa, a canid of somewhat hyena-like appearance, is quite distinct from any other canid group; Africa has a series of small, large-eared, foxlike animals, the fennecs, and one African form of similar appearance is notable in that, contrary to the carnivores' trend for cheek-tooth reduction, it may sport four molar teeth —more than any proper terrestrial placental mammal. But despite this diversity in habits and anatomical detail, all these forms are close to the common structural pattern of the family, and the "dogs" show much less variation than do the weasel and civet groups.

Raccoons.—The raccoons, together with a few relatives (almost all of them American animals) appear to be quite closely related to the dogs and descended from early members of that family. These likable little animals have persistently kept to the arboreal habitat of their early ancestors. They have, however, departed far from a carnivorous mode of life. One evidence of this is the fact that the shearing teeth can no longer shear; they have been modified into chewing teeth like the molars behind them. Apart from the raccoons, which range throughout most of the Americas, there are three tropical types: the cacomistle, rather raccoon-like in appearance but more slenderly built; the coati, distinguished by a long slender snout; and the kinkajou, with a smooth brownish fur and a prehensile tail. All are friendly little animals which are said to make delightful pets.

Sometimes included in the raccoon family, but perhaps not closely related, are two Asiatic animals from the mountains of southwestern China and eastern Tibet. One is the true panda, an animal of rather modest size (about two feet from snout to end of body) with a short head, reddish coat, and striped tail. Like the raccoon group, it has fairly good chewing ability and its diet is mainly herbivorous. Farther from the typical raccoon type is the animal popularly called the panda but more properly termed the giant panda or, better, the parti-colored bear. This is a rare animal, almost never seen in captivity; in fact, very few specimens have ever been collected. In general build it resembles a small bear; in coloration it is dominantly white but with black patches or stripes. Its teeth are still better adapted to chewing than other raccoon-like types, and in its native haunts bamboo shoots are said to be its favorite food. In general this curious creature splits the difference, structurally, between raccoons and bears, and it is perhaps most closely related to the latter.

Bears.—A final stage in the trend away from a carnivorous mode of life is seen in the bears, also members of the dog group of carnivores. These large forms have, in general, a mixed diet. The last molar tooth had, it would seem, already vanished before the bears drifted back toward a herbivorous type of diet; and teeth, like other structural features generally, when once gone never reappear. But in bears the lack of a full set of back teeth has been made up for by the great elongation of the two molars which are left; the two do the work of three. In the Miocene there were several large and rather heavily built dog types which seem to be intermediate between dogs and bears in structure.

Largest and most impressive of bears is a series of forms including the brown bears of Eurasia and the grizzly bears of western North America. The big brown bear once ranged widely over Europe but is now extinct except in the more remote regions; it extends over Asia south to the Himalayas and east to Kamchatka. The cave bear of the European Ice Age, contemporary with Stone Age man, was a large variant of the brown bear type. We need not, however, go to the fossil record to cite bears of impressive size. Grizzly bears are known to reach half a ton in weight, but of maximum size are the much larger Kodiak and other giant brown bears of Alaska—the largest land carnivores in the world—reaching as much as 1,500 pounds. Distinct from these big animals is the black bear, confined to North America but widespread there. A third northern type of bear is the large polar bear, with its appropriate white coat. A good swimmer, this form is (because of the nature of his country) much more of a strict carnivore, with fish and seals as the staples of his diet.

Apart from these familiar forms, there are four other living bears, tropical in distribution; all are mainly black in coloration, but all are quite distinct in nature from the American black bear. The Asiatic black bear is a

Giant bears. The largest of bears are related to the large brown bear of northern Eurasia, represented in the United States by the grizzly bear of the western mountains. The giant of the tribe is the Kadiak bear of the Aleutian Islands of Alaska (*left*), which is reported to reach a weight of 1,400 pounds. Still larger, however, was the great cave bear of the Pleistocene of Europe, illustrated *below*. (Kadiak bear photograph courtesy New York Zoölogical Society; cave bear from a mural by Charles R. Knight; photograph courtesy Chicago Natural History Museum.)

Pinnipeds—aquatic carnivores. In several instances mammalian flesh-eaters have taken to an aquatic life. A successful group is that of the Pinnipedia, including the seals, walrus, and their relatives. Above is shown a seal rookery—the Tolstoi rookery on the Pribiloff Islands, Alaska. This is the home of the fur seals of the northern Pacific. The photograph was taken at the height of the breeding season. Families in the background; idle bulls in the foreground. The fur seal and its close relative, the sea lion, are found only in the Pacific and may be distinguished from the common seals of the Atlantic by such features as the absence in the latter of any projecting ear. (Photograph courtesy American Museum of Natural History, New York.)

The walrus (*left*). A ponderous mollusk-eating member of the pinniped group, confined to cold northern seas. The upper canines are enormous tusks which appear to be useful in grubbing for oysters and mussels. (Photograph from a mounted group in the American Museum of Natural History, New York.)

The elephant seal (*Macrorhinus*), *below*, in which there is a curious development of a proboscis in the males. This animal is the largest of pinnipeds, attaining a length of 20 feet. It ranges from California to the Antarctic. The photograph (by Sir Douglas Mawson, courtesy New York Zoölogical Society) shows two bulls fighting on an Antarctic beach.

large and aggressive animal inhabiting the forests of the Himalayan and related mountain chains. The remaining forms are small. There is a Malayan animal, usually called the sun bear, and in India is the sloth bear, a relatively little fellow with large claws and shaggy hair. The teeth are quite small in correlation with a diet of insects, honey, and fruits. (This is "Baloo" of Kipling's *Jungle Book*.) There are no native bears in tropical Africa and but one in South America—a short-faced type, called the spectacled bear because of rings about his eyes which give him the learned appearance of wearing tortoise-shell spectacles.

AQUATIC CARNIVORES

A final group of carnivores is that of the purely marine types, the pinnipeds, including the various seals and the walrus. These are mainly fish-eating forms which very probably have descended from some primitive dog stock; their fossil history is poorly known. A curious feature is that the tail had seemingly become too feeble a structure in their land ancestors for it to resume any function in propulsion. The feet are webbed, and the hind legs (reduced in other marine mammals) have remained large and are turned back as a substitute for the missing tail. The group is one of temperate to cold waters of both Northern and Southern Hemispheres; except along the western coasts of the two Americas, where there are cold currents, they seldom venture far toward the Equator.

The pinnipeds include three families: eared seals, earless seals, and walruses. The earless seals, so-called because there is no projecting ear flap, are widespread in the north of both Atlantic and Pacific oceans and in southern oceans as well; the little harbor seal is a common form in both northern oceans. Otherwise purely marine, in Asia seals are found far inland in Lake Baikal, the Aral Sea, and the Caspian—a situation perhaps due to a former connection of these water bodies with the ocean at some time in the Pleistocene. In earless seals the limbs are useless on land, and the common seal comes ashore only very briefly for breeding purposes (they lead a properly monogamous family life). Most members of this group are small, but the elephant seal, of the Antarctic and the Pacific shores of America—named because of the development of a short proboscis in the males—is a giant form, in which the males may reach twenty feet in length.

The eared seals are confined to the Pacific, reaching both Arctic and Antarctic extremes. Most members of the family are termed sea lions; many are of rather good size. However, the California sea lion is relatively small; this animal, intelligent and amiable (as are most seals) is the most common seal in menageries and makes a good circus performer. In this group all four limbs can be used in locomotion on the beach, and members of the family stay ashore for considerable stretches of time at the annual breeding periods. The males are much larger than the females and are polygamous, the old bulls gathering—and defending—harems composed of half-a-dozen or more inmates.

In this family, too, are the fur seals, one in Antarctic waters, the other—a famous species—from the North Pacific. The northern fur seal breeds almost exclusively in the small Pribiloff Islands in the Bering Sea. The seals remain there from May to September, then take to the sea, journeying southeast in the northeast Pacific in the fall and returning up along the Canadian and Alaskan coast in the spring. When first discovered, the rookeries on the Pribiloffs were of fabulous size. Constant commercial hunting so endangered the existence of the species that only through international agreements has it been preserved.

The walruses, found throughout the Arctic from Atlantic to the Pacific and, with the seals, the main source of Eskimo livelihood, are enormous animals with breeding habits like those of the eared seals. Their diet, however, is primarily one of mollusks; their enormous upper canine tusks are useful in digging and gathering mussels, and their stout peglike cheek teeth deal effectively with mollusk shells.

CHAPTER 13

Hoofed Mammals

Quite in contrast to the carnivores just discussed are the hoofed mammals, or ungulates, of which several groups will be considered in the present chapter and the one which follows. By this term may be designated almost all the larger herbivorous mammals. The name, however, is not entirely a distinctive one, for while typical ungulates, such as the horse and the cow, have hoofs, the ungulate orders include a number of animals with well-developed claws and even such types as the purely aquatic sea cows. Nor do the ungulates form a single natural group, for the hoofed condition has undoubtedly been attained independently by various lines; and, strange as it may seem, a cow is, for example, probably as closely related to a lion as to a horse.

UNGULATE STRUCTURES

Despite the artificial nature of the ungulate assemblage, there are certain structural changes which have generally happened in the transformation of a primitive mammal of whatsoever group into a large herbivore—changes having chiefly to do with teeth and limbs.

Teeth.—The sharp, pointed cheek teeth of primitive mammals were unfitted for a purely vegetable diet and were not suitable organs for undertaking the thorough chewing which leaves, grain, or grass must undergo before passing into the stomach. In relation to this we find that in ungulates the cheek teeth have generally tended to become much enlarged and to develop a flattened grinding surface. In large forms and especially in animals which eat highly abrasive material such as grass (as, e.g., horses and cattle) the demands for increased grinding surface on the teeth are very great. In correlation with this the molars of many ungulates become high crowned so that the tooth may be ground down a very considerable distance before it is worn out.

There is, however, a curious fashion in which this high-crowned condition is attained in such a form as a horse or cow. The crown of a tooth is covered with a thin layer of a very hard material, the enamel; the tooth roots are fixed to the jaws by a connecting layer of porous bonelike material, the cement; but the main mass of the tooth is the material known as dentine. To make a high-crowned tooth one would imagine that nature would simply have built the dentine up into a high and solid pillar, covered by a hard film of enamel and kept the cement in its proper place at the roots.

Not so. Dentine is a relatively soft material, and once the hard enamel at the crown surface was worn off, the tooth would wear rapidly away if built in this fashion. Instead, each of the cusps on the original low crown grows up into a tall slender pillar. Each pillar, of course, is covered with enamel, and hence the amount of this hard material in the tooth is greatly increased. But merely having a series of separate tall cusps would give a weak structure. Nature has remedied this situation. Before the tooth breaks the gums, the cement material covers over the entire tooth and fills the interstices between the individual cusps, binding them together into a compact unit. And when wear occurs, it grinds down gradually through a complex structure of interwoven layers of cement, dentine, and hard enamel.

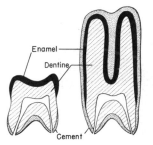

A cross-section of a low-crowned cheek tooth (*left*) and a high-crowned tooth (*right*), to show the set of folds of cement, enamel, and dentine.

Limbs.—In most ungulates a fast type of locomotion has developed. Speed is necessary for these harmless forms if they are to escape flesh-eating enemies, and long-distance transportation is essential for many ungulates which range from season to season over different feeding grounds. In primitive mammals the skeleton was a flexible one; in ungulates the joints of the various bones are so constructed as to be extremely efficient for straightaway forward motion but very poorly adapted for any other types of movement (we all know, for example, how difficult it is for a horse to regain its feet after it has fallen in an unaccustomed position on a slippery pavement). In running types the first joints of the limbs (humerus and femur) are short, giving a fast muscular drive, and the second segments long, swinging fore and aft over a wide angle. In addition, the bones which lie in the palm and sole are much elongated, and by thus running on its toes the animal adds a third functional segment to the limb. With further development of speed the toes themselves are lifted until they touch the ground only at their tips. With the result of attaining solid stance, hoofs ("ungulates") are developed. A herbivore in general no longer needs claws, and these horny structures become short and broad and surround the end joint of each toe.

With the lifting of the feet into the ungulate position, it is obvious that there will be a tendency for the short side toes to fail to reach the ground and cease to function. In many ungulate groups these side toes are reduced and the central ones strengthened. The "thumb" and big toe, which were short and primitively diverged from the others, usually disappeared early in ungulate history. Beyond this we find, as we shall see, two different types of toe reduction in hoofed mammals—one in which the axis of the foot lies through the middle toe, leading to the development of three- and one-toed types, such as the horses and rhinoceroses; a second in which the axis lies between the third and fourth toes, leading to the development of two-toed, "cloven-hoofed" types, such as the pigs, deer, and cattle.

The type of limb structure just described is that found in most small- or medium-sized ungulates. In large forms, such as elephants, a different series of problems is encountered, and the limb construction for the support of a heavy weight is quite a different one. The limbs are comparatively short, straight, and thick; the thigh is long, the shin short. The foot is broad; there is little tendency for loss of toes; these are short and stumpy and usually have a pad beneath them, forming a stout terminus for the pillar of the limb.

Speed is, of course, the best defense a harmless herbivore has against a carnivore. But, in addition, we find defensive weapons in many ungulate types. Some hoofed mammals, such as the swine or some of the small deerlike forms, have large stabbing canine teeth. Much more common, however, is the development of hornlike structures of one sort or another.

EARLY UNGULATES

Condylarths.—Hoofed mammals were unknown during the Age of Reptiles but made their appearance in considerable numbers even in the Paleocene, earliest of Tertiary epochs. Fairly representative of an early stage in ungulate history was *Phenacodus* of the Lower Eocene of North America and Europe. This animal reached a size rather considerable for the time, the largest species attaining the dimensions of a tapir. In general appearance it resembled some of the early carnivores, for the body and tail were long and the limbs fairly short and primitive in structure. There are, however, good signs that this form was already on the road toward higher ungulate conditions, for not only were the cheek teeth expanded for the chewing of vegetable food, but in the feet, while all the toes were still present, each toe was capped by a small hoof.

This interesting form was once believed by some to be the actual ancestor of many of the later hoofed mammals. This cannot be the case, for it is a bit too late in time (it was a contemporary, e.g., of the early horses) and was also somewhat too large to fit into the early ancestral stages of most later lines. But probably its smaller and less well-known Paleocene relatives of the order Condylarthra were close to the stem of some of the later great ungulate groups.

Uintatheres.—In many lines of mammalian development we may contrast "archaic" and "modernized" types. Thus, among carnivores, we have seen the development of archaic forms, the creodonts, which flourished, rapidly reached in many cases a very large size, and then disappeared quickly when placed in competition with other forms of more progressive structure. So among ungulates we find archaic and progressive forms. Some of the animals considered in the next chapter—the elephants and various extinct South American ungulates—are essentially archaic types which, however, managed to exist for a long period of time. Other large archaic forms which arose early in Tertiary times had less good fortune and soon died out. Of these very ancient forms the most spectacular were the uintatheres (order Dinocerata). These were characteristic of the Paleocene and Eocene. By the end of the latter period some of them reached the size of an elephant. These heavy types were not built for speed; the limbs were powerful, elephant-like columns; all the toes were retained to brace a presumably padded foot. It is probable that the uintatheres were swamp dwellers; but even so they needed protection against the larger contemporary flesh-eaters. Large upper canine tusks were developed as weapons at an early date, and in addition three pairs of hornlike structures projected from the top of the head. Uintatheres were exceedingly small-brained types, and their tooth construction also was too poor to stand anything but the softest of food. With the end of the Eocene this line became extinct.

ODD-TOED UNGULATES

In the Eocene more progressive ungulates were already well started on their careers. Most prominent of ungulates from the Eocene on were two groups—the perissodactyls, or odd-toed ungulates, and the artiodactyls, the cloven-hoofed or even-toed types; these will be treated in turn.

Toe reductions in odd-toed forms.—The perissodactyls include, among living hoofed mammals, the horses and their zebra and ass relatives, the

Early ungulates. *Above*, a restoration of *Phenacodus* of the early Eocene, a very primitive, hoofed mammal with five toes. *Below*, a scene in the Bridger Basin of southwestern Wyoming in Middle Eocene time. *Right*, uintatheres, specialized hoofed forms with three pairs of horns and small brains. *Left*, *Orohippus*, an early horse. (From paintings by Charles R. Knight, courtesy American Museum of Natural History, New York, and Chicago Natural History Museum, respectively.)

An archaic ungulate. In the Paleocene there had already developed prematurely large ungulates, of which *Barylambda*, pictured above, is representative. This clumsy and ponderous beast ran to a dozen feet in length. Such archaic forms were destined for rapid extinction with the advent of more progressive types. (Restoration, prepared under the direction of Bryan Patterson, from the Chicago Natural History Museum.)

The Malayan tapir, similar to the South American form on the next plate.

rhinoceroses, and the tapirs. In addition there were two interesting extinct families, the titanotheres and chalicotheres. Characteristic of this group is a type of foot symmetry with the axis through the middle toe and a tendency to reduce the toes from five to four or three and, finally, in horses, to one. In order to see how this type of toe reduction has worked out, try the experiment of placing your hand on a table in the flat position of the primitive mammal. Raise the wrist up off the table and the hand is in

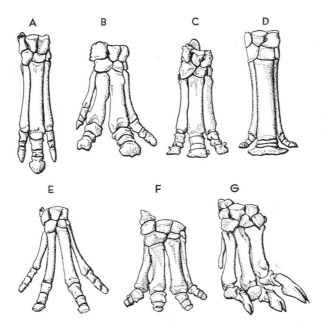

The right front feet of various odd-toed ungulates. *A*, an Oligocene running rhinoceros; *B*, a four-toed Oligocene true rhinoceros; *C*, a later (Miocene) three-toed rhinoceros; *D*, the gigantic rhinoceros *Baluchitherium*, with a pillar-like foot; *E*, a tapir; *F*, a titanothere; *G*, the chalicothere, *Moropus*. The "thumb" is absent in all. In some the "little finger" is present (*B, E, F, G*) but tends to be small and is lost in most rhinoceroses (*A, C, D*). This results in a three-toed condition, as in many of the (related) horses.

the position of a primitive hoofed mammal walking on its toes. As this is done, it is obvious that the thumb soon ceases to touch. If this were lost, your hand would be in the four-fingered condition seen in the front foot of the dawn horses or the living tapir. Raise the hand a bit more, and the little finger ceases to touch. An analogous three-toed stage is characteristic of most fossil horses and is still present in the rhinoceroses. And, finally, getting the hand straight up from the table, only the middle finger touches.

You have thus in a brief space of time repeated the essentials of the story of limb evolution in the horses from five toes to a one-toed condition.

Horses.—The oldest of horses, *Eohippus*, the dawn horse, appears at the beginning of the Eocene epoch; frequent remains of this tiny horse, no larger than a fox terrier, are found in deposits of that date in our western states. This form was already a slimly built little fellow with quite long legs; but while toe reduction had already begun, there were still four toes in front and three behind. The cheek teeth of the dawn horse were still low-crowned and incapable of coping with much hard food; presumably he dwelt in the forests, existing as a browser (living on leaves and soft vegetation) rather than a grazer (grass-eater). *Eohippus* seems surely the beginning of the horse line but is probably close to the stock from which have come the other odd-toed ungulates as well.

Eohippus and "Eohomo" (the "dawn horse" and the "dawn man"), a fanciful sketch made by Huxley, famous English zoölogist, on seeing the remains of the oldest horse. Needless to say the two were not contemporaries—by 50,000,000 years or so. (From Schuchert and Le Vene.)

Mesohippus, a typical Oligocene horse, was bigger, the size of a collie or larger. The "little finger" had been lost, resulting in a count of three toes on all the feet. By the Miocene the main line of horse evolution, leading to such forms as *Parahippus* and *Merychippus*, was acquiring high-crowned teeth, suggesting a grass diet and a change to plains life. This is also borne out by the feet, in which the two side toes, although still present, were becoming slim and short and probably did not normally touch the ground. A single toe is not well adapted to rough going but is an excellent structure on the hard surface of the plains. *Hipparion* and other forms which survived into the Pliocene were the size of a pony and lightly built, fast-running types. During the Pliocene epoch the side toes tended to disappear, and by the beginning of the Ice Age, horses of the modern genus *Equus*, with but one complete toe, were present on every continent except Australia.

Little *Eohippus* was present in Europe as well as North America, and a few aberrant descendants of his persisted for a time in the former continent. But otherwise the entire story of horse evolution right down to the development of modern one-toed types took place in North America, and during Tertiary times only a few types (such as *Hipparion*) wandered now and then into the Old World. Native horses were present in North America throughout the Pleistocene Ice Age. Subsequent to the last retreat of the ice, however, they became entirely extinct here. The horses are only one of a number of large animals which lived in America during a recent period and have since vanished from this continent. In addition,

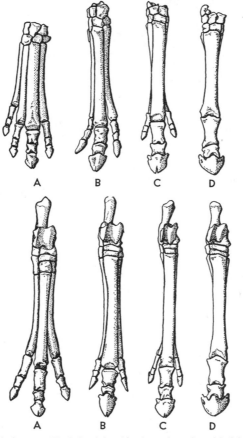

The evolution of the feet in horses. All of the right side; front feet *above*, hind feet *below*. The "thumb" has disappeared in even the earliest known horses. *A, Eohippus*, a primitive Eocene form with four toes in front and three behind; *B, Miohippus*, an Oligocene three-toed horse; *C, Merychippus*, a late Miocene form with reduced lateral toes; *D, Equus* of the Pleistocene and Recent. (*A* after Cope, *B, C*, after Osborn.)

ground sloths, glyptodonts, giant armadillos, camels, mastodons, and mammoths all inhabited the United States not so many thousands of years ago. What caused this mass extinction of large American animals (while not touching the smaller forms) we do not know. The plains of the

Americas are today quite well adapted for horse life, and, when reintroduced and allowed to run wild, horses (and asses) have flourished there
as feral forms. It was once thought that the vicissitudes of life in the Ice
Age might be responsible for the disappearance here of horses and other
native forms. But the evidence suggests that most of them survived
through all the glacial advances and gave up only when the ice retreated

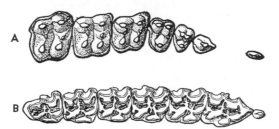

Crown view of the upper cheek teeth (molars and premolars) of *A, Eohippus; B,* the modern horse,
genus *Equus.* These are of the right side, with the front of the mouth to the right. The simple six-
cusped teeth of *Eohippus* have elongated and evolved into a very complicated pattern in the recent
forms. The three teeth to the left are the molars. These were already fairly well developed in the
early horses. In modern horses three of the more anterior teeth (premolars or "bicuspids") have
gained a form similar to that of the molars.

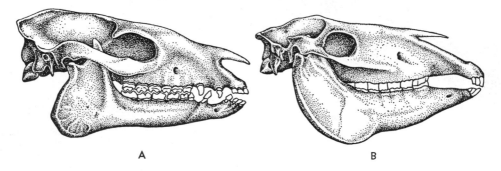

A   B

The skulls of *A, Eohippus; B, Equus,* a modern type of horse. In the course of horse evolution, with
the lengthening of the tooth row and the development of high-crowned teeth, the face has become
very much elongated proportionately and the jaws much deepened to accommodate the teeth.

for the last time and happier days were returning. The only change in
the American scene was the appearance of man. He is not directly responsible for their disappearance by hunting them down and exterminating them, although he may have killed a small fraction of the population.
But perhaps his appearance may in various ways have upset the balance
of nature and more or less indirectly led to their downfall.

In Eurasia, wild horses were very common in the Ice Age; today, except

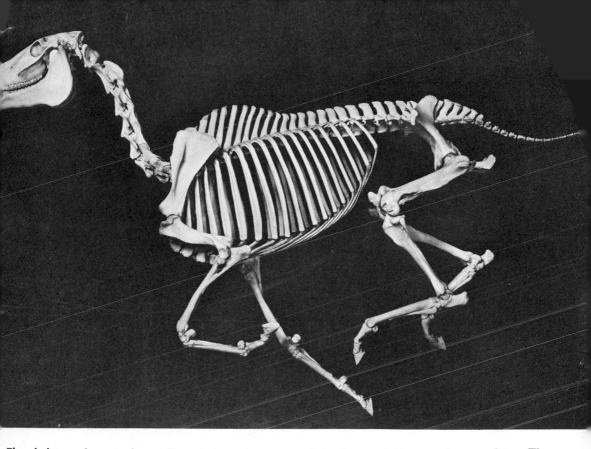

The skeleton of a race horse (Sysonby), to show some of the features of fast-running ungulates. The proximal joints of the legs are short, resulting in a powerful drive. The second joint is longer, which makes for a long fore-and-aft swing. The metapodials (bones of palm and sole) are much elongated, giving a third segment to the limb, and the number of toes is reduced. (From a mount by S. H. Chubb in the American Museum of Natural History, New York.)

Tapirs. These animals have specialized in relatively large size and heavy build plus the development of a protruding nose, but in teeth, feet, and forest-dwelling habits, they greatly resemble the earliest horses and other primitive odd-toed ungulates (perissodactyls). An earlier plate has shown the Malayan tapir; *at the left*, a South American form with its young. (Both tapir photographs courtesy New York Zoölogical Society.)

Horses, ancient, "medieval," and modern. (A series of paintings by Charles R. Knight.) *Top,* the dawn horse (*Eohippus*) of the early Eocene, about fox-terrier size. *Center, Mesohippus,* a three-toed horse of Oligocene days, about as large as a collie. *Bottom,* Przewalski's horse of the steppes of Asia, the only living wild species of horse. (Photographs courtesy American Museum of Natural History, New York.)

for some feral types, the horse appears to be represented only by a rare, shaggy-maned pony-like form from the steppes of central Asia, named "Przewalski's horse" after its discoverer. The tarpan, more common and widespread in the Asiatic wastes, is generally believed to be a feral—i.e., secondarily wild—form, descended from once domesticated progenitors. In Africa, the zebras, very similar to the true horses except for their striped skin, are still fairly abundant. It has been suggested that many of the Pleistocene fossil horses of both hemispheres may have been striped zebra-like equids rather than horses in the narrow sense. Wild asses are to be found over the more arid regions of western Asia and northern Africa. The Asiatic asses range from Syria eastward to Mongolia; the African form is more limited in distribution, essentially the regions of the Sudan and Ethiopia east of the Nile to the Red Sea. The ass appears to have been domesticated in the Near East at an early Neolithic date. Domestication of the horse seems to have been a later accomplishment by the tribes of the steppes of southern Russia and central Asia; when first mentioned in Mesopotamian history, it is termed "the ass of the mountains," carrying barbaric invaders from the north. Use of the horse (at first for drawing carts and war chariots, only later for riding) appears to have been responsible for the successful development of nomadic life in the Eurasian steppes and in great measure responsible for the successful invasion by barbaric Indo-European nomads of the more civilized regions to the south.

Tapirs.—A much less prominent group of odd-toed ungulates is that of the tapirs. Today their only living representatives are confined to the tropics of South and Central America and the Malay region. We have, in earlier chapters, noted similar cases of "discontinuous distribution" of animals in New and Old World tropics, and hence are prepared to find— as is the case—that in the Tertiary tapirs inhabited North America and Europe; the cold of the Pleistocene appears to have been responsible for their extermination in these northern latitudes. The tapirs, except for their somewhat larger size and stockier build and the development of a pendant snout, are in many respects quite close to the ancestral horse types. Like the earliest horses, the living tapirs have four toes in the front feet and three behind; like them, the teeth are low-crowned; and the tapirs today are persistently forest-dwelling browsers. These forms have departed least of all the odd-toed ungulates from the primitive stock of the order. The largest form is the Malay tapir, which may reach eight feet in length; the several American species are smaller. The Malay form is conspicuously colored, the front part of the body is black to brown, the rest white; the American forms are nearly uniformly brown. The tapirs frequent stream banks and often take to the water when pressed.

Rhinoceroses.—The rhinoceroses of the Old World tropics are the living remnants of a once-important group of odd-toed hoofed mammals. They are large, ungainly, three-toed creatures with one or two horns on the top of their long low skulls and a reputation for stupidity and bad temper. The horns are of a peculiar structure. They consist not of true horn or bone but of a bundle of hairs "glued" together into a compact mass. The incisor teeth in the front of the mouth are peculiarly modified, and lost entirely in living African species; mobile lips replace them in cropping vegetation.

Far different were the earliest rhinoceroses, which appeared in the Eocene as contemporaries of the early horses. These were comparatively small and slim "running rhinoceroses," seemingly much like their horse relatives in build, hornless, and at first with four toes still present on the front feet. In the following period there had evolved somewhat more normal three-toed rhinoceroses of larger size and heavier build. Horns, however, were slow to make their appearance and were absent in many early types.

Baluchitherium, an Oligocene hornless form recently excavated in central Asia, was perhaps the most remarkable of all rhinoceroses. This animal was some seventeen feet in height and was proportioned like a heavily built giraffe. It appears to have been the largest of all terrestrial mammals but was, of course, considerably smaller than some of the dinosaurs.

There were many variants among the older rhinoceroses; short-legged, water-living forms, for example, and the woolly rhinoceros of the Ice Age in Europe, whose warm shaggy pelt is made known to us not only by the drawings of cavemen but also through the fact that two of these forms have been found "pickled" entire in a Galician oil seep. Rhinoceroses were present in North America as well as in the Old World until the end of the Tertiary. Today they seem to be a dying group, close to extinction.

Living rhinoceroses, none abundant, are confined to the Old World tropics. Three forms are present in the oriental region. The great Indian rhinoceros, a grass-eater, bears a single broad-based horn above its nose; it is now found only in the plains of the Assam region. This rhinoceros has the most grotesque appearance of any member of the group for its thick skin, studded with warty protuberances, is divided into a number of areas by deep folds, giving the animal the appearance of being covered with riveted plates of armor. A small relative with similar build is the Javan rhinoceros, which actually ranges from Bengal through the Malay region

The zebras are African forms which, except for their striped skins, are almost indistinguishable from horses and asses. The species shown is Grant's zebra. (Photograph courtesy New York Zoölogical Society.)

The Indian rhinoceros. A young female in which the horn is not developed, but its swollen bony base is visible over the nose. The Indian and Javan rhinoceroses are one-horned and have a folded skin resembling armor; other living rhinos of Africa and Sumatra are two-horned and have a smoother skin. (Photograph courtesy New York Zoölogical Society.)

An extinct running rhinoceros, *Hyracodon* of the Oligocene. Early rhinoceroses were horn-less, often rather slimly built, and comparable to their cousins the horses in general appearance. (A restoration by Charles R. Knight; photograph courtesy American Museum of Natural History, New York.)

The largest land animal. *Baluchitherium*, a giant, hornless, fossil rhinoceros of the Oligocene of Asia, is the largest animal known which was a land dweller. It is, of course, exceeded in size by some whales, and some of the sauropod dinosaurs (as *Brontosaurus*) were larger but were obviously not true land forms. *Baluchitherium* specimens vary somewhat in size; the largest known individual appears to have been about 17 feet high at the shoulder and 27 feet in length. The skull was about 5 feet long but was (as may be seen from the restoration) rather small for the animal. (Restoration by Mrs. E. R. Fulda; photographs courtesy American Museum of Natural History, New York.)

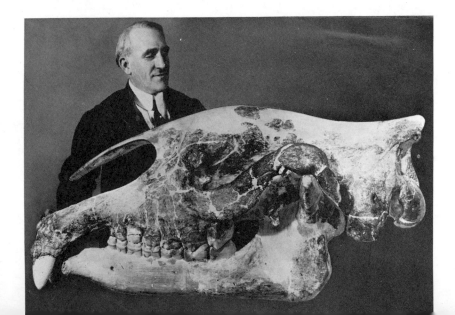

South Dakota in early Oligocene days. East of the Black Hills of South Dakota, in the region known as the Mauvaises Terres, or Big Bad Lands, the rocks and clays exposed by erosion have yielded a wealth of specimens which have made the region one of the most famous fossil-collecting regions of the world. Above are shown a few of the most common finds. The titanotheres in great measure deserve their name of "titanic mammals." Large individuals might have a length of about 15 feet, a height at the shoulder of 8 feet. They were related to the horses and rhinoceroses and tended to have the ponderous build of the latter. The peculiar horns, however, are quite unlike those of rhinoceroses and were massive bony structures. The titanotheres appear to have been unprogressive in both teeth and brains and, although very common at the time of this imaginary scene, vanished completely a short time later. *Right*, a pair of archaic carnivores (*Hyaenodon*). These may be suspected of having preyed on the titanothere herds by attacking the "calves." *Left*, specimens of a large land tortoise which frequently grew to a yard or more in length of shell; its remains are extremely common in the Bad Lands. (From a mural by Charles R. Knight; photograph courtesy Chicago Natural History Museum.)

The western plains in early Miocene days, about the middle of the Age of Mammals. In the foreground a group of giant hogs (*Dinohyus*), one of which is digging up roots for food (a habit suggested by the signs of wear on their teeth). These forms are but distantly related to the true hogs. An average individual would have had a height of about 6 feet. *Right,* a pair of chalicotheres (*Moropus*), ungainly odd-toed "ungulates," which despite their relationships to horses, etc., have powerful claws rather than hoofs. The claws indicate that they may have dug for tubers; on the other hand, the long legs and neck suggest that they browsed from tree branches as shown in the painting. *Left,* a pair of rhinoceroses (*Diceratherium*), peculiar in that the horns were arranged side by side on the nose. Behind them, contemporary small three-toed horses (*Parahippus*) and in the background a group of slender gazelle camels (*Stenomylus*). Remains of animals of this age are abundant in northwestern Nebraska, particularly at the famous Agate Springs quarry on the Niobrara River. (From a mural by Charles R. Knight; by permission of the Chicago Natural History Museum.)

to the larger East Indian islands. The Sumatran rhinoceros has the same range, except that it does not reach Java; it differs not merely in having a smoother skin but in having a small horn on the forehead in addition to that on the nose.

Two further rhinoceroses are present in southern and eastern Africa, both now much restricted in numbers. They are related forms and, like the Sumatran species, have a relatively smooth skin and two horns (the nasal horn may be long and sharply pointed). There are two species, termed, from their skin color, the black and the white rhinoceroses (the latter actually a rather dirty gray). These are big fellows, particularly the white species, which are the largest living land mammals except the elephants, reaching a length of fifteen feet and presumably having a live weight of several tons. The black rhinoceros is a browser; the white form, a grazer. Despite their awkward appearance, the rhinoceroses can travel rapidly, with a swinging trot which may break into a gallop; the white rhinoceros is said to be especially swift.

Titanotheres.—An interesting fossil group of odd-toed forms is that of the titanotheres ("titanic mammals"). These began in the Eocene with forms of somewhat horselike appearance. But there was a rapid tendency toward extremely large size and a ponderous elephantine build (four and three toes were retained on the front and hind limbs, respectively, of these forms to the end of their history). Slow of speed, these big herbivores seem to have been a source of prey for the larger carnivores of their day, and in relation to their needs for defense we find that in a number of distinct types of titanotheres paired bony horns appeared over the nasal region. These forms were almost entirely confined to North America. At the beginning of the Oligocene, giant titanotheres were the commonest of mammals in the western part of this country, and fossil remains of them are very abundant in the lower levels of South Dakota Bad Lands. But, following this climax in development of size and numbers of the group, they disappeared abruptly from the fossil record. A possible reason for extinction is the fact that the teeth of these forms were suitable only for the softest of plant food; a slight change in vegetation may have been enough to cause their undoing.

Chalicotheres.—A final group of odd-toed ungulates is that of the chalicotheres, now quite extinct also but not uncommon during the Tertiary. In general structure these forms were not dissimilar to the horses, while their teeth were quite like those of titanotheres. On both counts we must place them in the group of odd-toed hoofed mammals. But their feet are of such a character as to belie the name, for the toes terminated not in hoofs but in huge claws. So unexpected is such a feature in an ungulate group that it was long argued that we could not include them among the odd-toed forms. But belong here they do. Seemingly, the claws were an adaptation for digging out roots and tubers which may have formed a major part of their diet.

All in all, the odd-toed ungulates have not proved a success. In the early days of mammalian history they were numerous and widespread.

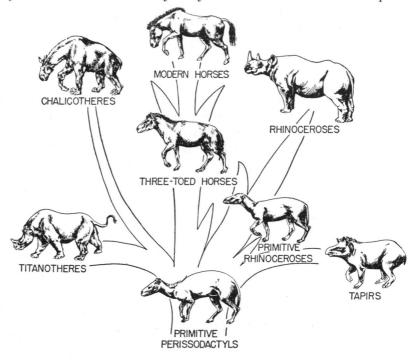

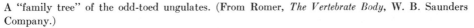

A "family tree" of the odd-toed ungulates. (From Romer, *The Vertebrate Body*, W. B. Saunders Company.)

But today two of the five groups are entirely extinct, and the tapirs and rhinoceroses are greatly reduced in numbers. The horse family has held its own; but even here wild forms are none too common, and the advance of machinery is making inroads on the numbers of our useful domestic horse.

EVEN-TOED UNGULATES

Much more successful, in modern times, have been the even-toed ungulates or artiodactyls. These forms were not as common as their odd-toed rivals in the earliest days of ungulate development, but they have become increasingly prominent during the course of the Age of Mammals and seem to be at the peak of their development at the present time. To this group belong the common large food animals and a host of wild types— pigs and peccaries, the hippopotami, camels, deer, giraffes, cattle, sheep, goats, and antelopes.

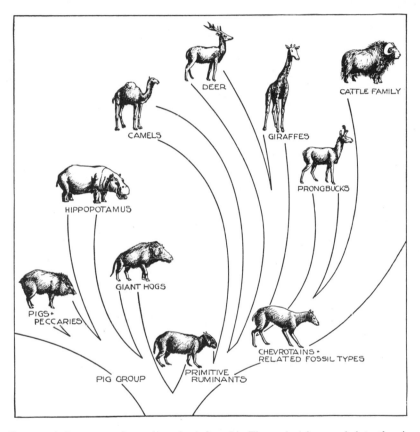

A family tree of the even-toed ungulates (artiodactyls). The major cleavage is into the pig group (*left*) and the cud-chewers, or ruminants (*right*). Among the latter the camels appear to have diverged at an early date.

Toe reduction in even-toed forms.—To gain an idea of the type of toe reduction encountered in the even-toed forms, try again the "experiment" of hand-lifting mentioned previously for odd-toed types. After the thumb has lifted, the hand is in a four-toed stage, as was the case in many early members of the present group and is still characteristic of the pigs and certain other living forms. If, with four fingers touching, these be shifted about until they rest comfortably on the table, it will be found that the two central ones (third and fourth) project somewhat in advance of the others, and, if the hand is lifted farther, these two alone touch, and the second and fifth are raised from the table. It is this type of toe reduction that has taken place in the higher even-toed forms; the cloven hoof of the sheep or cow in reality consists of two closely applied toes, the third and fourth.

Another characteristic of the artiodactyls which is even more distinctive is to be seen in the skeleton of the hind foot. In all mammals the proximal bones here are the heel bone (calcaneum), with its backward projection for

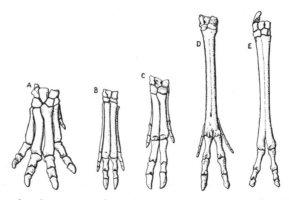

The right front feet of various even-toed ungulates. *A*, an oreodon; *B*, a primitive ruminant; *C*, a peccary; *D*, a primitive deer; *E*, a camel. The "thumb" was rapidly lost, and the foot tended to become four toed, with toes *2* and *3* more prominent. The lateral toes become reduced to "dew claws" or disappear, leaving the two center toes as a "cloven hoof." The two bones supporting these toes (metapodials) fuse in most cud-chewers to form a cannon bone.

attachment of the calf muscles, and a second bone, the astragalus. This forms the main connection between leg and foot, and its upper end is rounded, frequently with a pulley-like surface, to afford free movement of the foot on the leg. In artiodactyls alone among mammals, the lower end of the bone as well as the upper has a surface of this sort, rendering members of this group unusually agile in running and bounding movements of the hind legs.

The first even-toed types appeared in the early Eocene, but it was not until the close of that epoch that they became prominent members of animal society. From the first there seems to have been a distinct cleavage into two groups, the swine and their relatives, on the one hand, and the cud-chewers, or ruminants, on the other.

Piglike artiodactyls.—The pigs and their relatives are the most primitive of living even-toed forms in many ways. They are still four-toed animals (although the side toes are much reduced), and the limbs are little elongated; not much speed has developed. The diet of the pig is still of a rather primitive, omnivorous type, in contrast to their more advanced ruminant relatives; swine will eat anything from potatoes to rattlesnakes. The true pigs have large canine tusks; these usually curl outward and even upward at the side of the skull. The cheek teeth, in relation to their diet, are (like our own) low-crowned. A peculiar pig feature is the flat, disklike, inquisitive tip of the snout.

The most typical pigs are represented by the wild boar, still ranging from the wilder forest areas of Europe eastward to Asia and northern Africa and with close relatives in the oriental region and eastern Asia. The wild boar appears to be the species from which the domestic pig derives. Domestication appears to have taken place in southwestern Asia, and pig remains are common in Neolithic sites in this area. Curiously, this is the very region where the Jewish and Mohammedan dietary laws were formulated, forbidding the eating of pork. It is interesting to speculate as to whether this religious prohibition arose with any realization of the danger of infection by the trichina worm through eating imperfectly cooked swine flesh.

In Africa the bush pigs depart from the typical pig pattern to some degree, and in that continent is found the wart hog, a distinct pig type of hideous appearance; its name is due to the presence of large warty protuberances on the sides of the massive head. An aberrant form is the babirusa of Celebes. Here the tusks do not merely turn sideways but actually turn straight up and pierce through the top of the nose as a peculiar type of horn. This might be considered as a useful adaptation if the tusks stopped there, but they do not. They continue growing and curve back in a spiral in front of the eyes. Such a seemingly useless end product of a seemingly useful development constitutes one of the many puzzles with which the student of evolutionary processes must deal.

Closely related to the pigs are the peccaries, small and more lightly built forms in which (as a distinguishing feature) the tusks grow straight downward in normal fashion; the stomach is somewhat complex in build, as if foreshadowing the conditions seen in the ruminants (to be presently described). Peccaries passed their early history in North America, and even during the Ice Age they were numerous as far north as Michigan and Maryland. Today the peccaries are mainly found in South America, although one type ranges as far north as Texas. The peccaries are forest dwellers. The collared peccary, which reaches Texas, is a small and inoffensive form, which usually travels in family groups or small parties. A second species, ranging north only to Central America, is larger, relatively fierce, and often roams the country in large and dangerous herds.

A distant ally of the swine is the hippopotamus, a large water-living, piglike type now found only in tropical Africa but in the Pleistocene in regions as far removed as England and China. The hippopotamus is a large animal, with weights up to five tons or so, a barrel-like body, and

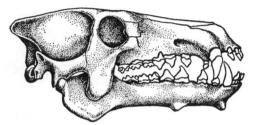

The skull of an Oligocene giant pig (*Archaeotherium*). The use (if any) of the bony knobs on the jaw and below the eye socket is unknown. This skull was about 1½ feet long; some were about twice this size. (After Scott.)

short but stout limbs. Its large and homely head has a broad muzzle of enormous gape, suitable for taking in large masses of the pulpy water plants on which it feeds. The animal is primarily a water dweller; its eyes, ears, and nose are situated high up on the skull so that the animal functions well when nearly fully submerged. The trunk and limbs are radically different from anything which one would expect in an aquatic form, giving warning to the paleontologist in his efforts to interpret the mode of life of extinct animals from their bodily build.

In the region of Liberia there is a distinct species of pygmy hippopotamus no larger than a pig when fully grown. In the Ice Age similar pygmy hippopotami lived on the Mediterranean islands—Sicily, Malta, and Cyprus. Island life has been seen to have an effect in selection for small size in various instances such as this, and we may note in passing that as contemporaries of these pygmy hippos on the Mediterranean islands there were pygmy elephants, the smallest of a size which would make it as reasonable a house pet as some of the larger races of dogs.

Hippopotami are unknown before very late Tertiary times, and their pedigree is uncertain. However, in middle Tertiary times there were abundant remains, mainly in Eurasia, of anthracotheres, which had a somewhat similar build except for a narrower snout. Possibly these were the ancestors of the hippopotami.

There are several additional extinct groups of animals which are allied to the swine. Most prominent of these fossil forms were the "giant hogs"

African rhinoceroses. The Indian rhinoceros is shown in an earlier plate; the two African species are figured here. The skin is relatively smooth, without the "armor-plate" effect of the Asiatic form, and two horns are present. The upper form is the black rhinoceros; the lower, the rare white rhinoceros, a larger form which may reach a height of $6\frac{1}{2}$ feet at the shoulder. The African forms are further distinguished by lacking incisor teeth, so that they pluck their vegetarian diet with the lips. Rhinoceroses are sometimes called "living fossils," the odd-toed group of mammals to which they belong having largely disappeared. (New York Zoölogical Society photographs.)

The wart hog, an ugly African relative of the pigs, with a large head, long muzzle, long tusks, and wartlike growths below and in front of the eyes. (New York Zoölogical Society photograph.)

Piglike ungulates. The swine are the characteristic members of a suborder of even-toed ungulates, the Suina, in which the cud-chewing habit of the ruminants is not developed and in which the feet have typically four well-developed toes.

In America the pigs are represented by the peccaries (*left, top*) not uncommon in the tropics and found as far north as Texas. They are distinguished by the fact that the upper tusks (canine teeth) are not curved upward as they are in the true pigs.

The babirusa (*center*) is a peculiar pig from the Celebes in which these upper tusks are highly specialized; they force their way upward through the roof of the snout and curve backward so far that their utility as weapons is much reduced.

The hippopotamus is a lubberly amphibious relative of the swine. In addition to the familiar large hippopotamus (seen asking for a peanut, *below*), there is a pygmy hippopotamus in Liberia. At the lower left is an adult female of this species and her offspring. (Photographs courtesy New York Zoölogical Society.)

(entelodonts) of the Oligocene and Miocene. These were animals of large size, one form being about as large as a buffalo. The giant hogs were rather more progressive than their porcine relatives in their limbs, which were fairly long and in which the toes had been reduced to two. Peculiar flanges of bone grew down from the sides of the skull and jaws, suggesting a repulsive appearance; that their dispositions were none too good is suggested by the fact that some fossil skulls show wounds apparently inflicted by the animal's own kind.

Primitive ruminants.—The piglike forms described above constitute one division of the even-toed ungulates; the other, and much more important, division of the group is that of the ruminants, or cud-chewers. These forms are, in contrast to typical swine, pure vegetarians. In all the living members of this group there is a complicated stomach, generally with four chambers. Food when first cropped enters, in a typical ruminant, the first two chambers; it is there reduced to a pulp. It is later, at leisure, returned to the mouth, the "cud" chewed, and then returned to further stomach compartments to continue its digestive travels. The side toes have tended to disappear fairly rapidly, and living members of the group have but two functional digits, although the lateral ones may be represented by vestigial

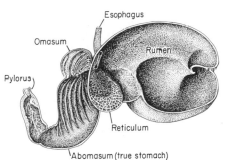

A section through the stomach of a cow, to show the various chambers. Food enters the first two chambers (rumen and reticulum); later, after the cud is returned to the mouth and chewed, it re-enters the stomach, proceeds through the other two chambers (omasum and abomasum), and then (through the pylorus) goes into the intestine.

"dew claws." In the pigs the two long bones of the foot which support the toes remain more or less separate; in ruminants the two are usually fused into a single structure termed a "cannon bone." In the earlier periods of the Tertiary there was a great abundance of primitive ruminants, ancestral as a group to later forms but most of them representing sterile side branches of the ruminant stock. Fairly characteristic of these early forms were the oreodonts, which, judging by the abundance of their remains in the White River Bad Lands of South Dakota, must have swarmed the western plains in enormous numbers in Oligocene times. The oreodonts are often called, for want of a better name, "ruminating swine," for the limbs were rather swinelike, but the teeth were similar to those of living cud-chewers. The body was stockily built, the limbs short and with four good toes. The teeth, on the other hand, show a pattern in which each cusp had a crescent-like shape—a feature repeated in all modern ruminants.

Camels.—Even while the oreodonts and other primitive cud-chewers flourished, higher ruminant groups were emerging. The earliest group to develop—and the one most distinct from other ruminants—is that of the camels and llamas. The camels, as we know them today, are Old World desert dwellers of large size, with a hump (or two) as a reserve food supply on their backs. There are but two divergent toes on the long limbs, bearing a pad beneath them for better support on soft sands. There are two quite distinct living species, both domesticated. The dromedary, the larger of the two, with but a single hump, is the basic element in the culture of the nomadic Arab and is widely used in the desert regions of southwest Asia and northern Africa. The bactrian camel, with two humps, shorter legs, a stockier build, and a more hairy coat, is better adapted to colder, rocky, and hilly country and is found throughout central Asia to western China and southern Siberia. Neither type is known in a wild state, and, except for a few Pleistocene remains, there are no fossil camels in the Old World.

The llamas and related South American forms are quite different in appearance; they are much smaller mountain and plains dwellers, lacking a hump and covered with a thick coat of wool. They are primarily natives of the Andean highlands. There are two wild species, the little vicuna, restricted essentially to the high mountains of Peru, and the larger guanaco, which ranges south to the islands of Tierra del Fuego and is common in the temperate plains of Patagonia. Few native American animals were tamed by the Indians, but the relatively highly cultured inhabitants of the central Andes had two domestic beasts derived from the guanaco. The true llama was primarily used as a beast of burden or a riding animal but also formed a source of meat and wool and is still important in the economy of the natives of the highland country. The alpaca is a considerably smaller animal, bred for the sake of its long fine wool.

Despite the differences in their appearance and habitats, camels and the llama group are closely related. The clue to their distribution lies in the fact that North America was their original home. The little early camels of the Oligocene and Miocene were remarkable for the rapidity with which the side toes were lost and the limbs lengthened. Presumably the hump is a very recent acquirement in camels; the older members of the

A chevrotain or "mouse deer"; tiny Old World tropical animals which are ruminants, but of a primitive sort, close to the ancestry of more advanced cud-chewers. The form shown here (*Tragulus*) is from the Malay region; a second genus is present in Africa. (This is a stuffed specimen; couldn't find a photograph of a live or better-looking one.)

Camels, although once natives of North America, are now inhabitants of the Old World. *Right*, the two humped bactrian camel; *lower right*, the dromedary of Arabia. (Photographs courtesy New York Zoölogical Society.)

The llama of the Andes (*below*) is a humpless member of the camel group, descended from North American ancestors. (Photograph courtesy American Museum of Natural History, New York.)

The annual replacement of the antlers of deer. A series of photographs of an American elk or wapiti (*Cervus canadensis*) at various seasons of the year. *Upper left*, winter: The antlers developed the previous year are still present. *Upper right*, early spring: The antlers have been lost by resorption of bone at the base (the winter coat is also being shed). *Lower left*, late spring: The new antlers are growing rapidly. *Lower right*, summer: The antlers are fully grown but still covered with skin (velvet), which will presently dry up and be rubbed off. (Photographs courtesy New York Zoölogical Society.)

The "Irish elk," named from the fact that the best remains are from Irish bog deposits. This extinct European deer had the most ponderous antlers of any known deer. Although these would seem to be a handicap, the animal was abundant in Europe in the Ice Age. (From a mural by Charles R. Knight; photograph courtesy Chicago Natural History Museum.)

family appear to have been plains and glade dwellers rather than desert types. Among the Tertiary camels of North America were a number of distinct types, one small form having the slender build of a gazelle, others developing not only long legs but a long neck similar to that of giraffes.

By Pleistocene times the llamas had gained their present home in South America, and true camels had migrated to Asia. Camels, however, persisted in the Southwest of this country until comparatively late times. Why they became extinct is, as in the case of the horses, a mystery, for camels were found to be well adapted to the southwestern deserts when reintroduced there by man.

Chevrotains.—The most progressive and successful artiodactyls, such as the deer and cattle families, are generally set off from the rest of the even-toed ungulates as the Pecora.[1] They are characterized by such features as the usual presence of horns or antlers, the complete fusion of the upper bones of the feet to form cannon bones, the incomplete nature of the little side toes, the full development of the complicated stomach described earlier, and the loss of the incisor teeth from the upper jaw (these animals do not bite off their grassy or leafy food but tear it off, gripping it between upper lips and lower teeth). But before describing

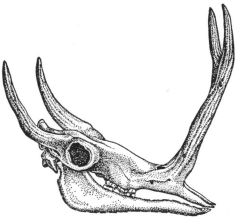

The skull of a fossil ruminant (*Synthetoceras*) from the Pliocene of America. Descended from the primitive mouse-deer stock, this form has acquired "horns," as have other advanced types, but has developed most prominently a huge fused pair above the nostrils. (After Stirton.)

the typical pecorans, we must mention some primitive relatives of theirs, the chevrotains, or mouse deer, of the Old World tropics. These are tiny, timid forest browsers, none much bigger than a hare or rabbit. They have no horns or antlers, but their sharp upper canine teeth are elongated as weapons, and they are more primitive than the pecorans proper in such technical features as the less complete fusion of the cannon bones and the retention of complete, if

---

[1] This name is merely a Latin one referring to cattle; our term "pecuniary" is also derived from it, owing to the fact that the older Romans, like the modern Kaffirs, measured their wealth by the number of cows they owned.

tiny, side toes. Several species are present in the Malay region, and another form survives on the west coast of Africa.

These mouse deer are living relics of a group which flourished greatly during Oligocene times. Some of these older forms were very similar to the chevrotains in size and structure. Others were beginning to branch off into various lines of specialization. Certain of these branches are extinct; others have led to the existing families of higher ruminants described below. In most cases there was a tendency toward increase in size; and while a number of early forms had canines as defensive weapons, most of the advanced types tended to develop horns of some sort or other. Of the extinct groups we may mention as an example one in which several pairs of horns started to develop along the length of the head and in which the dominant pair came to be, in *Synthetoceras*, over the nostrils rather than in the more usual position on the forehead.

Deer.—In the typical pecoran families to which we now finally arrive, the cannon bones in the feet are always long and perfectly formed, the side toes never more than tiny vestiges, the ruminant stomach fully perfected—in other words, these ungulates have reached the full development of the locomotor and feeding characters which have made them the dominant hoofed animals of modern times. There are four families, represented by the deer, giraffe, American prongbuck, and the cattle tribe; the first two types are browsers, the last two grass-eaters; three are basically Old World types, only the prongbuck being a native American. Except for some of the smaller deer, all have horns or antlers of some sort.

The deer, or cervids, have been a comparatively conservative group as far as habits are concerned, for (with low-crowned teeth) they have mainly remained browsers and have rarely left the forests, which appear to have been the original home of the higher artiodactyls. They are essentially animals of the Temperate Zone; some are present in southern Asia, but none have reached tropical Africa, while, on the other hand, they have extended their range into the tundras and barren lands of the Far North.

A most obvious feature of deer structure is the development of antlers. We sometimes speak of a deer's horns; but, properly speaking, a horn should be covered, as is a cow's, with actual horn—a hardened skin material identical in nature with our fingernails. In the case of the deer there is no such structure; the antler consists of bone alone. During the antler's growth it is covered by soft, furry skin—velvet—but this dies and is rubbed off when growth is completed. Antlers differ from horns in two other features. A true horn is a single structure, although it may be variously twisted or coiled, whereas in many deer the antlers are divided into a considerable number of branches, or tines. A still greater difference is the fact that horns are permanent structures, but antlers are shed. Every year the antler breaks off at the base, or burr, and then grows anew to larger size and greater complexity. This seems a very wasteful feature when we think, for example, of the great amount of food which must have gone yearly to built up the gigantic antlers of the great extinct "Irish elk" (with a spread of antlers of as much as eleven feet, whose remains are common in Irish bogs).

Members of the deer family are highly varied in such features as size, antler development, and range, and some of the more prominent forms may be briefly mentioned. If we are to use the word "deer" in any narrow sense, the forms to which this term most properly applies are the red deer of Europe and its allies. This is a large form, with well-developed antlers; a distinctive feature of this animal and its closer relatives is the fact that near its base each antler sends a stout branch forward over the forehead— a "brow tine" (lacking, for example, in the typical American deer species and in many other forms). The red deer has many close relatives in the Old World, including, for example, the deer of the Indian region—the sambur and axis deer among others—and further species in eastern Asia. A second common European form related to the red deer but of smaller size is the fallow deer, a native of the Mediterranean region but introduced into more northern countries. One member of this brow-tined group reached America—the large wapiti, often misnamed "elk," of the west.

The one European deer that lacks the brow tine is the roe deer, a form of small size with small and simple three-tined antlers. All American deer likewise lack this brow development and seem to form a special subcategory of the family. Largest of native American forms are the common Virginia deer of the eastern United States and the related white-tailed and mule deer of the west, with well-developed, multi-tined antlers. The Pampas deer of southern South America is somewhat smaller, with simpler antler structure; still smaller are a number of other South American forms in which the antlers may be but simple spikes.

Ranging into cold northern areas are two large and distinctive cervid types. The true elk of the Old World and the moose of the New (the two forms are very similar) are the largest members of the family. They are definitely cold-weather forms, inhabiting the most northerly forest regions of both continents. The antlers are unique in that they project out sideways from the skull roof before turning upward and expanding broadly, with a shape like the palm and fingers of a hand. Still more northerly in distribution are the reindeer of northermost Europe and Siberia and the practically identical caribou of northern Canada. To some degree these animals are found in the northern fringes of the forests, but they are adapted as well to living on the lichens and other small plants of the barren grounds north of the last trees. The reindeer is the only pecoran apart from the bovids to be domesticated. The large antlers have a typical brow tine, but they are placed unusually far back on the head; quite in contrast to typical cervids, the antlers here are well developed in females as well as males.

Even from what has been said above regarding antlers in various forms, one may gain the suggestion that antler development is correlated with body size, and that the larger deer tend to have disproportionately large antlers, the small forms small and simple ones. This is very true. If we look over the range of cervid types, we find a whole series of small forms with antlers small to absent. The little roe deer of Europe, already mentioned, has small horns with but three tines; the muntjacs of India and China, somewhat smaller, have short two-tined horns; the tiny tufted deer of China and the pudua of the Chilean mountains (the latter no larger than a hare) have only short little spikes for antlers; the small water deer of China have no antlers at all. Instead, however, we find that in some of these forms the males have long upper canines as fighting weapons.

Absence of antlers was, of course, a characteristic of the ancestral ruminants, as it still is of the chevrotains today. In some of these small deer, however, absence of antlers may be a secondary affair, due to their having become reduced to pygmy size, with an accompanying reduction of antlers, from ancestral deer of larger dimensions. Most workers, however, believe that one last type of deer to be mentioned is really primitive in lacking antlers. This is the musk deer, ranging from the Himalayan region north through central Asia. The name derives from a belly gland whose product is the musk of commerce. This is a form about as large as the roe deer but with no trace of antlers whatever (canines are well developed) and with some anatomical features suggesting that they are representative of a truly ancestral "antler-less" deer type.

Giraffes.—The giraffe is a representative of a further group of higher ruminants. This beast, too, is a browsing form; but the giraffe is a dweller in the savannahs rather than the woods and, with limited amounts of trees to graze upon, has evolved a mechanism for getting the most for his money. The long front legs and exceedingly long neck enable it to reach high branches inaccessible to ordinary animals. We have said that higher ruminants all have tended to develop horns of some sort, and the giraffe is no exception, for he bears small, skin-covered bony prongs on the top of his skull.

The giraffe's long neck is an interesting development; but it has not involved the addition of a single bone to the animal's skeleton. The neck of an ordinary mammal contains seven vertebrae; that of the giraffe contains exactly the same number, each one, of course, greatly elongated. On the other hand, we may note that in the whale, in which there is practically no neck and the head is buttressed right in front of the shoulders, there are again just seven neck vertebrae, but each one is exceedingly short. Reptiles, as we have seen, have a great degree of variation in their neck skeletons; mammals, for some unknown reason, almost never depart from the primitive number of neck vertebrae.

Fossil relatives of the giraffes, similar to the living form in most respects, but with short necks, have long been known from Pliocene deposits. It was therefore of considerable interest when there was discovered early in the present century in the forests of the Belgian Congo the okapi, a living, short-necked giraffe relative.

The prongbuck.—The giraffes are entirely Old World types; the one higher ruminant peculiar to North America is the prongbuck of our western plains. This creature has much of the build and habits of the antelopes of the Old World but is quite a distinct type, as is shown by its "horns."

These structures consist of a bony core which is never shed and is covered with horn. In these respects the prongbuck agrees with the cattle tribe. But, on the other hand, the prongbuck does shed the horny covering of the horn, and the horn is forked. These characters suggest the deer. To which group does the prongbuck, then, belong? To neither, really; it is a native product, end form of an independent, purely American line of ruminant development of which there are many fossil representatives. The cattle tribe, as noted below, are a highly developed series of grazing artiodactyls which developed in the Old World during the later part of the Tertiary. Few varieties of the cattle group ever reached America, and the prongbuck family developed in parallel fashion as dwellers in the American grasslands. Not improbably the entrance of the bison into America during the Pleistocene may have been responsible for reduction of the prongbuck family in numbers and variety. Once close to extinction, the prongbuck is now flourishing modestly in the more arid regions of the West.

Cattle and related types.—A final and the largest group of higher ruminants is that of the bovids, including the cattle and their numerous relatives—antelopes, sheep, goats, and other familiar forms. These are mainly plains dwellers, with high-crowned teeth capable, as is the case with horses, of utilizing grass as a food supply. In almost all members of the group true horns have developed, simple although often much-curved structures with a permanent bony core at the center and tipped with hollow horn. Bovids were comparatively late in their development, for there were but few of them in the Miocene, and it is only toward the end of the Tertiary that they became numerous. But once started on their careers they have swept all before them and are now, in open country of all sorts, the dominant hoofed mammals—indeed, almost the only ones.

We shall begin our account with mention of members of the family which are as different as can be from the familiar cattle types—the duikers, tiny animals of the African forest, which stand little over a foot in height and have very tiny horns, which may be almost hidden from view between their ears. These little fellows seem to stand in much the same relation to the larger bovids that some of the tiny small-horned or hornless deer do to their larger cervid relatives.

Next may be considered the gazelles and their relatives. The gazelles are small and very slender, graceful forms, mainly dwellers in the deserts of northern Africa and western Asia, generally of a concealing sandy color with ringed horns; these are usually gently curved in somewhat the shape of a lyre. Closely related are a number of other forms, also generally of small size, mainly African, such as the little dik-diks and klipspringer.

Related by technical characters to the gazelle group, according to students of the bovids, is the major Temperate Zone branch of the family, of which the sheep and goats are most characteristic. Connecting links between the group and more typical antelopes are furnished by the saiga antelope of the steppes of western Asia and a smaller relative from the high plateau of Tibet; these are the only Asiatic antelopes except for a few forms dwelling in the more hospitable climate of India. A readily identifiable if unimportant feature of the saiga is its peculiarly swollen muzzle. The remainder of the sheep-goat group are mainly dwellers in mountain country; the horns, generally ringed, usually curve backward, downward, and forward in spiral fashion. The sheep, of which there are a number of species, one even reaching western North America, are typically inhabitants of mountain meadows; the goats, of which there are a number of species from Asia and Europe, are, on the whole, high mountain

The giraffe is a ruminant related to both the deer and cattle groups but representing a family of its own. Like other members of the ruminants, it possesses "horns." These, however, are easily overlooked, for they are present only as a pair of short hair-covered processes. Its long neck, developed for browsing upon trees, has, strangely enough, only the regulation of 7 neck vertebrae of all mammals. This specimen, photographed in its native Africa, stood 17 feet high. (Photograph courtesy American Museum of Natural History, New York.)

The okapi, a short-necked relative of the giraffe. Such giraffids have been known only as fossils until a few decades ago when this living form was discovered in the Belgian Congo. (Photograph from mounted group, American Museum of Natural History, New York.)

The prong-buck or American "antelope." This western-plains animal is the only survivor of an American ruminant family common in fossil form. It is characterized by its peculiar "horns," which include a simple bony core which is not shed and a forked horny sheath which is shed annually. (Photograph courtesy New York Zoölogical Society.)

Bovids. The family *Bovidae* includes the cattle and related forms. Besides having in common high crowned teeth for eating grass, these ruminants possess true horns—that is, simple (although usually curved) bony projections covered by a horny sheath and never shed. The family is a group which developed in the Old World where, in addition to wild sheep, goats, and cattle, there is a great host of varied forms lumped together in popular usage as antelopes (the sable antelope of Africa is shown above at the left). Only three bovid types, illustrated here, reached America. *Upper right*, the sure-footed mountain goat (photographed in his native haunts in the northern Rockies by John M. Phillips). This form is not a true goat but belongs to a separate genus (*Oreamnos*). *Left center*, the mountain sheep which inhabits mountain meadows. To represent the bison, we have shown (*lower left*) not the familiar American species but the European one, now close to extinction. (Upper two photographs courtesy American Museum of Natural History, New York; lower photographs courtesy New York Zoological Society.)

forms. Both genera are represented in the Near East, where they appear to have made up, with cow and pig, Neolithic man's series of domesticated ungulates. Related to the sheep and goats but definitely distinct are the European chamois, the goral and serow of the Asian mountains, and the Rocky Mountain "goat." A final large and ungainly member of this group of cold-weather bovids is the musk-ox, with a circumpolar distribution in the barren Arctic.

Returning to the tropics, we may mention a group of "middle of the road" antelopes, not particularly related to either the sheep-goat group, on the one hand, or the cattle, on the other. Most are of good size, including in their number a variety of forms such as the oryx, addax, gnu, and hartebeest; they make up a fair proportion of the African antelope assemblage. Finally, several antelopes such as the kudu and eland of Africa and the nilghai of India, mainly large and heavily built animals with horns of a rather "twisted" appearance, show an approach to the cattle.

In the varied members of the cattle group we find forms of large size and powerful build, with horns which, in contrast to most antelopes, extend out laterally from a bony boss atop the head before curving upward. Some are tropical in natural range, but others (like the members of the sheep-goat group) are tolerant of chilly climates. The domestic cattle took their origin from a wild form which was widespread in Europe and Asia in Pleistocene times. Coexisting with this animal was a much larger wild ox, the urus ("ur-ochs" of the ancient Germans), which survived into medieval times; some of its blood may persist in a few preserved herds of European wild cattle today.

The yak of Tibet is a cattle type whose long hair aids it in surviving in the cold of the Tibetan plateau. The typical domestic cattle of India are a distinct species, the obvious differences being a conspicuous hump on the back and a large dewlap beneath the neck. In recent years these cattle, termed "brahmas," have been introduced into America where they have proved well adapted to resist the hot weather of the Gulf States. The Indian and Malay regions are the homes of a series of cattle-like forms, including the gaur, the little anoa of the Celebes, the gayal, the banteng, and the true buffalo. The first two are wild; the gaur is the largest of all the cattle tribe, the anoa the smallest. The gayal is semidomesticated in India, the banteng is domesticated in the Malay region; and the buffalo is a common Indian domestic form. Buffaloes of a different type are present in Africa; they are not, however, domesticated, and the Cape buffalo is an aggressive foe to the hunter.

A final member of the cattle group is the bison (often but incorrectly termed "buffalo"). In the Ice Age the bison ranged in great numbers over the European plain and during the Pleistocene made its way to North America, where it became likewise exceedingly abundant in the western grasslands. Man has, however, brought it close to extermination. By the opening of the present century there remained of the European form only a relatively few individuals inclosed in preserves, and the American form likewise has survived only in protected herds.

The history of the cattle-antelope group has been almost entirely confined to the Old World. It was there that these animals had their origin, and it is there that almost all the living wild forms are found today. Only four members of the family have successfully invaded North America—the bison, the mountain sheep and mountain "goat" of the Rockies, and, in the Far North, the musk ox. This lack of representation is seemingly due to the fact that the family developed late in the Age of Mammals. They were, for the most part, tropical plains dwellers, and only forms such as those above, which can stand cold climates or inhabit mountains, were able to make the (by then) difficult passage from Asia via Alaska.

The even-toed ungulates are, in contrast to the perissodactyls, a flourishing group. To what has their success been due? Perhaps not so much to their good teeth—although they are good—or to good brains—we are the animals that boast of that feature—or to good feet—although there are some clever mechanical adaptations in the limbs of this group. Perhaps (since it is the higher ruminants that compose the greater part of the modern artiodactyl population) the development of the stomach may have been the real cause of artiodactyl success.

Which are more important—brains or stomachs? Sometimes, when indigestion hits us, we may think of a good stomach as the really essential thing in life. But, when we sit down to a meal of beef or lamb or pork, we may reflect that brains do seem, on the whole, to have won.

Antelopes. These abundant, successful, and attractive Old-World tropical animals include four score or more species; on this page is given a sample. *Directly below*, the graceful desert-dwelling dama gazelle; *bottom*, the large, short-horned nilghai; *right, from the top down*, the tiny duikerbok, the white-bearded gnu, the kudu, with spirally-curved horns, and the waterbuck. (New York Zoölogical Society photographs.)

An Asiatic elephant, from a group in the American Museum of Natural History, New York, mounted by Louis Paul Jonas; a fine example of the art of the modern taxidermist. *Below*, a restoration of the American mastodon of the Pleistocene, a form in which the lower jaw and lower tusks have been reduced, very much as in the true elephants. (From a mural by Charles R. Knight; photograph courtesy Chicago Natural History Museum.)

CHAPTER 14

More Ungulates

While the odd- and even-toed groups are the most important of ungulate stocks, they are but two of a dozen or more orders, living and extinct, which are grouped as hoofed mammals. In the present chapter we shall consider two series of ungulates: first, the subungulates, including conies, elephants, and sea cows, and then a varied assemblage of forms, now extinct, which once inhabited South America.

SUBUNGULATES

Under the heading of subungulates are included the conies, small rodent-like creatures from Africa and Syria, and the proboscidians, including the elephants and their extinct relatives, and the sea cows. These make up a very diverse assemblage, a seemingly incongruous jumble of land and sea forms, large and small types; but early representatives of the varied groups show fundamental similarities which strongly suggest a common origin; and the fact that the earliest fossil forms are found in Africa suggests that that continent was their common ancestral home.

Conies.—Most primitive of subungulates in many respects are the conies, or dassies (*Hyrax*, etc.), of Africa and southwestern Asia. The word "coney" was an old English term for rabbit (Coney Island, e.g., means Rabbit Island) but has been used in the English Bible as the best equivalent for the animals now under discussion. The comparison is not inappropriate, for the dassies have chisel-like front teeth and gnawing habits similar to those of the rodents. The comparison would be better, however, with the woodchuck rather than the rabbit as regards general appearance and habits, for the ears are small and the hind legs are not greatly elongated. Most conies are ground dwellers in rocky country, but one group of tropical African species consists of tree dwellers.

But the rodent resemblances are superficial only. The cheek teeth are rather more like those of a miniature rhinoceros than anything else; there are little hoofs rather than claws, and the general build of these odd mammals is quite different from that of any common group; they have obviously had a long independent history. This conclusion is borne out by the fossil record. The oldest African land animals of the Age of Mammals are found in the Fayum region of Egypt, in beds of late Eocene and early Oligocene age. In these beds conies are already numerous and varied, one form being as large as a lion. They have never been discovered as fossils outside of Africa and the eastern Mediterranean region and apparently are a group characteristically African in origin.

Elephants.—The elephants are the only living representatives of the proboscidians, mammals with a trunk. These ponderous fellows are the largest living land creatures; of living animals only a few of the whales are larger, and in the past they have been exceeded in bulk only by some of the dinosaurs and one giant rhinoceros. Their great weight has resulted (as in all heavy land animals) in a pillar-like construction of the limbs, the straight legs with long proximal segments, ending in a short but broad, padded foot. The skull is very short and high. Of the originally numerous front teeth there remain but two upper incisors which extend forward as the two long curved tusks; above, the nose is extended to form the long flexible proboscis. The cheek teeth are of a very curious nature. All the molars of an ordinary primitive mammal are present at some stage of life and each one is a very large high-crowned structure with numerous cross-ridges which can undergo considerable wear. The jaw, however, is very short, and the elephants have evolved an odd type of tooth replacement whereby only four of the dozen molars are in place at once, one in each half of each jaw. The teeth are formed one after another in the back part of the jaws and gradually swing around into position as the preceding tooth is ground down to the roots and discarded.

There are but two types of living elephants, one in southern Asia, the other in Africa. In both the thick leathery skin is practically devoid of hair. In popular lore the elephant is noted for intelligence and long memory—"The elephant never forgets." The former attribute has been debated, but recent psychological tests do confirm beliefs as to the animal's retentive memory. Both forms typically are found in herds, inhabiting forested regions. The former is generally termed the Indian elephant, but its range extends through the wilder regions from India to the larger islands of the East Indies. It is readily distinguished from the African form by its relatively small ears and less domed forehead; structurally, its teeth are more complex and high-crowned. The Indian elephant is frequently captured and is not merely used for transportation but employed as a laborer in such activities as lumbering and construction work. In size, a

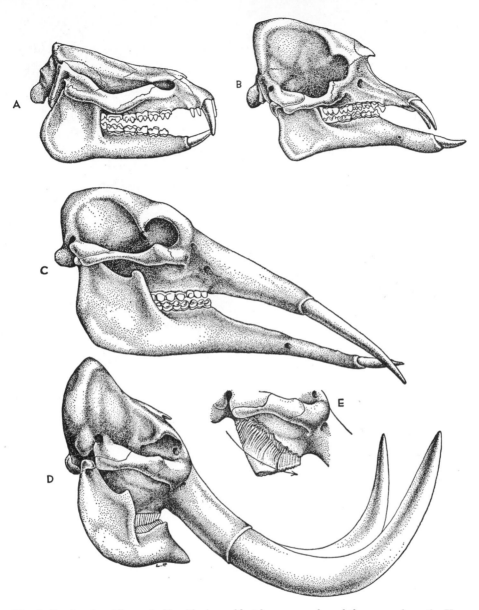

The skulls of proboscidians. *A*, *Moeritherium*, oldest known member of the group, from the Upper Eocene and Lower Oligocene of Egypt. The skull is fairly normally built; the upper and lower tusks are simple, somewhat elongated incisor teeth. *B*, *Phiomia*, an Oligocene mastodon with small upper and lower tusks; the jaws are elongated. *C*, *Trilophodon*, a Miocene long-jawed, four-tusked mastodon. In later mammoths and elephants, as the woolly mammoth (*D*), the lower jaw shortens and the lower tusks disappear. *E*, part of the side of the upper jaw of an elephant. The bone over the teeth is removed to show how the teeth swing down into place as earlier ones are ground down and used up. (*A*, *B*, after Andrews.)

typical form will run to a nine-foot shoulder height, but exceptional males may be a third again as large. The African elephant, with large ears, a very high skull, and relatively simple cheek teeth, was once found over nearly every part of that continent but is now relatively rare or absent in many regions. It is a somewhat larger animal than its Asiatic cousin. In modern times no attempt to domesticate the African form has been made. In ancient days, however, it was tamed in Barbary, and a number of these elephants accompanied the Carthaginian Hannibal in his invasion of Italy.

In the Pleistocene epoch just behind us, elephants inhabited all the northern continents and were present in great variety and abundance; to these extinct elephants the term "mammoth" is usually applied. Best known of these extinct forms is the northern woolly mammoth, a dweller in cold climates in both Eurasia and North America. Fossil remains of these forms are numerous, and in Siberia many woolly mammoths have been found preserved in natural cold storage, imbedded in the frozen tundra. So common are mammoth finds in Siberia that there is a flourishing local trade in fossil ivory, and specimens have been excavated with hair and flesh still present; some years ago guests at a scientific banquet in Russia were served with portions (very small) of mammoth steak!

In both Old and New worlds there were extinct mammoths of other types inhabiting temperate and tropical regions. These creatures, to judge by their remains, were quite common, and most are known to have lived beyond the time of recession of the Pleistocene ice sheets. Why they suddenly disappeared over most of their former range is as great a problem as that connected with the extinction in America of horses, camels, and other forms.

The oldest proboscidians.—The elephants are but a final stage in a long history of proboscidian evolution. The most primitive ancestor of the group is found in the same Fayum beds that contain the oldest remains of the conies. *Moeritherium* was no bigger than a good-sized hog and showed few definite elephantine characters. There was apparently no trunk, although there might have been a piglike type of snout; all the front teeth were present, although one upper and one lower pair were considerably larger than their neighbors; all the grinders were present in normal fashion and were quite simple in construction. In fact, so primitive was this beast, that it has been debated whether it really belongs to the elephant group. A careful analysis shows that a majority of its characteristics are in agreement with those of proboscidians, but it also had many features of the sea cows and some of the conies. *Moeritherium* was apparently close to

The banks of a Kansas river in the early Pliocene. Proboscidians, although not native to North America, had at that date arrived in this continent and are represented by a primitive mastodon (*Trilophodon*). In these forms both upper and lower jaws were long, so that there was probably but a short length of free trunk. The tusks, too, were short and were present in lower as well as upper jaws.

Left, representatives of a short-legged rhinoceros (*Teleoceras*) which appears to have had amphibious habits and rather resembled the hippopotamus in general proportions. Only a tiny horn appears to have been present. (From a mural by Charles R. Knight; by permission of the Chicago Natural History Museum.)

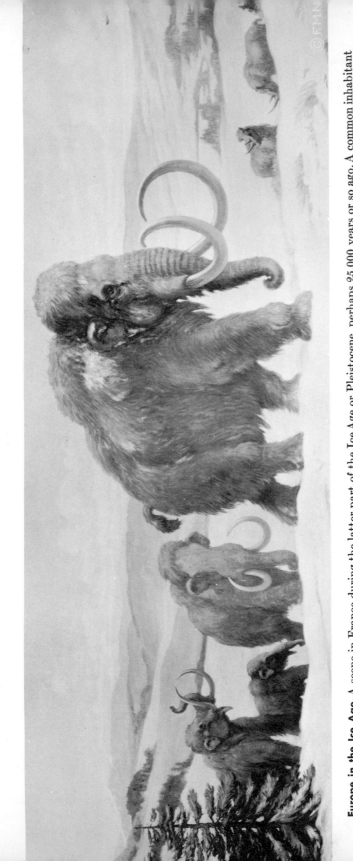

Europe in the Ice Age. A scene in France during the latter part of the Ice Age or Pleistocene, perhaps 25,000 years or so ago. A common inhabitant of all northern regions of both Eurasia and North America at that time was the woolly mammoth. This large elephant with shaggy hair and a domelike head is known not only from its bones but from paintings and engravings by cave men and from frozen corpses found in the Siberian tundra.

Right, the woolly rhinoceros of the colder regions of Eurasia at the time, a form well protected from the cold by its covering of thick hair, which contrasts strongly with the nearly naked skin of typical rhinoceroses. (From a mural by Charles R. Knight; photograph courtesy Chicago Natural History Museum.)

the common stem from which all three of these diverse groups have sprung.

Mastodons.—In the upper levels of these same Egyptian fossil beds are more progressive proboscidians, the earliest of the mastodons, which flourished from Oligocene to Pleistocene times. In these forms there were, to begin with, two pairs of long incisor teeth which gradually lengthened into tusks. All the older mastodons had tusks in the lower as well as the upper jaws. These tusks were at first straight, at the end of long bony jaws, and with a long nose tube which was probably flexible only at its tip.

Gradually, however, the jaws shortened, the lower tusks were reduced, the upper ones became free at their bases and much curved, and the trunk

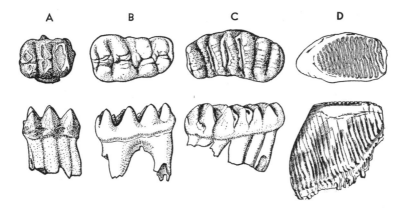

Teeth of mastodons (*A, B*), a primitive elephant (*C*), and the woolly mammoth (*D*), all much reduced. Crown views *above*, side views *below*. The typical mastodon tooth has but a few crests and is low crowned; the elephant tooth is high crowned, with numerous closely crowded transverse ridges.

became the highly flexible structure it is today; these features led to the head structure seen in the last of the mastodons and in the elephants. The elephants alone of animals other than man have a chin; but the elephant chin is not (as is ours) a new structure but merely the stump of the jaw which once bore the lower tusks.

The grinding teeth of the early mastodons were of simple construction—low-crowned, with a few cross-ridges, and all in place at the same time. But with increase in size or a trend toward grass rather than softer vegetation as a food supply, we find that the teeth (particularly in the line leading to the elephants) became higher crowned, increased the number of cross-ridges, and gradually tended toward the peculiar elephant type of tooth succession.

Although mastodons originated in Africa, they spread rapidly over the earth and in the Miocene and Pliocene were found all over the northern land masses and in the Pleistocene even reached South America. In the Old World they seem to have died out near the beginning of the Pleistocene; but in this country they survived until within ten or twenty thousand years of modern times.

Sirenians.—The living sea cows include only the dugong and manatee. These grotesque mammals are purely aquatic creatures which browse upon the vegetation of coastal waters but never come ashore; the hind legs are lost, the front ones have become steering flippers, and there is a well-developed transverse tail flap. The cheek teeth somewhat resemble those of primitive proboscidians, but the dentition is reduced and is supplemented by horny plates on palate and jaws and by a cropping mechanism formed by large fleshy lips. The name of the group recalls the alluring sirens of Greek legend. Even today in Aden one is invited in to see (at a price) a mermaid; but one should be forewarned, or the shock of seeing a rather ugly sea cow instead of a beauteous damsel may be too much.

The manatees are found on both coasts of the tropical part of the Atlantic Ocean. They are commonly found in shallow coastal waters, where there is an abundance of water vegetation, and frequently ascend the lower courses of such rivers as the Amazon and Orinoco. The dugong has its headquarters on the shores of the Indian Ocean; it reaches Australia to the east and Equatorial Africa to the southwest; it is not uncommon around Ceylon, but is most abundant in the Red Sea region.

A third sea cow—"Stellar's sea cow"—once existed in the North Pacific. A bit over two centuries ago the Danish explorer Bering conducted a Russian expedition eastward through and beyond the wild Kamchatka region to reach Alaska for the first time. On his return he was shipwrecked on the island and in the sea now bearing his name. Here his party was saved from starvation by finding enormous numbers of a gigantic sea cow, which was observed and described by his naturalist, Stellar. But within a few decades this spectacular animal, even then very restricted in its distribution, was exterminated by hunters.

These curious marine mammals are no relation to the whales (which are of flesh-eating origin), nor is there any reptilian parallel to them, except for the seaweed-eating lizards of the Galápagos. Now restricted, they were widespread in Tertiary seas. Most of the oldest sea-cow remains come from Egypt, and these forms show many features suggesting a common ancestry with the conies and elephants. It seems absurd to include these

A sea cow swimming. This figure shows some of the specializations of these ungainly sirenians—the paddle-like front limbs, the absence of the hind legs, the horizontal tail, rather comparable to that of a whale. The form shown is the American manatee. (Photograph courtesy New York Zoölogical Society.)

Conies—mother and young. The conies (*Hyrax*) are animals of Africa and Syria which in habits (and also in their relatively small size) invite comparison with the rodents. In their structure, however, they are clearly quite another sort of creature and represent an archaic African stock of hoofed mammals from which both proboscidians and sirenians have arisen. (Photograph courtesy New York Zoölogical Society.)

African elephants (*below*). The African elephant belongs to a different genus from the Asiatic form. There are various anatomical differences; a readily observable feature is the contrast of the enormous ears of the African species with the more modest ones of its Indian cousin. (New York Zoölogical Society photograph.)

Rodents, or gnawing animals. Characteristic of this order are the two pairs of continuously growing, chisellike gnawing teeth well seen in the "portrait" below. A few representative forms are illustrated on this page. *Left*, a Brazilian tree porcupine, peculiar in the development of a prehensile tail. *Below this*, the familiar beaver. *Lower left*, a colony of prairie dogs, related to the squirrels and woodchucks. *Lower right*, the giant of all rodents, the capybara of South America (a relative of the guinea pigs) which grows to the size of a hog. (Photographs courtesy New York Zoölogical Society.)

marine and almost limbless animals in a group of "hoofed" mammals, but the resemblances in the older types are unmistakable.

Our fossil record in Africa is very incomplete, but as far as it goes it suggests that in the earliest part of the Age of Mammals that continent was for millions of years separated from the rest of the world. During that time there developed a peculiar native stock of archaic herbivorous animals which gradually diversified into the conies (as the least specialized representatives), the great proboscidians, and the ancestral sea cows. When, at about the end of the Eocene, interconnections with Asia were established and other animal types could enter Africa, the conies merely held their own, but the proboscidians migrated outward in triumph. The sea cows, too, migrated but, already far along in aquatic adaptations, radiated outward along marine rather than terrestrial routes. The main evolutionary center for the mammals has been the Northern Hemisphere; but Africa has been the point of origin of the odd subungulate types,

SOUTH AMERICAN UNGULATES

The southern continents appear to have been isolated regions for part or all of the Age of Mammals. Australia has remained disconnected and has evolved its own peculiar marsupial fauna; Africa was separated long enough to initiate the development of the subungulates but later became faunally similar to the northern continents. The history of South America is somewhat intermediate in nature. It was isolated at the dawn of the Age of Mammals and appears to have remained so until the close of the Tertiary, when connections were renewed. During its long period of isolation many curious mammalian types developed there which are now for the most part extinct. In a previous section we mentioned the flesh-eating marsupials of that continent; in the following chapter we shall speak of the curious edentates that developed here. Here we shall touch very briefly upon the development in South America of a number of groups of ungulates which are now entirely extinct but which flourished there during Tertiary times.

When South America became separated from North America, there were but few placental mammals which had then reached that continent. There were some forms ancestral to the armadillos and other edentates, perhaps some primitive monkeys and some rodents. The only other placental mammals which had penetrated were a series of primitive ungulates—small herbivorous creatures comparable to the archaic ungulate types of northern continents. No higher hoofed mammals, no true even- or odd-toed

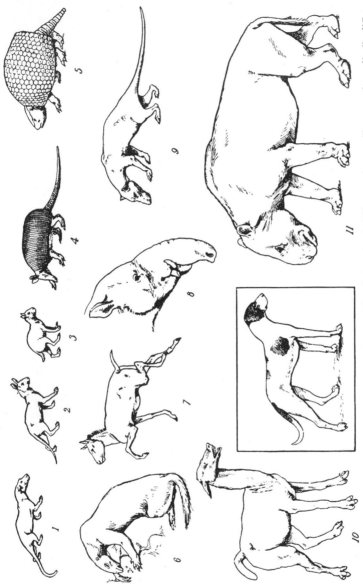

Some representative members of the mammalian fauna of South America in Miocene times to illustrate the radically different nature of the groups found there from those of other continents. A pointer dog is shown to give the scale. Representatives of most of these types survived until the Ice Age, when they had become much larger in size. The only flesh-eaters were marsupials, represented by numbers *1* and *9*, which were, respectively, forms rather comparable to weasel and fox. Rodents were absent earlier in South America, but, by this time, forms related to the guinea pig had gained entrance (*3*). The peculiar South American edentates were already present and diversified. Besides armadillos (*4*), large ground sloths (*6*) and the armored glyptodonts (*5*) were present. The remainder of the fauna consisted of ungulates of peculiar sorts. A group known as litopterns had produced such forms as a good imitation of a one-toed horse (*7*) and the three-toed form (*10*) of larger size and rather camel-like appearance. Another group of hoofed animals, known as notoungulates, had developed into a great variety of types ranging from little typotheres (*2*), which were rather rabbit-like in their adaptations, through the peculiar astrapotheres (*8*), with an elephant-like proboscis, to big toxodonts (*11*) which can be described only by comparing them to a blend of woodchuck and rhinoceros. (From Scott, *Land Mammals in the Western Hemisphere*, by permission of the Macmillan Co., publishers.)

forms, were able to enter until in the Pleistocene the land connection at Panama had become re-established. In the meantime the archaic "old settlers" had the field to themselves and, during the Tertiary, developed into a bewildering variety of hoofed types. It is difficult to describe them without going into a maze of technical details, for all of them are now extinct, and we cannot fairly compare them with existing groups from other continents. In an accompanying figure is shown a representative selection of Miocene mammals—both of these ungulate types and their native contemporaries; the later, Pleistocene representatives of these groups were, on the average, of much larger size.

Conspicuous and numerous among the ungulates were the toxodonts, or "bow-tooths," most of which tended to grow to large size. In general build they seem to have been somewhat like a cross between a rhinoceros and a hippopotamus. Closely related to them but very different in appearance were the little typotheres, with seemingly rodent-like habits and even, in some cases, long hopping hind legs which make a comparison with rabbits not inappropriate.

Another group (the litopterns) not particularly related to the last were South American parallels to the odd-toed ungulates of other continents. These forms were fairly orthodox ungulates and tended to reduce the toes to three. One group paralleled the horses very closely; the two lateral toes dwindled and vanished as in horses, leaving only splints to represent them. These are the only animals in the world except horses that ever attained a monodactyl condition.

Most of these South American ungulates, free from the attack of placental carnivores and free from the competition of other ungulate groups, flourished during the Tertiary and were still present in nearly full array in the Pleistocene. But then came their abrupt downfall. Land connections with North America became established. Sabertooths, wolves, and other carnivores against which they had no adequate means of defense entered the continent and feasted upon them; horses, tapirs, deer, and llamas, with more efficient ungulate adaptations, competed with these archaic hoofed mammals for feeding grounds. Between these upper and nether millstones the old South American ungulates disintegrated. Not one of this vast host of Tertiary forms has survived.

Their disappearance is the most spectacular of the many cases of extinction of animal types in the Pleistocene. In many other orders various forms became extinct or had their ranges much restricted; but here a whole section of several orders of the world's long-established mammalian population was absolutely wiped out.

CHAPTER 15

A Diversity of Mammals

An account of the varied mammalian forms remaining for consideration before our own group, the primates, is reached would, if at all adequate, require a volume in itself. In this chapter we can give merely a brief sketch of the other major types: of the rodents, most flourishing of all mammal types; of the hares and rabbits; of the bats, the only flying mammals; of the whales and porpoises, the most highly specialized of aquatic forms; of the curious edentates, so-called "toothless" mammals.

RODENTS

Among the rodents, or gnawing animals, are included the squirrels, beavers, rats and mice, porcupines, guinea pigs, and hosts of less-familiar forms. The rodents are without question the most successful of all living mammals. In the number of different types included in the order they exceed all other mammalian groups combined; they are found in almost every inhabitable land area of the globe and seem to thrive under almost any conditions. Their range of adaptations is a wide one. A majority of rodents are terrestrial and often burrowing types. No purely aquatic forms have developed, although the beaver and muskrat have progressed far in this direction. Others, such as the squirrels, are arboreal; and while there are no truly flying rodents, the "flying" squirrels have gone far along this evolutionary path. In their general bodily build the rodents are little specialized. It is their gnawing and chewing adaptations that particularly characterize the group.

The rodents are primarily vegetarian in their habits. There is an efficient series of grinding teeth in the cheeks. In the front of the mouth, separated from the cheek teeth by a wide gap, are the curving chisel-like gnawing teeth. Of these highly modified incisors there are but four—a pair in both upper and lower jaws. Subject to constant and hard use, the roots of these teeth grow continually and have their bases far back in the skull and jaw. So relentless is their growth that if, through accident or deformity, one tooth fails to meet its mate and be worn off in the normal fashion, it may keep growing outward, curve back into the mouth, prevent the jaws from closing, and eventually cause death by starvation. The cheek teeth are generally complex and efficient grinders, and they too frequently have a constant growth from open roots. Most mammals chew as man does, by an essentially straight up-and-down motion of the lower jaw. Rodents, in contrast, are able to really grind their food by a fore-and-aft jaw motion. The jaw muscles which effect this motion are powerful and complex; it is primarily through study of the variations in these muscles and the skull structures associated with them that we are able to classify them and arrange most members of this large and varied group into three suborders, named for typical representatives—the squirrels, the South American rodents, and the rats and mice.

Squirrel relatives.—Most primitive of the rodent groups is that typified by the squirrels and hence termed scientifically the Sciuromorpha; the earliest well-known rodents from the Eocene were essentially squirrel-like in structure. A very primitive rodent of this group is the sewellel, or mountain "beaver," of the Rockies of the northwestern United States and

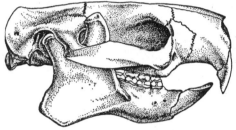

The skull of a rodent to show the elongate, chisel-like incisors, widely separated from the cheek teeth to form the dental structure characteristic of the order as a whole.

British Columbia. Apart from the fact that they are partly aquatic, there is nothing beaver-like about these shy and stockily built animals, which are about a foot long, with a stub tail. More advanced and typical of this first suborder are the widespread squirrels, found on every continent except Australia. The common squirrel of Europe is reddish-brown in color, with a white breast. The same animal extends eastward through temperate Asia, where the coat is generally duller in color, and is represented in North America by the equally common gray squirrel; there are related Central and South American squirrels. The smaller American red squirrel is a very distinct form. In the Old World tropics there is an abundance of squirrels of different sorts, some related to the red squirrels. The flying squirrels, mainly found in tropical Asia but with Temperate Zone representatives in both Europe and North America, have developed a skin membrane between front and hind legs, with which they do not truly fly but do glide outward and downward from one tree to another. In Africa there is present a second group of "flying" squirrel-like rodents with similar adaptations but with structural differences indicating that their development of good gliding membranes occurred quite independently of that of the typical flying squirrels.

Most squirrels are arboreal, but there have developed ground-living, burrowing, side branches. One such group is African in habitat; a second and more familiar one inhabits the Northern Temperate regions. Small and attractive members of the latter group are the little chipmunks which are mainly North American but are also present in Siberia. Somewhat larger are the ground squirrels, likewise mainly American but extending through Asia to Central Europe; in America they are commonly called gophers; in Europe, susliks. Still larger ground types have developed. In the American West the prairie dog, a stub-tailed and chubby animal, was once exceedingly abundant, its "towns" of closely spaced burrows often covering areas many miles in extent; it is now much reduced in numbers. The largest of the tribe is the stoutly built animal known in North America as the woodchuck or ground hog, in Europe as the marmot. The European marmots are found in mountain regions or in the colder plains of the northeast; the common American woodchuck, however, lives in lowland plains and fields as far south as the latitude of the Carolinas.

Beavers, pocket gophers, and kangaroo rats are forms generally included in the squirrel suborder in a broad sense, but the relationship is at best a distant one (and some recent workers deny that there is any special relationship of any kind). The beavers, notable for their industry and aquatic habits, were once present in forested regions over the entire Holarctic region. The beavers are skilled foresters, felling trees to eat the bark, and excellent builders of dams to form ponds in which they construct their homes—"lodges" stoutly built of sticks and mud, with underwater entrances. To their misfortune, the beavers wear a valuable fur and as a result have been nearly exterminated over much of their former range. The quest for beaver produced a romantic chapter in the winning of the American West, where, in the early part of the last century, there developed the hardy breed of mountain men who braved the dangers of the wilds and hostile Indians in search of beaver pelts. Except for some South American forms, the living beavers are the largest of all rodents. In the Pleistocene there were still larger beavers in both Europe and North America, reaching the size of a half-grown black bear.

The pocket gophers are a peculiar North American group of rodents which owe their names to the development of large hair-lined pockets within their cheeks for food storage. They are subterranean animals, which dig long tunnels in search of the roots on which they feed, throwing up mounds of fresh earth at frequent intervals. Having the same peculiar cheek pockets but very different in every other regard are the kangaroo rats and their allies, likewise a North American group. As the name implies, these little animals have elongated hind feet and a hopping gait. This type of locomotion has evolved quite independently a number of times within the rodent order. Other hopping types are found among the ratlike forms, but even here we may mention one further hopping form possibly related to the squirrel group—the spring haas ("jumping hare") of South Africa, similar in size and general appearance to a hare but with a bushy tail and a rather squirrel-like skull and dentition.

South American rodents.—The second of the three "classical" rodent groups is essentially South American. During the Tertiary isolation of that continent, there developed among the rodents—as among monkeys and ungulates—a whole series of peculiar native forms whose relationship to one another is certified by diagnostic features of the skull and the jaw musculature. A characteristic member of this assemblage is the familiar guinea pig, but most members of the suborder are forms quite unfamiliar to readers living in the northern continents. The guinea pig is one of the few animals domesticated by the Indians of either America. The wild relatives of this fertile little creature are the cavies, of which there are a number of species of varied size and habits; together with the related agoutis and pacas, mainly forest dwellers, these terrestrial forms range widely over South America. The chinchillas, known for their highly prized soft fur are small, rather squirrel-like, burrowing forms from the Andes; the related viscacha is a form from the open Pampas. Rather ratlike in appearance are the degu of the Andes and the burrowing tuco-tucos. The beaver-like coypu is of larger size, reaching a length of two feet. In the West Indies there is a native type, the hutia, similar to the last but a forest dweller. The giant of all rodents is the carpincho (*Hydrochoerus*), an amphibious form of eastern South America, which reaches a weight of nearly one hundred pounds and, as its name ("water hog") indicates, has rather the appearance of a member of the pig group.

One family of the suborder is of wider distribution—that of the New World porcupines. These are arboreal rodents, which live not only in their native South America (where one form has, like a number of natives of the region, a prehensile tail) but have very successfully invaded North America to become common animals of cold Temperate Zone forests.

We have gone on the assumption that the guinea-pig group (technically, the suborder Caviamorpha) developed in isolation in South America. But there is here a confusing situation. In Africa there are a number of rodents which have internal structural features comparable to those of South American rodents. Most are unfamiliar forms—the cane rats, rock rats, gundi, and African mole rats (if you have ever heard of any of them, you are a most unusual reader). However, this series includes the Old World porcupines, common in Africa and extending their range into southern Asia as well. These large rodents differ markedly from the New World porcupines in many details of structure and habit—they are, for example, terrestrial rather than tree-lovers—but like the other African rodents just mentioned, they have important anatomical characters found elsewhere only in the New World caviamorphs. To account for this similarity, various workers have assumed a land bridge by which members of this rodent group (and some stream-dwelling fishes as well) could have crossed the South Atlantic. But—as usual with hypothetical land bridges—the question arises as to why, if these animals could cross the bounding waves in this fashion, could not such animals as carnivores and typical hoofed mammals have done so as well? Recent workers are inclined to the opinion that the structural resemblances are due to parallel developments among rodents in the two continents, and the superficially similar prickliness of the two porcupine groups is pure coincidence. But the problem is an interesting—and somewhat embarrassing—one.

Rats and mice.—The third major group of rodents is that of the rats and mice—technically, the suborder Myomorpha. The family of typical rats and mice, in a broad sense, is in many ways the most successful of all mammalian groups. Even on a conservative basis there can be counted about two hundred distinct genera of rats and mice, and the number of individuals is presumably an astronomical figure. By many workers on rodents the family is split into two major components of approximately equal size. One consists of all the rats and mice of the New World, including the common meadow mouse, white-footed mouse, and so on, as well as such large forms as the muskrat. Members of this group are the only rodents that have invaded South America to compete with the forms native to that continent. This type of rat and mouse is a relatively primitive one, primarily characteristic of the Northern Temperate Zone. Probably in earlier times this type was as abundant in the Old World as it is now in the New. Today, however, the group is less common in the Eastern Hemisphere, although represented by a variety of specialized types such as the voles, lemmings, hamsters, and hopping desert types, the gerbils or sand rats. The more typical, "progressive" rats and mice are a more recently evolved group, which are primarily tropical Asiatic and African forms. None have reached the Americas (except where introduced) nor Madagascar, where the more primitive types still flourish, but members of the group are the only land mammals of any sort to reach Australia. The domestic mouse, presumably of southern Asiatic origin, has followed man the world around, and the omnivorous true rat has spread over the entire world, making the major jumps on shipboard. There are two distinct types of this unwelcome animal; both apparently are of Asiatic origin. The black rat, perhaps from India, was the first of the two to become widespread and seems to have reached Europe by the fourteenth century. Later to appear was the brown rat—often called the Norway rat but probably of Chinese origin—which became widespread in the eighteenth century. The brown rat is the larger and more aggressive of the two and has displaced the black from most areas.

Belonging to the general rat-mouse suborder, but very distinct from its more typical members, are the dormice[1] of Eurasia and Africa. These are attractive little animals, squirrel-like in appearance and habits but with diagnostic mouse-like features in skull and teeth. A final series of mouse-rat relatives includes still another group of hopping rodents, of which the prominent members are the jerboas of the steppes and deserts of Asia and northern Africa—extremely long-legged hopping types in which the hind feet are three-toed in somewhat birdlike fashion.

Hibernation.—We have noted previously the fact that various fishes and reptiles "hole up" during one season or another when environmental conditions are unfavorable and remain in a state of partly suspended animation until better days return. Among Northern Temperate Zone mammals, and most especially the rodents, a suspension of activity during the winter months, when temperatures are low and food scarce, is not uncommon. Ordinarily one speaks of any animal as hibernating if it becomes torpid and inactive for part or all of the winter. But students of the subject use the word "hibernation" in a more restricted sense. Such animals as bears, badgers, or raccoons, they say, are merely having a good sleep but are otherwise functioning normally; true hibernation, as seen in such animals as ground squirrels, woodchucks, hamsters, and bats, involves major physiological changes.

---

[1] They have nothing to do with doors; the name refers to their somnolence. They undergo a very long period of hibernation; being nocturnal, they also spend their summer days in sleep.

Most notable is a sharp drop in temperature. Mammals generally maintain a constantly high body temperature, but during hibernation temperature control is practically abandoned; internal temperatures may drop to practically the same as those outside and even approach the freezing point. Respiration is very slow; a hibernating woodchuck, for example, may take only a dozen breaths or so an hour. The heart beats at only a small fraction of its normal rate. The nervous system is almost inactive, and the animal responds only to the strongest stimuli. Animals about to hibernate usually have a good store of fat in their bodies; this must suffice to furnish fuel for their low metabolic activities during the entire hibernating period, and by spring a large fraction of the body weight has been lost. In general the onset of cold weather initiates the shift to the hibernating condition, which ceases with the coming of spring weather. A fair amount of the facts of hibernation are known; but we know little of the mechanisms which initiate and control the hibernating state.

HARES AND RABBITS

In the earlier days of zoölogical study the hares and rabbits were included in the order Rodentia on the grounds that, like squirrels or rats, they have chisel-like incisor teeth. Many decades ago, however, it was realized that except in this one regard there were practically no similarities between the two groups. And even in the teeth there are differences, for, whereas true rodents have but a single pair of chisels above, opposite a pair below, the hares and rabbits have two upper pairs and one lower pair. In consequence these forms have for many years been placed in a separate order Lagomorpha. But fixed ideas are hard to change. Even today one not uncommonly finds, for example, in the medical or physiological literature, a paper in which the author states that such-and-such is characteristic of "the rodent" when, come to find out, the subject of his work has been a rabbit!

For the most part the members of the group (barring some domestic varieties) are remarkably uniform in structure and appearance. The hind limbs are long, although there is less specialized elongation of the hind foot for the hopping gait than in such true rodents as the jerboas. The long ears are characteristic, and the back and sides of the body are generally of a gray to reddish-brown color, while the underside of the upturned stub tail is pure white. Several inhabitants of colder climates change to white fur in the winter season.

Most members of the order are placed in three genera. The hares, relatively large forms which do not construct burrows, are widespread in Eurasia and Africa and in North America, where the jack rabbit is a familiar representative. The common Old World rabbit, a smaller form which shelters in a burrow, appears to have inhabited the Mediterranean region originally but has been introduced and flourishes in more northern European countries now. The American rabbits, including the cottontail, are members of a distinct genus which also reaches South America. Hares and rabbits, although few in number of genera and species nevertheless are a successful group, naturally present in every continent except Australia. The disastrous spread of rabbits when introduced into Australia by man is a familiar story. Currently, however, a virus disease has, at least for a time, reduced their numbers there and in various European countries.

The order contains one animal which is neither hare nor rabbit but a form definitely less specialized. This is the pika, or tail-less "hare," of which there are several species, especially characteristic of the Himalayas but ranging widely from eastern Europe through northern Asia to the northern Rocky Mountains. These are shy little mountain dwellers, with short ears but hind legs of normal length, which appear to represent a primitive stage in lagomorph evolution.

BATS

Only in the bats has true flight been developed by mammals. As in the pterosaurs (and in contrast with birds), the wings are formed by webs of skin; but instead of being supported by a single elongated finger, as in flying reptiles, nearly the whole hand is involved. The thumb, a clutching organ, is free and clawed; the other fingers are all utilized in the support of the wing membrane and usually have lost their originally clawed end joints. In having the wing expanse broken by the long digits, the bat has evolved a more flexible and less easily damaged wing than that of the pterosaurs. The wing membrane extends backward to attach to the hind leg and, further, spans the hind end of the body between the legs, usually attaching midway to the tail. The hind legs are relatively weak and peculiarly posed and are of little use in walking, so that a bat is nearly helpless on the ground. All its active life is spent on the wing; at rest, during the day, it perches on a limb or cave wall upside down, hanging, with folded wings, by the hind legs. In most cases the food consists of insects caught in the air. The majority of bats are tropical, but there are common Temperate Zone forms as well. During the winter some northern bats migrate; many, however, hibernate, frequently in large clusters on the walls of sheltered caves.

Bats are skilled nocturnal flyers; their ability to avoid obstacles even in pitch darkness is well known, and it is obvious that they must depend for aerial navigation upon senses other than sight. Tactile sense appears to be highly developed, and there are in many cases grotesque fleshy outgrowths about the nose and ears which appear to lodge delicate sensory structures. Hearing, however, has recently been proved to be the major reliance of the bat in avoiding obstacles. The flying bat appears to be a silent creature. But this is far from the case. The use of refined acoustic methods shows that the flying bat continuously emits a series of shrill cries which are so high pitched as to be inaudible to the human ear. The highest tones which

Full face and profile view of the head of a leaf-chinned bat of the West Indies. Nose and ears have grotesque sensory outgrowths which almost completely obscure the normal physiognomy. (From Anthony.)

we can hear are waves of about 20,000 vibrations per second; the bat's cries may have a frequency of about 50,000 vibrations and are thus two and a half times higher. The bat ear, however, is capable of hearing them, and the bat appears to avoid obstacles by noting the echoes reflected back from them. This is an auditory system analogous to the radar system developed during the last war for spotting the position of objects by their reflection of radio waves.

The familiar bats of Europe and North America are among the least specialized members of the order, lacking the curious facial "ornaments" of many types but feeding, as do most, on insects caught on the wing. Similar forms are widespread over almost the entire world, but in the tropics there is an abundance of bats of other sorts, of which we shall mention only a few. As a curious variant in dietary habits, the hare-nosed bat of the American tropics may be noted, for this bat feeds in great measure on small fish which it scoops up in flight from the surface of the water. Also peculiar in diet are the members of a varied family of bats found only in South America and often called "vampires"—which most of them are not. These are in part characterized by the development of leaf-shaped fleshy structures surrounding the nose (similar to the outgrowths found in some Old World bats). Most of these South American forms have shifted from an insect diet to fruit-eating. In this they run parallel to a very distinct group of large bats, the flying foxes, which are described

Primitive gnawing animals. The sewellel or "mountain beaver" (*above, left*) and the pika (*above, right*), a tiny rabbit relative, with short legs and small ears. Both are from the Rocky Mountain region.

The spotted cavy or paca (*left*) a common (and typical) South American rodent.

An African porcupine (*below, left*) which resembles those of the New World.

The Indian fruit bat, hanging from its perch (*below*). (New York Zoölogical Society photographs.)

Blackfish stranded on Cape Cod. This is a small whale (maximum length 28 feet; genus *Globicephala*) related to the dolphins and porpoises. It travels in schools which often become helplessly beached. (Photograph courtesy American Museum of Natural History, New York.)

Archaic whales. In the Eocene are found the oldest and most primitive whales, usually known as zeuglodonts. Although already aquatic, they show many features suggesting descent from archaic land carnivores. (From a mural by Charles R. Knight; photograph courtesy Chicago Natural History Museum.)

Unborn young of a pygmy sperm whale. A fetus, close to the stage of birth, removed from the body of the mother, a pygmy sperm whale (*Kogia*) stranded on Staten Island, New York. It is perfectly formed except for the umbilical cord still attached to the navel. The two-foot rule gives the size. The sperm whales are toothed forms in which the expanded snout carries a "case" full of oil and waxy spermaceti. (Photograph courtesy American Museum of Natural History, New York.)

Right whale (*Balaena*). These whalebone whales were considered the "right" whales for oil as well as whalebone by the old New Bedford and Nantucket whalers. Once common in both Atlantic and Pacific, they are now relatively rare. Length about 60 feet. The model photographed here shows well the great strainers of whalebone suspended from the upper jaw and the tiny eye at the corner of the mouth. (Photograph courtesy American Museum of Natural History, New York.)

below. But while most have modified their diet in this way, two small tropical American forms have taken up a very different mode of life—that of true vampires. Various bats have been accused of this and other crimes, but only here is there definite proof of blood-sucking among the Chiroptera. The operation is neatly performed. Chisel-like upper incisor teeth slice off a small area of skin, and with little pain or inconvenience to their prey, they suck blood from the wound.

So far our discussion of bats has been limited to forms familiar in northern regions and to their tropical relatives. There exists, however, a second and very different group of bats confined to the Old World tropics—the flying foxes, or fruit bats. These form a distinct suborder, termed the Megachiroptera ("big bats") in contrast to the familiar little members of the Microchiroptera. The name is well deserved, for they are of relatively large size, one having a five-foot wing spread. The popular name is likewise appropriate, since most have a well-developed, foxlike muzzle, without the facial peculiarities of many of the members of the other group. The wings are well developed, but, although these animals are able flyers, they lack the "radar" navigation system of their smaller cousins. Such aids are not so greatly needed in this group, for their diet consists of fruits, and flight is necessary only as a means of travel between the trees, where they rest by day—commonly in large swarms—and the forests or orchards, where they forage at night.

Presumably the fruit-eaters are a side branch, and the stem of the order lies among the insect-eaters. We may be sure that the bats arose from the primitive placental insectivores which had a similar diet. Fortune has preserved for us specimens of bats from the Eocene in which the flight structure was nearly as completely developed as today. Farther back, however, we lack knowledge of transitional forms which must have existed at the dawn of the Tertiary.

Although surely not a bat ancestor, the "flying lemur" of the East Indies illustrates an intermediate stage in the development of flight. As writers on natural history have often remarked, the name is hardly appropriate, for the animal is not a lemur nor does it fly. It does, however, have a very large fold of skin extending back from its arms by means of which it can "plane" for a considerable distance. Probably the earliest bats began with such gliding apparatus.

WHALES AND PORPOISES

The cetaceans—the whales and porpoises—constitute the largest and most important group of mammals that has turned to an aquatic life and that best adapted to an existence in the water. They have become so completely divorced from their former land life that they are helpless if stranded on a beach. Only in their need for air-breathing do they show any marked functional reminiscence of their previous terrestrial existence.

The cetaceans have reassumed the torpedo-like streamlined shape of primitive aquatic vertebrates; the body, however, is thick and rounded in section, and hence the main propulsive force (unlike that of the typically slimmer fish) is confined to the tail fin alone. As in other aquatic mammals, the tail has failed to resume its original fin structure; horizontal flukes, supported by fibrous tissue (similar to those seen in the sea cows), supply the motive power. Many whales have redeveloped a fishlike dorsal fin, as the ichthyosaurs did. As in the sirenians, the hind legs are lost (except for internal rudiments), and the front limbs are steering flippers. Parallel again to the marine reptiles, extra joints may be found in the toes. Hair has been abandoned as a covering and may be absolutely lacking on the skin of the adult whale; a thick layer of fat—the blubber—affords insulation against the constant cold of sea waters and is, further, a reservoir of food and possibly (as a result of metabolic processes) of water as well.

The skull is large, and the jaws are long; together they include as much as a third of the total body length in some whales. The teeth in modern whales are simple peglike structures. In many forms there has been a considerable increase over the original forty-four; in others, teeth have been abandoned completely. There is little or no sense of smell, and the eyes are small. But, although structurally the ear seems to be somewhat reduced, hearing is present and may be acute. Research in underwater sound during and after the last war indicates that whales (particularly the white whales) can communicate with one another in a whale "language" composed of a considerable variety of sounds.

Whale and porpoise brains appear to be of an advanced type. Psychological study of cetaceans is, of course, difficult, but there is evidence that the dolphins, at least, rate very high among non-human mammals in mental ability.

Some of the larger whales, such as the sperm whale, dive to great depths in pursuit of food and may remain submerged for periods of time said to be as much as forty-five minutes. It is obvious that these habits must involve major physiological specializations. The whales are adapted to withstand rapid changes between normal surface conditions and the immense pressures encountered in the depths. Without such adaptations they would suffer, as does a man under similar circumstances, from the "bends" (pain caused by liberation of gas bubbles in the blood when pressure is decreased too suddenly). Further, there must be, and are, adaptations for taking in abnormally large supplies of oxygen before a prolonged dive. The lungs are large and can be greatly expanded, and among the body tissues there are large networks of vessels which can be filled with oxygen-carrying blood.

In relation to breathing problems, the nostrils, originally of course at the front end of the snout, have moved backward over the top of the head until they open directly upward as the blowhole; as may be imagined (and

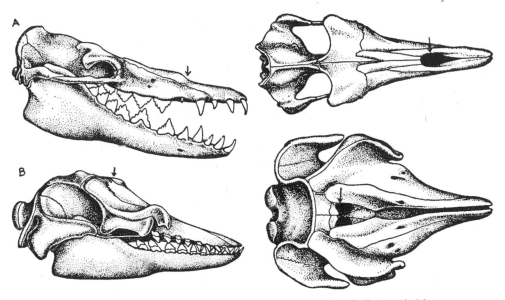

The evolution of the blowhole in whales. *A*, side and top views of the skull of a primitive cetacean, *Zeuglodon*. The external nostrils (marked by an arrow) were in a primitive position near the front end of the skull. In *B*, a typical modern cetacean, the nostrils and the bones surrounding them have shifted to the top of the skull.

as may be seen from a figure of a typical cetacean skull), this has caused major modifications in cranial architecture. The "spouting" of the whales is the expiration of air through this peculiar type of nose when the top of the head comes to the surface; the vapor seen is due to the condensation of water when the column of warm moist air, blown from the lungs, strikes the cooler atmosphere.

Toothed whales.—Living whales are arrayed in two major groups—the toothed whales and the whalebone whales. The former group is much more numerous; teeth are generally retained, and fish and squid are the common articles of diet. The numerous varieties of porpoises and dolphins are close to the common stem of the group. Typically there is a good battery of peglike teeth. Porpoise-like cetaceans became common in mid-Tertiary times; in their very long snouts and in some further technical respects, the river dolphins of today are regarded as closer to these older types than are their marine cousins. Such fresh-water forms are found in the Amazon and La Plata rivers in South America, in the Ganges and Indus rivers of India, and in a single lake far inland in China. As might reasonably be expected from the relatively small bodies of water in which they navigate, these dolphins are small, from five to eight feet in total length. The Indian dolphin is quite aberrant in that it is blind, probing the mud of the river bottoms for concealed fishes and fresh-water shrimps.

Among the numerous marine members of this group of smaller cetaceans there is considerable variation. The term "porpoise" is generally applied to the *smaller* members of the family which are short-snouted. The familiar and playful Atlantic porpoise never exceeds six feet in length, and there are a number of other species of similar size. The term "dolphin" properly belongs to larger forms, with pronounced noses. For example, along the Atlantic coast of North America the animal frequently called a "porpoise" is actually the common bottle-nosed dolphin, a larger animal (up to twelve feet in length). The dolphin of the ancients is a type with worldwide distribution but is especially abundant in the Mediterranean. It averages eight feet or so in adult length and has a prominent beak.

Two interesting members of the family inhabit cold northern seas. The "white whale," or beluga, is unique in its pure-white coloration, but since it seldom reaches a length of more than fourteen to sixteen feet, it hardly deserves to be called a whale. Schools of this gregarious cetacean may be found as far south as the St. Lawrence River and the northern part of the Norway coast. Related and similar in many ways is the narwhal, notable for the presence—in the male only—of a greatly elongated and spirally twisted tusk, formed by the elongation of an incisor tooth. Possibly this structure may be used for dueling purposes by males in the mating season. (This is pure speculation, but no other reasonable explanation has ever been given.)

Two final members of the dolphin family are often called "whales" with greater justice, if size be the criterion. The blackfish, or pilot whale, a feeder on cuttlefish, is found in large herds in the North Atlantic; it is black in color, as its name suggests, and has a rounded, swollen forehead; it may grow to as much as twenty-eight feet. Similar in size, but very different in other respects, is the "killer whale"; packs of this black and white spotted form may be found in every ocean region. This is an aggressive and vicious animal; it feeds to some extent on salmon and other fishes, but sea birds and seals are its favorite food; it preys upon other members of the porpoise-dolphin group; still further, a killer pack does not hesitate to attack even the very largest of whales, many times their own size.

The remaining toothed cetaceans are unquestionably whales. They are few in numbers of genera and species. Three types may be distinguished—the beaked, bottle-nosed, and sperm whales. The beaked whales get their name from the presence of a distinct projecting beak comparable to that of a dolphin. The greatest length they attain is about twenty-six feet, and some species are much smaller. Although technically toothed whales, they are, in fact, almost toothless, for the only teeth in the adult are usually a pair in the lower jaw, and even these may be much reduced. These forms are of little commercial importance. The bottle-nosed whales, somewhat larger in size, also have beaks, but this feature is less prominent because of the presence above it (as in the similarly named dolphin) of a bulging forehead. The teeth are reduced in a fashion similar to that of their beaked-whale relatives.

Largest of all toothed whales is the great sperm whale, or cachalot. Reports of individuals eighty feet or more in length appear to be on a par with fishermen's stories, but authenticated measurements of sixty feet or so testify to animals of considerable bulk. Curiously, however, this animal has a tiny relative, the pygmy sperm whale, which resembles it in every feature except size, for this little animal when mature is seldom more than a dozen feet long.

In body proportions the great sperm whale is very different from all other cetaceans because of its enormous head, bounded below by a short and slender lower jaw. The size of the head has nothing to do with the skeleton; the front and upper parts of this massive structure are occupied by tissues surrounding a great reservoir of oil, called "spermaceti"—as much as fifteen barrels of it in a large individual. This is presumably a valuable store of reserve food for the whale—and made the sperm whale a valuable catch for the whalers of the pre-petroleum days. Apart from the spermacetic reservoir, the structure of the sperm whale is fairly similar to that of the beaked and bottle-nosed forms, and even this reservoir has a smaller counterpart in the latter form. As in those whales, the dentition is much reduced, although the sperm whale retains a row of small lower teeth. Tooth reduction may be related to diet, for the sperm whale, like its beaked and bottle-nosed relatives, feeds on squids and cuttlefish.

The sperm whale is also the source of ambergris, a lightweight material offensive in odor when fresh but valuable as a perfume base; masses of it are frequently found floating in the sea or cast up on the shore. The material originates in the whale's intestine; horny cuttlefish beaks are frequently found embedded in it, and it has been suggested that they form the nucleus for its formation. (Note to fortune-seekers: specimens of supposed ambergris brought to museums for identification usually turn out to be sewage!)

Whalebone whales.—Sharply marked off from the toothed forms are the whalebone whales. Here teeth have been lost entirely. The huge mouth is filled with a series of sheets of whalebone, fringed with hairs at the edge and hanging down from the roof of the mouth in parallel rows like the leaves of a book. This whalebone is composed of hardened horny skin. In the roof of the mouth of the dog, for example, may be seen crosswise ridges of skin; whalebone is formed by the elaboration of such a series of skin ridges.

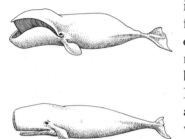

The right whale (*above*) and sperm whale (*below*); both have large heads but that of the right whale is mainly occupied by the large mouth, filled with whalebone plates, while that of the sperm whale consists mainly of the great spermaceti sac, lying above the upper jaws.

These whalebone whales live upon plankton, small animal organisms found in sea water, particularly a type of tiny shrimp; water, passing through the mouth, is strained through the whalebone filter, and the edible material deposited is licked off by the tongue. It would be impossible for this type of whale to swallow any large object (such as Jonah!); although the mouth opening may be huge, the gullet does not exceed nine inches in diameter. It is paradoxical that these eaters of tiny food particles should include in their number the largest animals of any kind that have ever existed.

Best known of whalebone whales are the "right whales." These whales get their name simply from the fact that in early days this was the right kind of whale to hunt because, in addition to a good yield of blubber, its whalebone was of excellent quality and, unlike some other whales of the group, the animals would not sink when killed. Now rare in most areas, the right whales once ranged the world over. There are a number of species; a pygmy member of the group is only twenty feet long, but typical right whales generally run to about sixty feet in length as adults. The head is

A finback, hauled ashore on the Norwegian coast. The finback (*Balaenoptera physalus*) is a very large whalebone whale reaching a maximum of 80 feet in length; the closely related blue whale reaches 100 feet and is the largest of all known animals. These and other closely related whales, known collectively as rorquals, are the major prey of the modern whalers. (Photograph courtesy American Museum of Natural History, New York.)

A bat on the wing. This small insect-eating bat is seen in an instantaneous photograph made with the the stroboscopic camera by Professor H. E. Edgerton of Massachusetts Institute of Technology. The structure of the outspread wings is well displayed. Note the extra membrane area in front of the elbow and the membrane between the hind legs. The first finger has a free claw for clutching purposes.

The skeleton of a whale (*Ziphius*), to show the extreme specializations of the cetaceans. The hind legs have been lost, the front legs shortened into steering paddles. The seven neck vertebrae are extremely shortened, and the head is also much modified. (Photograph courtesy American Museum of Natural History, New York.)

Pangolins (*below*). These dwellers in the Old World tropics are among the oddest of mammals. They are unique in their complete covering of overlapping horny scales. In the absence of teeth and the presence of a long wormlike tongue and powerful claws, they resemble the South American anteaters, shown on the next plate, but there is no evidence of true relationship. (New York Zoölogical Society photograph.)

large and blunt, with about the proportions of that of a sperm whale. But on even superficial inspection, it can be seen that it is built in very different fashion. In the sperm whale the head proper is enormous, the lower jaw short and slender. Here the head appears to consist mainly of lower jaws, the line of the long mouth-opening curving up, back, and then down, ending close to the small eye. This build of the jaws is necessary to give a cover over the enormous series of plates of whalebone, several hundred in number, hanging down from the roof of the mouth.

The rorquals, or fin whales, are today the most important animals in the whale fisheries; they were formerly neglected because the whalebone was of poor quality and smaller in amount, and—more important in earlier times—because the carcass was hard to deal with since it tended to sink to the bottom. The head is broad but much more slender in side view than that of a right whale due to the lesser depth of the whalebone "battery"; the flippers, too, are more slender and tapering than in the right whales. The "lesser" rorqual runs to but thirty feet in length; on the other hand, the great blue whale, or sulfur-bottom, a member of this group, is the largest animal in the world, reaching a maximum length of nearly one hundred feet and (judging by the known weight of smaller individuals) a weight close to two hundred tons. Related to the rorquals is the humpback whale, a fifty-foot animal with a stout build and very long flippers. Quite distinct from either right whales or rorquals is the "California" gray whale, a relatively slender-headed animal that never reaches more than forty-five feet in length. It inhabits the North Pacific, generally spending the summer in the Arctic and descending along the American and Siberian coasts in the winter.

Archaic whales.—Little is known of either the toothed or whalebone whale groups before the Miocene epoch. In the Eocene, however, archaic whales, the zeuglodonts, were already common in the seas. These were, in some cases, of considerable size (seventy feet is the apparent maximum), with a body comparatively longer but slimmer than later whales; the proportions were those which modern imagination ascribes to the sea serpent. These forms were considerably closer to their land ancestors than the typical whales. In the skull, for example, the nostrils were still near the front of the head, and the teeth were very much like those of primitive flesh-eating mammals. Not improbably the whales came from early creodont carnivores which gradually took up a fish-eating existence in the way that otters and seals later did; but we have no fossil record of the early stages in the transition from land to water.

Whaling.—Whale-hunting has furnished some of the most romantic chapters in the story of man's attempts to gain a living from the sea. Although there are current attempts to popularize whale steaks, the meat has been little used for food, but flesh and bone are useful for fertilizer, if nothing more. Whalebone, a strong but pliable material, was formerly very important (and used notably for corsets) but became less valuable in post-Victorian days, due to the development of substitute materials and to changes in fashions. The thick coat of blubber produces oil, which was formerly widely used for illumination and other purposes and is still of value despite the rise of the petroleum industry.

Presumably, stranded whales were utilized in ancient times, and pursuit of whales by small boats from the shores was not uncommon in the Middle Ages. The Basques of northern Spain, from at least the twelfth century, regularly hunted the right whales of the Bay of Biscay, and presently sailors of other western European nations joined in the chase. From off-shore fishing, the industry developed into one in which sailing vessels engaged, the sailors rowing off from the vessels in small boats to spear or harpoon the whales. As the Bay of Biscay whale population decreased, the center of activity moved north to pursue the even more abundant right whales of the Arctic.

In America the settlers on the New England coast, hard pressed to make a living from their rocky or sandy soil, early began to hunt whales—particularly the sperm whale—in coastal waters, and during the 1700's they began long voyages in search of their prey. Meanwhile, the best days of the Arctic right-whale fisheries were ended, and European whalers, too, undertook longer voyages—to the South Atlantic, the Antarctic, and, rounding the Cape, into the Pacific and far up into the cold northern reaches of that ocean. In this stage of the industry, particularly during the early half of the last century, American vessels predominated in number. As many as seven hundred or more vessels would sail annually from New Bedford, Nantucket, and other New England ports, often for voyages of several years' duration.

But in the latter part of the 1800's there was a great decline in whaling. The right whales had been hunted close to extinction; the sperm whales were much reduced in number. Profitable whales were hard to come by, and both oil and whalebone had decreased in value.

The present century has seen a revival of whaling, primarily among the Norwegians. But it is now big business, not adventure. No longer is a whale harpooned by hand by men in a small wave-tossed rowboat. The harpoon is shot from a gun, from the deck of a power vessel; further, this vessel does not do the job of cutting up the whale in the water and "rendering" the oil on its own deck; the whale is towed to a mother "factory" vessel; it is pulled intact into this ship and "processed" in efficient, business-like fashion. With these commercial improvements, whaling has gained a new lease on life, particularly in the Antarctic where whales—notably the rorquals, little hunted in early days—still abound. So successful have the new methods been that it has become necessary to establish an international commission to regulate whaling and attempt to prevent too great a depletion.

EDENTATES

South America is the main seat of an odd series of mammalian types usually known as the edentates ("toothless" mammals), including anteaters, tree sloths, and armadillos. The name is misleading, for, while the anteaters are toothless, the others have a considerable number of teeth; these are of a degenerate structure, however, being simply blocks of dentine, lacking the enamel covering found in other mammals; and teeth are never present in the front of the jaws. The brain is relatively small and of an archaic nature. In all members of the order the limbs and girdles are of odd construction, and in most there are large claws; these features suggest that, although many are terrestrial, the ancestors of the group were not improbably tree dwellers. An additional feature of the group is the presence of a peculiar type of articulation between the vertebrae.

The South American anteaters are essentially feeders not on ordinary ants but upon termites, the "white ants" numerous in all tropical regions. The best-known form is the great anteater, a spectacular-looking fellow with long hair and a long bushy tail, which reaches a length of six feet. The skull is peculiarly developed for its insectivorous diet; the snout is long and slim, and the jaws, in the absence of teeth, are quite feeble. The mouth is merely a small terminal opening from which a long sticky tongue may be extended to gather up the insect food. The front feet are armed with long, heavy, curved claws, which can dig ably into a termite nest. These claws make ordinary walking difficult, and in the front feet the claws are turned inward, the weight resting on the outer side of the knuckles. This animal is terrestrial; there are, however, two smaller forms, one no bigger than a rat, which are arboreal, have shorter snouts, and prehensile tails. (Curiously, prehensile tails in mammals are almost exclusively a South American patent. They are present in the opossums, native there; in the kinkajou among carnivores; in a Brazilian porcupine.

In monkeys, also, such a "fifth hand" is present only in South American forms.)

The tree sloths of the South American forests are among the oddest of mammalian types. They are small, stub-tailed, nocturnal forms, with a growth in their hairy covering of tiny plants (algae), which gives them a greenish tinge matching that of the surrounding foliage. They are clumsy arboreal animals, with the limbs—particularly the front ones—greatly lengthened; they spend much of their time slothfully hanging upside down from branches and holding on by their long, curved claws, two or three in number. These leaf-eating animals have teeth, but these comprise only four or five pairs of cylindrical pegs in either cheek. The tree sloths show an interesting anatomical anomaly. In mammals the number of vertebrae in the neck is almost invariably seven; the giraffe (with its long neck) and the whale (with no visible neck at all) have the same number. The only animals which violate this rule are one of the sirenians and the two species of tree sloths—of the latter, one has six; the other, nine or ten.

The third living South American edentate group is that of the armadillos, the only living mammals with a bony armor. As in the sloths, there are peglike cheek teeth; here, however, they are numerous and there may be as many as twenty-five teeth in the cheek series. Not more than seven are present in a jaw ramus in any ordinary mammal; except for cetaceans, the sloths are the only mammals which transgress this upper limit. The most interesting armadillo development, however, is their armor—numerous rows of bony plates, covered with horn, over the back and sides of the body and even over the top of the head. In the more primitive members of the group the armor forms a uniform series of movable bands the length of the body; in more advanced types the shield consists of solid partitions over the hip and shoulder regions, with movable hoops between. Armadillos are omnivorous terrestrial forms, with stout claws which make them excellent burrowers. In addition to a basic diet of worms and grubs and some vegetable food, they will eat birds' eggs, birds or other small animals, and carrion. One form—the nine-banded armadillo—extends northward to the United States. Until recently it was confined there to a small area in southwestern Texas. Within the last two decades it has, however, spread northward across this large state into Oklahoma and Arkansas; it has crossed the Mississippi and is seen occasionally from Florida to South Carolina.

Besides these three living groups, South America gave rise to two other and more spectacular edentate types, both extinct—the great ground sloths and the glyptodonts.

Edentates. The major group of toothless mammals is of South American origin. It includes five main types: anteaters, armadillo, tree sloths, and the extinct ground sloths and glyptodonts.

Right, representatives of the three living types. Most prominent feature of the large anteater, *Myrmecophagus*, is the long tubular snout, from which may emerge a very slender, sticky tongue, an ant-catching device. Among the various armadillos, the two illustrated in the second figure are a giant South American form (*Priodon*) and the familiar nine-banded form (*Dasypus*). The tree sloths suspend themselves from boughs by their sharply curved claws. The specimen shown below is a two-toed form (*Choloepus*); a three-clawed form also exists.

The aardvark, "earth pig," of South Africa (*below, right*), is generally classed as an edentate despite the presence of teeth (rather poorly constructed). He is not closely related to the South American forms, although his adaptations for ant-eating are similar to the anteaters. (All photographs courtesy New York Zoölogical Society.)

Giant South American edentates of the Pleistocene. Giant mammals were present at the time of the Ice Age in all continents. Characteristic contributions of South America were the ground sloths and glyptodonts illustrated here. The ground sloths were related to the living tree sloths and probably were also leaf-eaters. They had abandoned the trees for the ground, however, and grew to large size; *Megatherium*, shown above, was the size of a modern elephant. The glyptodonts were related to the armadillos but had a solid domed shell of bone rather than the more flexible armor of their surviving cousins. Note the tail, a bone-sheathed club with a spiked end. These edentates were South American natives, but several genera of ground sloths invaded North America in the Pleistocene, and glyptodonts reached the region of the Gulf of Mexico. (From a mural by Charles R. Knight; photograph courtesy Chicago Natural History Museum.)

The ground sloths developed in South America in Tertiary times and reached their peak in the Pleistocene, when a number of genera were present in North America as well. One form reached the size of an elephant, and others were as large as an ox. These forms were undoubtedly related to the tree sloths but were ponderous ground dwellers. As in tree sloths and anteaters, the feet were armed with good claws; this caused a clumsy type of foot construction, which was complicated by the heavy bulk of these animals. The gait was obviously a shuffling one; the front feet rested more or less on the knuckles, the hind legs bore down on the outer edge of the broad feet in the manner of the modern giant anteater. These ground sloths appear to have been able to rear up on their hind legs in bear fashion to crop leaves from the branches of trees. The anatomy of these animals suggests that, like their small living cousins, the ancestral forms were dwellers in the trees, descending after an increase in bulk.

Related to the armadillos were the extinct glyptodonts. They were covered with armor which formed a solid, domed carapace covering the entire trunk. Not only was the head protected by bony plates, but the tail also was incased in a bony sheath, sometimes bearing spikes. These glyptodonts, some of them ten or more feet in length, were a close mammalian parallel to the armored dinosaurs of the late Cretaceous.

A few small forms, seemingly related to the ancestry of the South American edentates, were present in North America in the Paleocene and Eocene but soon disappeared here. The ancestral edentates were marooned in South America during the Tertiary and there evolved into the three living and two extinct types described above. The archaic ungulates of South America were rapidly exterminated, as we have seen, with the re-establishment of North American connections at the close of Tertiary times. The edentates, however, were of sterner stuff. They not only held their own territory during the Pleistocene but, going against the tide of migration, invaded North America; large glyptodonts are found in the Pleistocene of the Gulf coast, numerous ground sloths penetrated as far as Pennsylvania and Illinois, and even today, we have noted, an armadillo ranges well into North America.

At the end of the Pleistocene the glyptodonts and ground sloths became extinct. The latter appear to have lingered late; in the southwestern United States there is positive evidence of their contemporaneity with man, and in Patagonia one appears to have been imprisoned in a cave used as a "stall."

In the Old World tropics are two types of termite-eaters exhibiting features similar to South American edentates. The aardvark of Africa is a grotesque, long-eared beast which has well-developed ant-eating adaptations in the long snout, slim whiplike tongue, and powerful claws; but there the resemblance to American anteaters stops. There are peglike teeth of peculiar construction, and the animal in general resembles nothing else at all. The term "earth pig" is not particularly appropriate, but there is nothing with which the animal can be compared.

The pangolins of southern Asia and Africa have, again, termite-eating adaptations in the slim snout, long tongue, toothless mouth, and sharp claws. They are armored but not by bone; instead, there are large overlapping horny scales, giving these creatures somewhat the appearance of animated pine cones.

Neither aardvark nor pangolin has fundamental features indicating real relationship to the true American edentates. We know little of their fossil history; presumably they represent isolated types which have taken up termite-eating as a vocation and have, in relation to this, tended to develop similar adaptive features.

CHAPTER 16

Primates: Lemurs, Monkeys, Apes–to Man

The order of primates is a group of particular interest to us, for it includes not only the lemurs, monkeys, and great apes but man himself; in the study of primate evolution we are studying our own kin, climbing toward the summit of our own family tree. We shall, therefore, treat the primates more fully than we have a number of other mammalian groups.

To the paleontologist the primates are a source of considerable difficulty. We have well-documented pedigrees for various mammalian types such as horses, camels, and dogs. In the case of our own ancestors we should like to present an equally adequate fossil history. Unfortunately, however, fossil remains of primates are more rare than is the case in any other large group of mammals. The reasons for this paucity of material are fairly obvious. Primates are, for the most part, tree dwellers; rock deposits in which fossil vertebrates are to be found are not normally formed in forested regions. Again, primates are mostly dwellers in the tropics, whereas most of the known Tertiary fossil beds are in what are today zones of temperate climate. It is only in Eocene times, when these regions (as shown by the vegetation) knew tropical conditions, that primates are found to any extent in the fossiliferous deposits of Europe and North America.

The living primates, for our purposes, may be divided primarily into three groups: (1) the lemurs, or prosimians, typically small four-footed forms of rather squirrel-like appearance, found today in the Old World tropics and present in the Eocene in Eurasia and North America; (2) *Tarsius*, a curious small, hopping, ratlike creature from the East Indies and its fossil relatives, occupying an intermediate position between lemurs and higher primates; (3) the anthropoids,[1] comprising the South American

[1] The manlike (anthropoid) apes are sometimes spoken of simply as anthropoids; but this loose usage should be avoided; the term "Anthropoidea" properly includes monkeys as well.

monkeys, the more advanced monkeys of the Old World, the great man-like apes, and man.

ARBOREAL ADAPTATIONS

Primates are essentially arboreal, only a few forms (such as the baboons and man) having taken up a life on the ground. There is evidence, we have

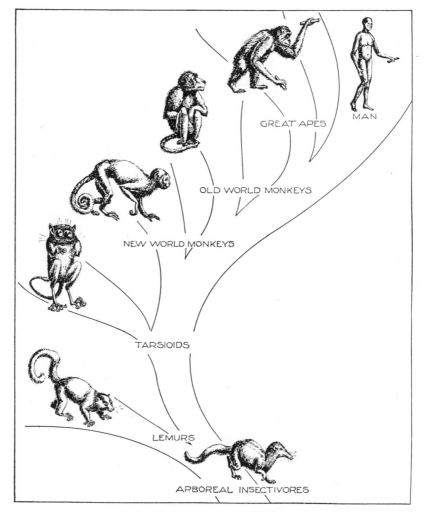

MAN

GREAT APES

OLD WORLD MONKEYS

NEW WORLD MONKEYS

TARSIOIDS

LEMURS

ARBOREAL INSECTIVORES

A simplified family tree of the primates

noted, suggesting that the primitive placentals were tree dwellers to begin with; the primates have merely continued on this ancestral track. Arboreal life is apparently responsible for much of the progressive development of primate characteristics; and, although man is not a tree dweller, this life of his ancestors has left its mark deeply upon him and is perhaps in great measure responsible for the attainment of his present estate.

Limbs.—Locomotion in the tree has left the skeleton of the primates in a condition much closer to that of the primitive placentals than is the case in most groups. Flexibility is necessary for climbing trees, and there is none of the restriction of limb movement to one plane found in ungulate groups. In contrast with such arboreal types as the squirrels, in which climbing is facilitated by digging the claws into the bark, the primate hold

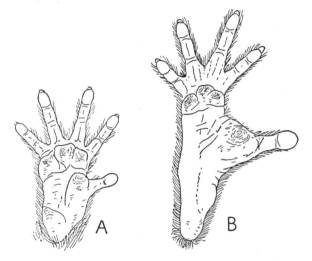

Hand (*A*) and foot (*B*) of an Old World monkey; viewed from the underside. Note the short thumb and the opposable big toe and long toes of the hind foot. In man, in contrast, the thumb is highly developed, while the big toe (once opposable) has, with upright posture, shifted forward parallel to the other toes. (From Pocock.)

is generally accomplished by grasping boughs or twigs. The primitive claws have been transformed into flat nails serving as a protection to the finger tips, although in many cases (particularly in lemurs and marmosets) there is a retention of or return to a clawlike structure.

In primitive mammals thumb and big toe were apparently somewhat divergent and opposable to the other digits. This grasping characteristic has in general been retained and emphasized in primates. In most lemurs, monkeys, and apes it is the big toe which is the most highly developed and divergent. The thumb is sometimes reduced, or even absent, in forms which habitually progressed by swinging from branch to branch, the other four fingers here hooking over the bough and the thumb being not only useless but actually in the way. Otherwise there has been almost no reduction of toes; the primates are in this respect more conservative than most mammalian groups.

Four-footed locomotion is the rule among primates, bipedal tendencies becoming apparent for the most part only among the higher members of the order. The hands, however, while primarily possessing their grasping adaptation in relation to locomotion, are well adapted (in contrast to most mammals) for grasping food and other objects; and even among the lemurs there is a tendency toward a sitting posture and the release of the front feet from supporting the body.

The primitively long tail is retained in almost all the lower primates and is apparently of considerable use in balancing. In many South American monkeys it has developed into a prehensile "fifth hand." In a number of cases, however, the tail is reduced or absent, and there is no external tail in the manlike apes and man.

Teeth.—An omnivorous diet appears to have been general in early primates, and similar food habits characterize many living types (although there has been a strong herbivorous trend). The dentition is, hence, usually less specialized than in most mammalian groups; the molar teeth are low-crowned with rounded cusps. In diet primates are not dissimilar to the swine, and the cheek teeth of these two groups are fairly similar; indeed the teeth of a fossil native pig from Nebraska were for a time mistakenly identified as those of a manlike American ape. In all primates, including man, the number of incisor teeth has been reduced from the original three to two in each half of each jaw, and there has also been a trend toward reduction in the number of premolar ("bicuspid") teeth from four to two, so that the higher monkeys and man have but thirty-two teeth instead of the forty-four present in our primitive placental ancestors. Primates are in general short-jawed and, consequently, short-faced types. In most of them the canine teeth are rather long and make effective biting weapons. In man they have ceased to project above the general level of the tooth row, but they still develop massive roots and many readers may have noticed the fact that, in consequence, they are slow to break the surface in a child's dentition.

Sense organs.—Arboreal life has had a profound effect upon the sense organs. A ground-dwelling mammal is in great measure dependent upon smell for his knowledge of things about him, and his olfactory sense is highly developed, while sight is usually comparatively poor. The reverse is true in these arboreal types, and in man, descended from them. Your dog recognizes you by smell; you recognize him by sight. For locomotion in the tree good eyesight is essential, and throughout the primate group there has been a progressive series of advances in the visual apparatus. Even in the lemurs the eyes are large, and there is a tendency to rotate them forward from their primitive lateral position. This results in a situation where parts of the two fields of vision overlap; in *Tarsius* and the higher primates the two fields are identical. In vertebrates below the mammals the eyes appear to furnish the brain with two separate visual images. In mammals where the fields overlap, there typically arises the stereoscopic type of vision, well developed in the higher primates. In this condition there is a sorting-out of the nerve fibers running from eye to brain, so that the brain-pictures formed by the two eyes coincide, and the minute variation between the two pictures gives the effect of depth. In addition, the monkeys and all higher types possess in the middle of each retina a specialized region, the "central pit" (fovea centralis), in which detail is much more clearly perceived.

With this improvement in vision goes a corresponding degeneration of the sense of smell. Even in the lemurs the olfactory sense is reduced, and in the manlike apes and man it is probably in the most rudimentary state to be found in any terrestrial placental. This loss is perhaps regrettable. We have no idea of the delicate nuances of smell which may be evident to the aesthetic senses of many other mammals. Had we their sensitivity to smell odoriferous art forms would presumably be added to the visual and auditory types, and composers would perhaps write olfactory symphonies; we might attend "the smellies" as well as movies and concerts.

Brain.—The brain, as well as the senses, has been profoundly influenced by arboreal life. Locomotion in the trees requires great agility and muscular co-ordination. This co-ordination in itself demands development of the brain centers; and it is of interest that most of the higher mental faculties are apparently developed in an area alongside the motor centers of the brain. Again, the development of good eyesight rendered possible for primates a far wider acquisition of knowledge of their environment than is possible for forms which depend upon smell. Perhaps still more important in the evolution of primate mentality has been the development of the grasping hand as a sensory aid in the examination of objects.

With these potential advantages to be gained from any trend toward increased mental ability, it is not to be wondered at that selective action has resulted, in monkeys, apes, and man in the development of large brains and especially of greatly extended areas of the gray matter of the cerebral hemispheres in which the higher intellectual faculties reside. Primates all have relatively big brains. The cerebral hemispheres have grown backward and downward in higher anthropoids so that they almost completely cover the rest of the brain from view, and the area of gray matter is greatly increased by a very complex infolding of the surface.

Skull.—These changes in dentition, sense organs, and brain have been associated with great changes in the primate head. Most lemurs have an elongate skull of primitive appearance, with a long face and a long, low braincase. But with the reduction of the sense of smell and the concurrent abbreviation of the tooth row, the muzzle in most primates has shortened considerably, the face sloping sharply downward toward the mouth. While the face has been reduced, the braincase has, on the contrary, expanded greatly in higher primates to accommodate the ever growing brain. In primitive mammals there was no bone separating the eye socket from the temple region. In a number of mammalian types a superficial bar has been built between these two openings. Such a bar is present in all primates from the very first. But, in addition, monkeys, apes, and men have built up underneath this a solid partition between eye and temple. This is unique among animals. Presumably it gives a better attachment and support for the eyes in their rotated position and also prevents the jaw muscles pushed out sideways by the expanding brain, from crowding forward into the eye sockets.

LEMURS

Primitive members of the primate order are the lemurs, found today in tropical Asia and Africa, more especially on the island of Madagascar. An ordinary Madagascar lemur of today is a fairly small arboreal animal, nocturnal in habits, with a bushy, hairy covering which contrasts with the rather sparse coat of most higher primates. The limbs are moderately long, the ears pointed, and the eyes are directed more laterally than forward. The lemur face may undergo some shortening, but it is in most cases a comparatively long and foxlike muzzle. The tail, usually long, is never a grasping organ. Thumb and big toe are always widely separable from the other digits; the big toe is especially well developed and has a flat nail in all forms, whereas the other digits are variable in their covering. Typical lemurs have a clawlike nail on the second toe (for scratching purposes) but normal nails on the other digits. Their name is derived from classical Roman days, when the souls of the dead, haunting the night in similar ghostly fashion, were termed lemurs.

Lemurs. *Upper left*, a restoration of the Eocene fossil lemur *Notharctus*, which may be close to the ancestry of modern forms. *Upper right*, a typical lemur of the type common today on the island of Madagascar. This is the most primitive living form, with a foxlike nose and eyes directed somewhat laterally. *Right*, a slender loris, an advanced type of lemur from southeastern Asia; this specimen was captured in Siam. Note the relatively short face and also the peculiar foot structure. Lemurs have hand-like feet with the thumb or big toe opposable to the other digits. *Bottom*, a family of galagos; parents and twin offspring. The galagos, or "bush babies," are inhabitants of continental Africa. In these advanced lemurs the eyes, as can be seen, are large and turned well forward in the fashion of higher primates, and the nose is much less prominent than in the typical lemurs. (*Notharctus* from a painting by F. L. Jaques under the direction of W. K. Gregory, photograph courtesy American Museum of Natural History, New York; loris photograph by American Primate Expedition, courtesy Sherwood Washburn; galagos photograph courtesy F. deL. Lowther.)

Tarsius. Seldom have these delicate little oriental primates been successfully "exported" from their native country. This one lived more than a year at Yale University. The photograph shows the large eyes, reduced nose, and the specialized pads on the toes. (Photograph courtesy John F. Fulton, Yale University.)

An Eocene tarsioid (*left*). A restoration of an early tarsius-like form. (From Scott, *Land Mammals in the Western Hemisphere*, by permission of the Macmillan Co., publishers.)

As well as the more typical lemurs, Madagascar houses a number of relatives of varied size and build. The typical lemurs are about cat-size. The mouse-lemurs, as the name suggests, are tiny forms, notable for an elongation of the foot bones which allow a hopping gait. The indris and sifakas are, on the other hand, much larger, the former having a trunk length of two feet. They are ungainly-looking creatures, with long limbs and, in the indris, with only a stub tail. Oddest of Madagascar lemurs is the little animal called (because of its cry) the "aye-aye." The front teeth are developed as chisel-like structures very similar to those of rodents (a feature also developed in some early fossil lemurs), and the toes are long and slender. A favorite food is said to be a type of grub found in tree bark, and an especially long third finger is said to be used in scooping this delicacy out of its hole. Still further lemur types, now extinct, were present in Madagascar in Pleistocene days, some of rather monkey-like appearance; one large form (which may have been seen, before extinction, by an early explorer) appears to have looked much like a four-footed caricature of a man.

The reason for the survival of these primitive forms on the island of Madagascar is seemingly the fact that this area has been long separated from the mainland of Africa, and as a consequence but few flesh-eaters (as we have noted) have been able to enter this region; members of the civet family are the lemurs' only enemies. On the mainland, on the contrary, lemurs are extremely scarce, and there are only a very few living forms. All are more advanced in structure, with, for example, shorter faces and eyes turned more forward, thus paralleling to some extent the higher primate groups. On continental Africa are found the galagos, or "bush babies," and the related potto—quite attractive but timid little nocturnal animals. Like some of the Madagascar lemurs and like *Tarsius* (to be described presently) the hind feet are somewhat elongated for hopping purposes. In India and the East are the lorises, of two types. The slow loris, as its name indicates, is a slow and deliberate creature, with a stocky build and stub tail, about the size of a cat. The slender loris is a much smaller form, with long slim limbs but equally lethargic habits.

In the Eocene, lemurs not unlike the living Madagascar types were common in both Europe and North America. Beyond the Eocene, however, they vanish from the fossil record of the northern regions, which presumably were already becoming a bit too chilly for tropical forms.

The lemurs and their kin are not far removed in many characters from the insectivore stock from which all placental mammals have come. We noted in a previous chapter that the oriental tree shrews appear to be intermediate in character, and many workers would, not unreasonably, place them, too, in the primate order.

TARSIUS

Much higher in the scale of primate evolution than any living lemur is *Tarsius*, a small nocturnal tree dweller from the East Indies. He has a peculiar ratlike tail and long hind legs adapted for a hopping gait. These are specialized features in which *Tarsius* is off the main evolutionary line of primates. But the rest of its structures are curiously intermediate between lemur and monkey types.

The brain is large, and the braincase is short and rounded. The visual apparatus is much advanced over that of the lemurs. The eyes are exceptionally large and are turned completely forward from the primitive lateral position, with the orbits close together above the nose. In this creature the two fields of vision are identical, and recent work indicates that stereoscopic vision is well developed.

While vision has thus advanced, the sense of smell has been proportionately reduced. The foxlike snout of the typical lemur has disappeared, and the nose has shrunken to a small nubbin tucked in between and below the eyes. There is little projection of the muzzle (the tooth row is comparatively short), and we have here the beginning of the type of face found in monkeys, apes, and man.

An interesting feature in which *Tarsius* shows affinities with the higher primates is the fact that the placental connections between mother and young, instead of being distributed, as in lemurs, all about the surface of the sac containing the embryo, are concentrated in one discoidal area and shed at birth just as in monkeys and man.

Tarsius is not a monkey, but it is far above the lemur level. The living form is too specialized in itself to be a true intermediate type, but it seems not unreasonable to regard it as an offshoot of a group transitional from lemurs to monkeys. In the fossil record, however, the story is a blurred one. In Paleocene and Eocene strata we find numerous remains of a varied assortment of small animals which are certainly primates below the monkey level. Some, we have noted, appear to be ancestral to typical lemurs, some definitely *Tarsius* ancestors; others do not fit well into either category, in default of knowledge of soft parts of the body. Apparently the early primates were a flourishing group of little tree dwellers which spread out into a number of lines. Most were short-lived; modern lemurs and

Tarsius represent two levels of evolution of these early forms in the rise of primates toward the monkey level.

MONKEYS

With the monkeys we reach the third and highest main division of the primate group, that of the suborder Anthropoidea, or manlike creatures in a broad sense, including monkeys, the great apes, and man.

Monkey characters.—In all these forms we have a very distinct general advance. The eyes were already large and placed far forward in *Tarsius*, but, in addition, in even the lowest of monkeys the special area (fovea centralis) in the center of the eye making for clear perception of detail is developed; the monkey's eye is on a par with our own. With this increase in vision we find that smell in anthropoid primates is unimportant and the snout usually much reduced.

The monkey is normally a four-footed walker, but there is a great tendency for an upright sitting posture and the consequent freeing of the hands for the manipulation of food and other objects. The brain is comparatively large in all members of the group, and the much-swollen braincase gives the skull a rather human shape. The freeing of the hands and the perfection of the visual apparatus undoubtedly give much greater scope to the higher mental status found in even the lowest monkeys. We speak proverbially of "monkey curiosity." This monkey trait is probably due in great measure to the fact that the monkey has good eyesight; he can see more to be curious about. Then, too, he has a great advantage in a means of satisfying his curiosity not possible in most animals; he can handle the objects seen.

We speak often, with respect, of scientific research. But, after all, what is research but the same old urge of the monkey to handle some new or curious thing? It may well be on a higher plane, but it is the same monkey curiosity which has made our simian relatives the pests of the jungle.

The anthropoids are readily divisible into two groups. On the one hand, are the South American monkeys (platyrrhines); on the other, are Old World monkeys, the great apes, and man (catarrhines).

New World monkeys.—The South American forms are generally known as platyrrhines, or "flat-nosed" monkeys, owing to the fact that the nostrils are widely separated, opening more sideways than forward or downward. If we wish a further technical character we may note that in these New World monkeys there are three premolar teeth in each half of each jaw, whereas in all the Old World forms, as in man, but two such teeth are present.

The smaller forms among the South American monkeys are the marmosets. These little fellows are squirrel-like in general appearance, with thick fur and a bushy tail, and are, except for the short face, rather like the lemurs superficially. Some writers have suggested that the marmosets are really primitive monkeys close to the lemur ancestors. Others, however, believe that their seeming primitiveness is due to specialization and degeneration. As an example of marmoset specialization may be mentioned the fact that the last molar has disappeared. In man this molar, the "wisdom tooth," is degenerate and often fails to function; but with its total loss the marmosets have progressed farther than we have in this one regard.

Much more typical monkeys are also present in South and Central America; of these the organ-grinder's monkey, the capuchin, is a typical representative. (Its name comes from the fancied resemblance of a crest of hair atop its skull to a monk's cowl.) In a majority of these forms the tail is not only long but has developed into a prehensile organ of great utility. Some members of this group, the spider monkeys, are very clever acrobats with extremely long limbs and no thumbs at all. An interesting side branch is that of the howler monkeys. These have a large, bony, resonating chamber in the throat; by means of the sounds emanating from it, night is made mournful in the South American forests. In contrast to the lemurs, the South American monkeys are almost all active in the daytime; and in contrast with some Old World cousins, all are good arboreal forms, seldom descending to the ground.

Little is known of the fossil history of these American monkeys, although there are some remains from middle to late Cenozoic beds in South America. They are as advanced as Old World monkeys in certain regards, more primitive in others. At first sight, one would tend to say that they form an intermediate stage in the evolution of monkeys from lower primates. But the facts of geography are against this. Quite surely they never lived anywhere but South America, separated by a broad ocean from the Old World, where (as we shall see) the evolution of higher monkeys and apes began at an early time. There are no fossil remains of monkeys of any sort in North America. Hence it is probable that monkeys evolved from higher, *Tarsius*-like lemuroids twice, in parallel fashion—once in South America; once, more successfully, in Eurasia and Africa.

Old World monkeys.—Grouped as the catarrhines are the monkeys of Asia and Africa, the manlike apes, and man. The name indicates one of the many features in which they are more advanced than the South American monkeys—the fact that the nostrils are closer together and open forward and down. There is a rather general tendency toward an increase in size, which culminates in the chimpanzee, man, and gorilla. Although primitively arboreal, some of the higher or more specialized types, as man and the baboons, tend toward a terrestrial life. The tail is often shortened or absent and never prehensile. Primitively four-footed, some have tended to an arm-swinging type of locomotion in which the body is held erect; man has attained an erect posture. Although progression on the ground is usually on all fours, the long arms bring the back into a slanting position, and sitting upright is the common catarrhine pose when not in motion. The thumb is usually well developed and opposable, and the big toe universally

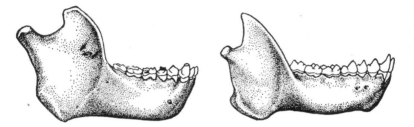

Left, the jaw of *Parapithecus* of the Oligocene of Egypt, oldest known monkey. Length of original about 1½ inches. *Right,* the jaw of *Propliopithecus,* oldest known manlike ape, from the same beds. Length of original about 2¾ inches. (After Schlosser.)

so except where secondarily modified in man. The hairy covering is always thin and the face naked.

The brain is large. The face, primitively short, tends to elongate in some instances in correlation with a tendency toward a vegetable diet and a consequent increase in tooth size. There are only two premolar teeth in each jaw half; we have finally arrived at the stage of reduction still found in man, with only thirty-two teeth instead of the original forty-four.

Anthropoids date back to the beginning of the Oligocene epoch in the Old World. In fossil beds of that age in the Fayum district of Egypt have been found two jaws and other fragments which are sufficient to show that small primitive Old World anthropoids had evolved by that time. At least one of the jaws, as noted below, even suggests the faint beginnings of the great-ape line; other specimens, however, are more primitive or truly monkey-like in nature.

The Old World monkeys include a great variety of forms, with a wide range in structures and habits. All are, in terrestrial locomotion, four-footed types walking flat on palm and sole. Correlated with the common sitting posture is the universal presence in these monkeys of hardened skin areas, sitting pads (callosities), on the buttocks. Two groups of living Old World monkeys are easily distinguishable today: one group has cheek pouches for the storage of food; the other has a complicated stomach, somewhat comparable to that of ruminants and with the same purpose— the digestion of vegetable fodder.

To the cheek-pouched group belong most of the small African tree dwellers seen frequently in our zoos—green monkey, mona monkey, etc.— often brightly colored and sometimes referred to collectively as the guenon monkeys. The mangabeys, easily recognized in a zoo by their white eyelids, are West African forms, somewhat less agile and transitional in technical features to the macaques. These last forms include the Barbary ape, present on the rock of Gibraltar—the only monkey inhabiting any part of Europe today. Typically, however, the hardy macaques are inhabitants of southern Asia, where they abound in numbers and variety; the rhesus monkey, frequently used in medical research, is a familiar representative. The macaques are less arboreal in habits than the common African monkeys. The tail is frequently shortened, and the face is somewhat longer, for the grinding teeth have tended to lengthen with development of better chewing power. Superficially this facial elongation—still more emphasized in the baboons—gives somewhat the appearance of the primitive mammal snout. It is, however, a secondary development, as witnessed by the fact that the nostrils, which had retreated backward with shortening of the face in primitive monkeys, open out above the snout, rather than having their primitive position high above its tip.

From the macaques several intermediate types lead to the African baboons and their hideous cousins, the drills and mandrills. These largest of monkeys have become as purely ground dwellers as has man; but they have become four-footed ground types, walking flat on palm and sole. In relation to their herbivorous mode of life, the tooth row and muzzle have become even more elongated. The baboons tend to operate in formidable bands which may be of considerable size.

The second Old World monkey group, that of forms with a complicated stomach for better digestion of their purely vegetable diet and without cheek pouches, has as central types the langurs of southern Asia, such as the sacred hanuman monkey of India. All these forms are good arboreal types, showing none of the ground-dwelling tendencies of the cheek-pouched monkeys. An interesting variant in this group is the proboscis monkey with, in old males particularly, a long pendant nose which outdoes that of man in its development and is said to be attractive to the ladies of the species. African representatives of this group of vegetarians are the guereza monkeys, with manes of long hair, often strikingly colored, extending down the flanks. Reduction of the thumb, seen in several primate types here reaches an extreme for Old World forms; there is never more than a tiny vestige of this generally useful digit.

Monkey society.—Man is not only an intelligent animal but a social animal as well. So, to a considerable degree, are his monkey cousins. In various groups of vertebrates there is some tendency for numbers of individuals to aggregate as herds or flocks or swarms, usually rather loosely organized; and there are also numerous situations in which care of the young keeps parents and offspring together for at least a short season. Such tendencies are strongly emphasized in higher primates, very probably due initially to the fact that here breeding is not seasonal but may occur at any time, and that young primates have

Head of an adult male proboscis monkey (*Nasalis larvatus*). (From Pocock.)

a longer period of infancy. In consequence the need for care of the young is essentially continuous and some sort of sustained family or group organization is necessary.

The great apes are difficult to study in the wild, and we know relatively little about them except for the gibbons (who enjoy a strictly monogamous family life). But monkey habits are well known in a number of cases. The psychologist Carpenter, for example, has studied the howler monkeys in the American tropical forests. These forms travel about the forests in fairly large bands and in an amicable fashion. To be sure, some individuals are dominant, but everyone knows his or her place in society and generally keeps it. Rather shocking to us is the fact that there are no permanent marriages; instead there is a general and friendly promiscuity. When traveling through the tree tops, the old males scout out the path for the tribe, and if there is an argument with another group over territory, the two groups of males wage the battle. But even this is conducted without bloodshed. They howl; and the group that howls the loudest wins.

Baboons and macaques are often kept successfully in zoos in large groups and their behavior has been studied by Zuckermann and others.

The results are somewhat embarrassing, for they present a picture including the elements of human conduct but in the raw, lacking the veneer—and inhibitions—generally seen in human society. In a baboon colony the question of dominance—"who is who" in power, from the most muscular males down—is constantly in the picture. A fight may be necessary to establish a baboon's position in the "peck order." Once established, the individual acts with propriety as becomes his station; he may bully those below him but submit to bullying by his peers. Here there is marriage of an essentially permanent nature. This consists of each of the more dominant males corraling as many females as he can and keeping other males away from a harem over which he rules. This is a society and one of some complexity in the interrelationships of the individuals which compose it. But it is hardly a source of pride to think that this may have been the type of social setup out of which human societies eventually emerged.

MANLIKE APES

But the Old World monkey groups, interesting as they are in showing how monkeys can vary in structure and adaptations, have little to do with our own history. Our ancestors came from a third line of monkey evolution—one leading to the development of the anthropoid apes. This line diverged from that of the modern monkeys at a very early date, for in the same Egyptian beds where are found the earliest known monkeys occurs the jaw of an animal which is definitely a small ancestral anthropoid ape.

Structure of the great apes.—The anthropoid or manlike ape group includes four living types: the gibbon, the orangutan, the chimpanzee, and the gorilla. They range in size from the comparatively small gibbons, through the chimpanzees (a bit smaller than a man but with powerful muscular strength), through the orangs (of essentially human size), to gorillas (several times as heavy as man). The skeleton is rather close to the human type. The chest is broad and flattened, with a broad breastbone, in contrast to the thin, deep chest of monkeys and of mammals in general. The hands are fairly similar to those of man, but the fingers are generally quite long in comparison with the thumb, and the arms are elongate; the legs, on the contrary, are short. The foot is still an excellent grasping structure, with the toes long and the big toe opposable to the others in handlike fashion.

These features, particularly those of the arms, seem to be associated with the great-ape type of locomotion. Except for the gorilla, all the great apes clamber about the trees to some extent, with all four limbs being

Monkeys. Sharply divisible into South American and Old World groups, the former are the more primitive. Two South American forms are illustrated above, a pygmy marmoset (*Callithrix*) and the black spider monkey. The marmosets are the smallest of all monkeys and, except for their faces, look much like squirrels. More "normal" New World monkeys are the capuchin monkey, favored by organ grinders, and the spider monkeys (*Ateles*), agile acrobats which parallel the gibbons in the use of their long arms.

At the right of the page are several examples of the Old World monkey group. *Top*, the Hanuman monkey, the sacred langur of India. *Next below*, the black guenon (*Cercopithecus*), typical of the African arboreal monkeys. *Bottom*, the mandrill, peculiarly pigmented, long snouted, and short tailed, an extreme example of the ground-dwelling baboon type. (All photographs courtesy New York Zoölogical Society.)

Brachiation is the term applied to the arm-over-arm type of locomotion highly developed in great apes. This type of swinging through the trees is comparable to the use of "traveling rings" in a gymnasium. In such progression the body is held, of course, erect. Long-continued retention of this type of locomotion may be associated with the development of relatively long arms and reduction of the thumb in great apes. In the four views above, a gibbon is seen swinging his way along a limb in the Siamese forests. *Upper left,* the beginning of a swing to a new hold, reached in the third figure; in the fourth, a second swing has begun. (Photographs from American Primate Expedition, courtesy Harold J. Coolidge.)

In full flight. An enlargement of a telephoto moving-picture "shot" of a gibbon in the middle of a long leap in the treetops in Siam. (Photograph by American Primate Expedition, courtesy Harold J. Coolidge.)

Portrait of an acrobat. A gibbon, captured in Siam, of the species seen in action on the opposite page. (Photograph by American Primate Expedition, courtesy Harold J. Coolidge.)

An adolescent female chimpanzee, Meshie, reared in an American family. (Photograph courtesy Harry C. Raven.)

Orangutans, young and old. *Below*, *left*, a young specimen, showing the concave profile characteristic of this East Indian ape. Note the small thumb and with an opposable big toe on the hand-like foot. *Right*, an old male, showing the characteristic development of a ring of calloused skin around the face and a skin fold below the chin. (Photographs courtesy New York Zoölogical Society.)

An adult male highland gorilla (*below*, *right*) shot in the Kivu district of the Belgian Congo. This animal weighed more than the three scientists, shown in the photograph, put together. (Photograph courtesy Harry C. Raven.)

Chimpanzee life. Extensive studies on the behavior and mentality of chimpanzees, are being made by the Yerkes Laboratories of Primate Biology at Orange Park, Florida. On this page are shown photographs of characteristic activities.

Upper left, a six-year old female chimpanzee using a "chimpomat" (variety of automat) to get food as reward for success in an experiment. (Courtesy of Drs. J. B. Wolfe and H. W. Nissen.)

Upper right, half-grown chimpanzees working co-operatively to get food by drawing a box toward them. (Courtesy of Drs. M. P. Crawford and H. W. Nissen.)

Left, a chimpanzee family: mother at left with two-weeks-old infant clinging to her while she busily grooms the father. (Courtesy of Dr. R. M. Yerkes.)

Lower left, a full-grown female chimpanzee using a push-button drinking fountain to get water to wash her hands. She is pressing the button with one foot while catching the water with her cupped hands. (Courtesy of Dr. R. M. Yerkes; all photographs courtesy of Yerkes Laboratories of Primate Biology.)

used. But with increasing weight, four-footed running along a single branch becomes difficult. The weight is better distributed if the ape rests his feet on one branch and grasps a second overlying one with his hands. From this it is but a step to the type of locomotion in which these forms, particularly the gibbons and orang, are adept. These apes swing by the arms from bough to bough much as one swings along the traveling rings in a gymnasium, a type of locomotion technically termed brachiation. With this feature is associated the trend in apes, noted above, for long arms, long, hooking fingers, and relative reduction of the thumb (which is, in this method of locomotion, more of a nuisance than a help). The feet are so definitely adapted to a grasping function that most great apes cannot walk flat on their soles but must support their weight on the outer side of the foot.

It will be noted that in swinging by the arms the body is necessarily held erect. Except in the gibbons there is little tendency for the use of the erect posture when on the ground. But the front limbs of all these apes being much longer than the hind, the body is necessarily tilted up considerably in front even in four-footed walking. Erect posture, an essentially human character, thus has its beginnings among these tree-dwelling types.

The brain is large, especially in the gorillas, but its growth has not kept up proportionately with increase in bodily bulk, and its relatively small size is in great measure responsible for the rather ferocious appearance of such a form as the gorilla, with its low forehead and beetling brow ridges. There is, as in most mammals, no chin, the jaw sloping away under the lips. The grinding teeth tend to be rather heavy and elongate, giving more of a projecting face than in man, and the canine teeth are prominent.

Gibbons. Smallest and most primitive of anthropoid apes are the gibbons (including hoolock and siamang) of the Malay region. The ordinary gibbon does not exceed three feet or so in standing posture. Alone among living anthropoids these small apes customarily walk erect when on the ground, with the extremely long arms used as balancers. But this type of locomotion is comparatively rare, for the gibbon is a highly developed arboreal acrobat capable of feats which no circus artist could attempt to imitate. The gibbons are also notable for vocal abilities; only the South American howling monkeys can successfully compete with them in effective broadcasting. The brain cavity has a capacity of about 90 cc. In a modern European man the comparable figure is about 1,500 cc. The gibbon is, unquestionably, a much less intelligent creature than the other, higher apes; but in considering brain size, we must take into account also the fact that we are dealing with a much smaller creature.

Orangutan.—The orangutan, the "wild man" of Borneo and Sumatra, is a red-haired beast considerably larger than the gibbon, adult males reaching nearly five feet in height. The orang brain is very much larger than that of the gibbon, reaching a maximum of over 500 cc. in capacity, a figure more legitimately compared to man's average of 1,500 cc., since man and orang are in the same size-class. There is no development of heavy brow ridges and the consequent ferocity of appearance which we get in the gorillas and to a lesser extent in the chimpanzees; the eyes are set close together above the deeply concave nasal region, giving a type of face quite unlike that of other apes. The orang is, again, a good arboreal type, with

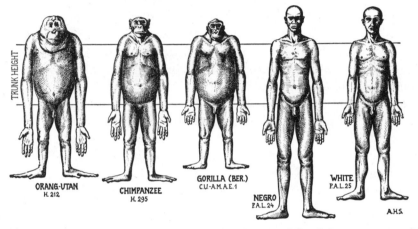

Exact semidiagrammatic front views of the four largest primates at fully adult age, constructed from detailed measurements on actual specimens, drawn without the hair, all reduced to the same trunk height. The proportionate differences between man and the great apes lie mainly in the construction of head and neck, breadth of trunk, and leg length. Note particularly that, while we think of the apes as long-armed, this is not the case with chimpanzee or gorilla; the legs have shortened, but the arms are not proportionately long. (From A. H. Schultz.)

long arms which can reach the ankles in the erect position; the orang and the remaining great apes, however, go about the trees in a more leisurely and less spectacular fashion than the gibbons.

This ape, while obviously above the gibbon in station, is probably far off the line leading to the two higher manlike apes and man, and but little is known of its fossil history.

Higher manlike apes—chimpanzee and gorilla.—Thus far our primate history has with few exceptions been one of increasing adaptation to arboreal life. But our present knowledge, small as it is, leads us to believe that by Miocene times a new tendency was already beginning, the first traces of a return from the trees to the ground. Three living creatures are descendants of the ancestral forms in which this trend first appeared. Two forms, the chimpanzee and the gorilla, have not gone far with this process of terrestrial adaptation, nor have they met with great success; a third form, man, has succeeded.

The great African apes, the chimpanzees and gorillas, are quite similar to each other in many respects and are obviously closely related. Both these living types are less pronounced in their arboreal adaptations than the lower apes and monkeys. The chimpanzee is more at home on the ground than are the gibbons and orangs but is still essentially a tree dweller. The gorilla, on the other hand, is (particularly as regards the highland type from central Africa) essentially a ground dweller and takes to the trees but little. In both forms the arms are much shorter than in the lower manlike apes but are nevertheless quite long as compared with the legs—this in contrast with human proportions. Both are large animals, the chimpanzee reaching as much as five feet in height, the male gorilla, six feet; the chimp rarely exceeds 125 pounds in the wild, but the more massive gorilla, it is said, sometimes reaches a weight of six hundred pounds in an old male.

Brains are comparatively well developed in these great apes. That in an exceptional gorilla may attain a maximum of 630 cc.; this is halfway to the average of certain human races. And intelligence, too, is increasing. Recently there has been considerable experimental work done by psychologists in testing the mentality of these forms, particularly the chimpanzee. They seem to show the rudiments of a human type of intelligence. There is a good memory, so that chimpanzees are generally much better than the average mammal at learning by trial and error such problems as which one of a choice of symbols or procedures will give a desired reward (usually food). But, more than that, there seem to be the beginnings of reasoning power as we use it. For example, instead of merely jumping blindly and vainly for food suspended out of reach, the chimpanzee may, of its own volition, pile up boxes to stand on or fit together sticks with which to knock the food down.

The chimpanzee and gorilla thus have tended toward ground-dwelling (although not as bipeds) and toward a human type of mentality. These are features which would be expected in human ancestors. Neither of these living apes, obviously, is a direct ancestor of man, but the evidence suggests strongly that we are here dealing with members of a group of animals from which man might well have descended.

But the evidence presented by these living apes leaves us with a puzzling problem with regard to the history of locomotion in the evolution of man. All these forms are brachiators, and it is not unreasonable to think that man's arboreal ancestors went in for at least a bit of arm-swinging before they descended from the trees. But the lower the ape, the more highly specialized the brachiating habit; the more like man, the less are the limbs specialized for acrobatics. Did the human ancestor first tend toward the limb build (and habits) of a gibbon and then revert to more human proportions? This seems unlikely, and some able workers have, perhaps in reaction, gone to the opposite extreme and advocated a descent of man from a primitive anthropoid along lines quite distinct from those leading to his vulgar living cousins.

Fossil apes.—What does the fossil record give us by way of an answer to this surmise? The answer must, at the present time, unfortunately be a limited and qualified one, and until recently was extremely limited. We have long known fragmentary remains of rather advanced anthropoid apes from the Miocene and Pliocene rocks of the Old World, chiefly in the Siwalik Hills of India, where we find fragmentary remains of large primates. To most of them has been applied the scientific name of *Dryopithecus* (the "oak-ape") so called because of the presence of oak leaves in the deposits from which the first remains of this form were obtained. The finds consisted almost entirely of teeth and jaws. This was disappointing; but, fortunately, teeth are the most characteristic parts of any animal, and these teeth and jaws have been carefully studied. Of these fragments we can say that they possess features found today in but three mammals—the chimpanzee, the gorilla, and man—but in default of any knowledge of the face, braincase, or body proportions and structure, we can go no further.

In recent years, however, some light has been shed on the problem by materials from East Africa. There, in the region of Victoria Nyanza, have been discovered fossil beds of early Miocene age—later, that is, than the Fayum beds that have the oldest anthropoids but definitely earlier than those with the tantalizing fragments of the "oak-apes." Some of the primate material discovered pertained, as might be expected, to ordinary (if extinct) monkeys and a gibbon relative. More interesting were a number of specimens of a more advanced great ape, named by its describer *Proconsul* (in honor of Consul, a performing chimpanzee of his acquaintance). There is some variation in size among the specimens of this form, but on the whole they indicate animals of considerably smaller size than the larger apes of today; the best-preserved skull, for example, is only about five inches long. The skull and jaws are definitely of an advanced ape type but

A mountain gorilla, M'bongo, who was a longtime resident at the San Diego Zoo. He was about 13 years old when the photographs were taken and weighed 602 pounds. The upper picture shows well the quadrupedal pose of gorillas. The crest atop the head is characteristic of the mountain gorilla of Central Africa. Like the much lower crest in the gorillas of the western lowlands and the chimp's eyebrow ridges, these crests serve as attachment for the muscles which support the massive lower jaw and chewing apparatus. (Photographs courtesy Belle J. Benchley, Zoölogical Society of San Diego.)

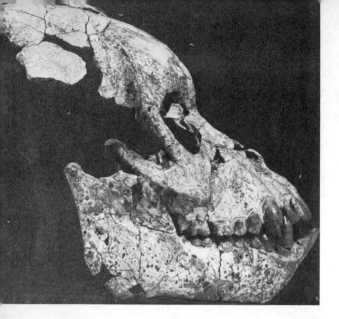

Fossil apes. *Above*, side and front views of the skull of a fossil great ape, *Proconsul*, from the early Miocene of East Africa (after Le Gros Clark). *Below*, *Oreopithecus*, a small form from the Italian Pliocene, which shows features suggestive of human relationships. This is currently being intensively studied by Dr. Hürzeler of Basel; only the parts in light color are well preserved. *Right*, *Dryopithecus* ("oak ape"), a relatively large form of which fragmentary remains, such as those figured, have long been known from the late Tertiary of India and Europe.

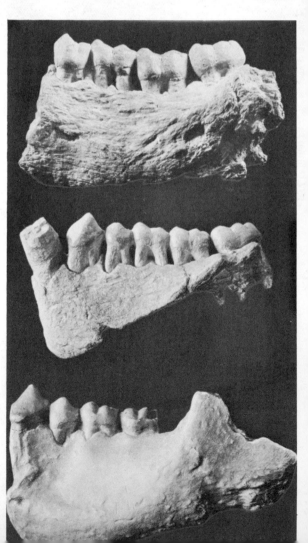

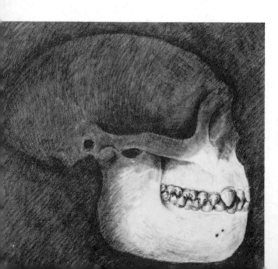

quite unspecialized; such an animal might have been ancestral to the modern chimpanzee, but, more than this, there is little in the animal's structure to debar it from ancestry to the gorilla or even man. In fact, in the absence of certain specializations of skull and teeth which are found in chimpanzee and gorilla but never developed (it would seem) in the human line, little *Proconsul* seems even closer to an expected ancestor of man than to living great apes.

What of the problem of the history of locomotion, which concerned us in the discussion above? Regrettably, little of the skeleton of the body and limbs has been so far discovered. Those bones which we do have indicate that the animal was slenderly built and obviously active and agile. Further, there is enough material of the arms to show that although *Proconsul* may have brachiated to some extent, he had not become markedly specialized in this direction. Still further, there is some evidence from the material that he may have progressed to some degree on the ground and was not purely a tree dweller. This is in agreement with geological evidence suggesting that *Proconsul*, unlike the modern chimpanzee and gorilla, did not live in a dense tropical forest but in an environment rather like that found in much of East Africa today—a savanna country, with patches of woodland separated by grassy prairies.

How and why man's arboreal ancestors left the trees for the ground has been often discussed. Did they descend voluntarily, seeking advantages to be found in life on the ground? Or were they driven out onto the earth by competition with primate cousins? A third alternative is suggested by geological evidence indicating that in middle Tertiary times grasslands increased greatly at the expense of forest areas; ancestral men may have been forced to become ground dwellers because the forest, so to speak, disappeared from under them. A fourth idea is one not unrelated to the last, suggested by the conditions thought to be present in the Miocene landscape of East Africa. Even if *Proconsul* was primarily a tree dweller, he could not have traveled far without crossing open plains. He may have evolved potentialities as a ground walker so that he could live successfully in the trees (a seeming paradox, parallel to that which we have already cited in the case of the tetrapod ancestors, namely, that they became able to walk on legs so that they could stay in the water).

FROM APE TO MAN

Human versus ape characters.—The gulf between the living great apes and a modern man at first sight seems to us a huge one, but, as a matter of fact, the anatomical features which distinguish men from apes are comparatively few. Bone for bone, muscle for muscle, organ for organ, almost every feature of the ape is repeated in the human body. The differences are almost entirely differences in proportions and relations of parts; the structures are almost identical.

The differences are related mainly to locomotor habits and brain growth. Among the higher apes we have seen trends toward erect posture and toward terrestrial life; man alone has carried out these tendencies to their logical conclusion and has become a ground-dwelling biped. His arms, not used for swinging from the branches, are comparatively short. The evidence of *Proconsul* indicates, we have seen, that our ancestors probably never had the disproportionately long arms and short legs of the extreme brachiators. The human hand, except for its somewhat greater flexibility, is of a primitive primate type, while the living great apes have tended to elongate the fingers, except the thumb, for hooking over branches.

A greater difference between man and other primates lies in the foot. In the great apes, as in primates generally, the foot, like the hand, is an efficient grasping structure. But this is an awkward affair for ground life, particularly in a biped where the feet must bear the entire weight of the body. In man the toes have shortened, the big toe has been brought into line with the others, and the heel bone is expanded for a prop at the back (this last feature was already partially developed in the great apes). In the evolution of erect posture the backbone has become more sinuously curved than in the apes, with the result of bringing the trunk and head directly above the hips and centering the weight over the legs.

In the great apes there is more of a muzzle than in man, for the teeth are more powerful and longer; the canines, too, project. In man the teeth are less powerfully developed, and the face is shorter; the nose has become more prominent, and the chin has developed in modern types.

Outstanding, of course, has been the development of the brain. In modern human races the average brain size ranges, roughly, from about 1,200 to 1,500 cc., or, on the average, two to three times that of the great apes. This difference has, of course, been responsible for many changes in the proportions of the skull and the general appearance of the head, the reduction of the heavy eyebrow ridges, and the development in their place of a high forehead. Brains have been responsible for giving man, otherwise a rather feeble creature, the place in the world which he now occupies. But, except in size and the great development of certain areas which are presumed to be the seat of the higher mental faculties, every feature of the human brain finds its parallel in the brain of the great apes.

Man has gone far and, we trust, may go still farther along new lines of evolution. But in his every feature—brain, sense organs, limbs—he is a product of primate evolutionary trends and owes, in his high estate, much to his arboreal ancestry, to features developed by his Tertiary forefathers for life in the trees.

South African man-apes.—The differences between man and living apes are, as we have just seen, rather trivial and would not concern us were we dealing with some other group of animals. But human evolution is a matter of personal concern, and we naturally desire in this case to close the evolutionary gap. Discoveries of fossil forms in South Africa, which are structurally intermediate between apes and primitive men, have in recent years gone far toward satisfying this desire.

Some three decades ago limestone cliffs at Taungs in the western Transvaal were being quarried for cement-making. In the deposits were partially filled caves containing remains of fossil Pleistocene mammals (notably baboons). However, one skull, sent for study to Professor Dart, an anatomist at Johannesburg, was startlingly different, for it was that of an anthropoid different from any living or fossil form. To this specimen the name *Australopithecus* ("southern ape") was given. Part of the braincase had been lost, but this was rather an advantage than otherwise, for its removal exposed a cast of much of the brain. The individual was a youngster, the stage of eruption of the teeth corresponding to a human child of six or seven. The youth of the specimen was an unfortunate feature, for the differences between a young ape and the human infant are much less obvious than differences between the adults. The creature was certainly below normal human mental standards, for estimates of brain size indicated that the adult brain would have been much closer in volume to that of the great apes than to that of man. However, the youngster's skull shows marked differences from those of chimpanzees and gorillas of similar age. The teeth, particularly, show many human characters, although they are of quite large size for the jaws (a feature not found in either typical great apes or typical man). It was hence claimed that *Australopithecus* was a very advanced ape type approaching, although not reaching, the human level.

This claim appears to be substantiated by later discoveries of adult skulls from other quarries in the Transvaal by the veteran paleontologist Broom, his successor Robinson, and further work by Dart himself. From four other localities have been recovered remains—fortunately adult remains—of advanced anthropoids related to the original form, although differing to a fair degree among themselves (and rather unnecessarily given a number of different generic names).

Nearly all the remains are those of skulls and jaws, none entire but among them giving us a nearly complete knowledge of these structures. The adult skull is very similar to the structure predicted from that of the child; it is, of course, much more apelike, with a face projecting to a considerable degree and with a low forehead and prominent brow ridges. Nevertheless, it is much less apelike than that of a gorilla or chimpanzee and a bit more toward the human side. The muzzle is less projecting than in the living great apes; the brow ridges are much less prominent than in the higher apes; and while the forehead is low, it is much more domed above and the whole braincase, in rather human fashion, is set higher above the face than in great apes. Particularly interesting is the rear of the skull. In four-footed primates, and even in great apes, the back of the skull is a flattened area to which are attached the muscles holding the skull straight on to the backbone in a more or less horizontal position. Here the braincase slopes smoothly down behind; the muscles were attached to the underside of the skull. This fact, and a related shift in the position of the condyles which join the skull with the backbone, proves that the skull was set, much as in man, at right angles to the summit of the backbone and that (as already deduced from the imperfect infant skull) the creature stood erect, or nearly so.

In skulls sufficiently well preserved to allow the size of the brain to be estimated, the figures do not run above 700 cc. or so—that is, far below any human standard and no bigger than those for large gorillas. But relative to the size of the creature, the brain is proportionately large, for most (not all) of these man-apes appear to have been small, with a height of only four feet or so—about the size of modern pygmies.

The teeth are large for the size of the skull and jaw but, as in the child's skull first discovered, are closer to the human than to the ape pattern. The canine teeth are stout but wear down to the general level of the tooth row, rather than remaining as projecting "tusks" as in many apes. Particularly significant is the shape of the tooth row. In apes the teeth in either upper or lower jaws are arranged (as shown in our figure) to form three sides of a square, the canines placed at the corners. In all men they are arranged in an even curve; and this is the case in the "southern apes" as well.

Of the rest of the skeleton our knowledge is very incomplete. But such bones as we have are much more of the human pattern than of the ape

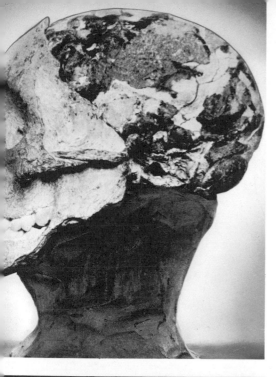

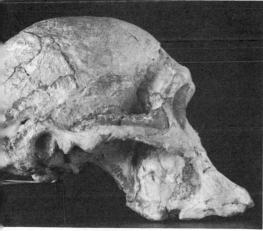

South African man-apes. The first *Australopithecus* ("southern ape") find, in 1925, was the skull of a child (*upper left*), followed by later findings of adult specimens. *Left*, an adult *Australopithecus; left below, Paranthropus*, a related form; *below*, an australopithecine skull and jaw, restored. *Above*, home life as it may have been among these man-apes. (Juvenile skull after Professor Raymond Dart; adult skulls from Dr. J. T. Robinson; restored skull after Dr. Robert Broom; drawing from *Life*, by Michael Ramus.)

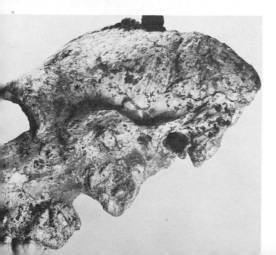

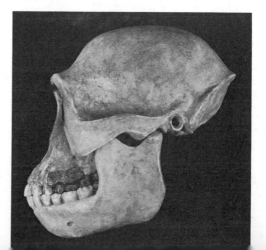

Pithecanthropus, the Java apeman. The original remains were found on the banks of the Solo River, near the village of Trinil. The site is shown at the left. Most of the excavation was made by a later expedition which found, however, no further human bones. Next below (*left*) is a view of the original skullcap seen from the upper surface. *Lower right*, a restoration by Dr. J. H. McGregor of the skull. The dark line from brow ridges to the back of the skull marks the lower limit of the original specimen. Below this point, reconstruction was once hypothetical. New and better braincase finds, together with considerable portions of upper and lower jaws, now enable us to picture the entire skull with considerable confidence. This skull is closely comparable to *Sinanthropus*, pictured over-page. *Lower left*, a bust in which the flesh and skin have been carefully restored by McGregor to give an idea of the appearance of *Pithecanthropus* in life. (Site after Selenka and Blanckenhorn; other photographs courtesy J. H. McGregor.)

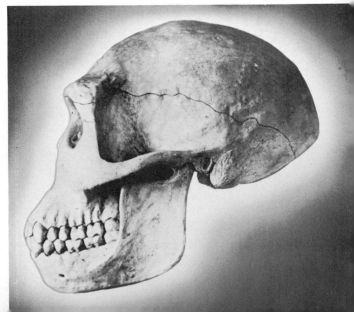

type and confirm our impression that we are dealing with an erect ground dweller and not an arboreal ape. Particularly significant are the hip bones, the pelvic girdle. In all other primates besides man the upper bone of this structure (the ilium) is relatively narrow. In man this bone is greatly broadened, in relation to muscular needs for erect posture; in *Australopithecus* the bone is nearly as broad as in a true man.

Were these "southern apes" purely southern, or is the fact that our specimens are all from South Africa merely due to accidents of preservation and discovery? There is some suggestion that they were actually widespread in the Old World at an early Pleistocene stage; and the indications of this also help in solving a seemingly unrelated puzzle.

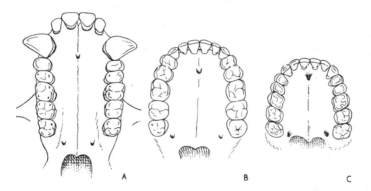

A palatal view of the upper tooth-row of *A*, a chimpanzee, *B*, an australopithecine, and *C*, a modern man (Australian), to show the similarity of the rounded curve of the australopithecine tooth-row to that of man and the strong contrast with that of a modern great ape. (From LeGros Clark.)

In the 1930's a Dutch scientist, von Koenigswald, discovered in a Chinese merchant's shop some primate teeth obtained from a Pleistocene deposit. These were teeth either of an advanced ape or a primitive man, but of enormous size, so that if the teeth had the same proportions to the rest of the skeleton that is usual in apes and man, their former owner must have been a giant among primates, far larger than a gorilla. The same worker also discovered in Java a fragment of a jaw which appeared to be that of a very primitive man, also with very large teeth. Dr. von Koenigswald (and the teeth) presently disappeared into a Japanese prison camp in Java. During the war, before he (and the teeth) re-emerged, a distinguished student of fossil man, Professor Weidenreich, proposed the hypothesis that this material indicated that man had come from giant-ape ancestors. But we have noted that in australopithecines the teeth are disproportionately large for the animal. Very probably these Asiatic finds (and a similar find from East Africa) belong to members—or descendants— of that man-ape group, and the bodies to which these teeth belonged may not have been of anything like the size that good Dr. Weidenreich imagined.

What sort of life did these man-apes lead? They were definitely not tree dwellers, for the country in which they dwelt appears to have been then, as today, essentially an open one. They were, at least in part, car- nivorous forms, for the caves in which they lived have yielded numerous remains of animals which appear to have been used for food supply; particularly abundant are baboon skulls, broken and with the brains removed. (A repulsive note: the known skulls of the man-apes them- selves appear to have under- gone similar treatment, per- haps for similar reasons.) Al- though there is no trace of their having manufactured tools of any sort, it is possible that they may have used animal thigh bones to bash the heads of their prey, and in one site there is a suggestion of the use of fire.

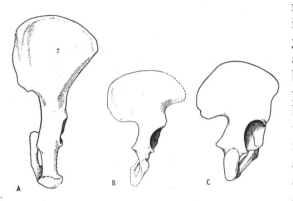

The pelvic girdles of *A*, gorilla, *B*, australopithecine, and *C*, man, to show the manlike broadening of the upper bone (ilium) in the australopithecine, in contrast with the narrower great-ape ilium. (From LeGros Clark.)

What is the evolutionary position of these "southern apes"? I have here included them in a chapter dealing with apes rather than one concerned with men and have called them "man-apes" rather than "apemen." But in fact, placing them in either category would be equally proper (or im- proper), for, as we have seen, they structurally split the difference between primitive man and the type of true ape from which man might be ex- pected to have originated. But does this mean that they are actually hu- man ancestors? This is debatable. The age of the South African specimens is still not too well settled. Probably all are to be regarded as Pleistocene, but some may be well along in that period. If so, they are too late in time to be ancestors of man, for primitive men and implements made by them are known far back in the Ice Age. The extreme opposite point of view,

held by some, is that these man-apes are not at all related to the human
line and simply represent a sterile side line of advanced apes which paralleled man in certain features. But the human resemblances are so numerous that this is difficult to believe. A reasonable point of view, based on
current evidence, is that the australopithecines may well represent a stage
in the ascent of man, but that part if not all of the known South African
forms may be surviving relicts, later in time than the actual members of
the group from which man arose.

CHAPTER 17

Human Origins

The history of man is, quite naturally, of particular interest to most of us; and, quite naturally, we would wish to inquire into it in greater detail than in the case of other animal lines. This eagerness for details regarding human history is, however, somewhat embarrassing to the paleontologist. As with the primate group in general, ancient remains of man are rare. It seems probable that the center of our early evolutionary history lay in Asia or Africa, regions which are comparatively poorly explored and which have yielded few remains. The human story is hence very fragmentary and inadequate at the present time. Our knowledge is constantly increasing; many of our most important fossil finds have been discovered within the past few decades, and, doubtless, with the continued advance of exploration and research other interesting finds will continue to be made. For the present, however, we can merely describe the isolated remains so far discovered and mention the theories as to lines of descent which they suggest.

THE ICE AGE

Human evolution is essentially a Pleistocene story. Advanced manlike apes are known in Miocene and Pliocene rocks, and types transitional to man were presumably in existence in the latter epoch. These are, however, quite unknown as fossils, and we have at present no human remains back of the Pleistocene, which covers (at the most) only the last million years of earth history.

This epoch was marked by the occurrence of a vast amount of glaciation. Great sheets of ice formed in the northern regions of Europe and North America and, moving southward, covered large portions of those continents with great glaciers, while smaller glacial areas formed about the higher mountain ranges of the Temperate and even Tropical regions (and a southern ice cap, which persists today, covered Antarctica). Such conditions (which might easily recur if the average temperature of our present Temperate Zones fell but a few degrees) unquestionably exerted a profound influence over the life of the northern continents. As the glaciers moved southward, regions with warm climates changed to temperate areas, to pine forests, to barren tundras, and, finally, if in the glacial path, became deeply covered with slowly moving masses of ice; with the retreat of the ice cap this sequence was reversed.

Pleistocene animal life.—Under the influence of conditions of this sort, we might expect vast migrations of animals and the extinction of forms unable to adjust themselves to new environmental conditions. We might expect, too, that changed conditions, with a premium placed upon adaptations to meet them, would stimulate evolutionary development. The Ice Age may have been a potent influence in the advance of man.

Among the animals with which our Pleistocene ancestors contended were most of the living forms and, in addition, many types now entirely extinct or much restricted in their distribution. In Eurasia there were great elephants—mammoths—of several types, the woölly mammoth of the north and types adapted to warmer conditions in the south. Rhinoceroses were abundant in Europe, and great herds of horses were present on the steppes which covered much of that continent. Remains of giant bears and lions are plentiful in European cave deposits. The reindeer ranged at times south to southern France, and, on the other hand, the hippopotamus reached England during the warmer parts of the period.

In North America, too, there were many creatures now extinct. Here, also, were various types of mammoths, as well as the mastodon, a more primitive proboscidian. Camels and horses roamed the American plains. Giant, short-snouted bears, lions, and great saber-toothed cats were among the most conspicuous of carnivores. Great ground sloths, one as large as an elephant, were numerous, and in the south were present the glyptodonts, giant cousins of the armadillos.

The glacial areas.—We must not overemphasize the extent of the Pleistocene glaciation. It was almost entirely confined to the northern Arctic and Temperate Zones, and even here vast regions were untouched. In North America the line of the Ohio and Missouri rivers marks out, roughly, the outer limits of the advance of the glaciers arising in arctic Canada. In Europe, Scandinavia was the center of a glacial area which covered most of Great Britain, northern Germany, and northern Russia. The Alps formed a smaller center of activity from which glaciers descended the mountain slopes some distance into the surrounding areas, and other European mountains formed minor centers. Much of northern and central Asia appears to have been unglaciated, although with a cold climate, and large areas of Europe and North America were never touched by the ice caps.

The glacial sequence.—When glaciation was first seriously studied, somewhat over a century ago, it was thought that there had been a single advance and retreat of the ice. This is now, however, known not to have been the case; there were wide fluctuations in the extent of the glaciated areas, and, between successive advances of the ice, regions in the glacial areas have had climates seemingly warmer than they have today.

In America glacial studies have resulted in the clear distinction of four successive glacial maxima, between which for long periods of time—tens or even hundreds of thousands of years—there were long interglacial periods with temperate to warm climates in the northeastern United States and southern Canada. In Europe studies of the Alpine glacial areas have furnished considerable evidence that there, too, there were four times of maximum glaciation, with three interglacial periods. These four maxima have been given names derived from Alpine localities in which the deposits formed at that time are well preserved—Gunz, Mindel, Riss, and Würm, names chosen partly because they follow one another in alphabetical as in chronological sequence. The Gunz and Mindel glaciations may have occurred close together in time in the early Pleistocene, with but a short interval between them. The Riss and Würm may have been similarly associated toward the end of the Pleistocene. A long, warm, major interglacial period may have occurred between the two great times of ice activity. These ideas have been incorporated in the accompanying diagram.

Dating human remains.—Debatable as this subject of the number of glaciations is, it is the only "timetable" we have for most of the Pleistocene, and it is important to attempt to tie our knowledge of human history into this glacial sequence. This, however, can be done only in a limited way, and the evidence is none too clear. If human remains were actually found in glacial deposits, their position could be determined with comparative ease; but the edge of a glacier is not the place that man would seek as a comfortable habitation, and such finds are negligible. Finds in caves often occur in stratified fashion, thus giving us a sequence for fossil remains; but these cave deposits usually cover but a limited period of time. The advance and retreat of the ice caps affected climatic conditions over large areas and caused great fluctuations in the flow of streams far from the glaciated areas. These variable conditions resulted in the development of terraces representing old valley levels along stream channels, and the sequence of remains frequently found in such terraces may be determined. Associated with glacial advances and retreats there seem to have been great fluctuations of the coast lines. In a number of regions are found

coastal terraces, indicating that the shore once lay at higher levels, sometimes several hundred feet above the present strand.

In tropical and subtropical regions the glaciers cannot, of course, give us
direct evidence of the geologic age of human finds. Indirectly, however,
they are of aid. Changes in the glaciated regions presumably influence

The Northern Hemisphere on a north polar projection, to show the extent of the Pleistocene glaciation (glaciated areas in black). The probable maximum extent of pack-ice in the North Atlantic and
North Pacific is indicated. The glaciated area included mainly large areas of North America and
northern Europe, with various outliers in the Alps, Pyrenees, and other mountain chains. Much
of western and central Europe was free of ice, even at the times of extreme ice advances, but obviously
had a cold climate at such times. Part of Alaksa was never ice-covered, and, except at the extremes
of glaciation, a migration route (if a chilly one) presumably extended southward near the Rockies.
(After Flint.)

climatic conditions in far distant regions. In correlation with this, workers
in the tropics find strong evidence of a sequence consisting of two wet or
"pluvial" periods separated by a dry "interpluvial." The wet stages cor-
respond, it would seem, to times of glacial advances to the north, the early

pluvial including Gunz and Mindel times, with little trace of the short first interglacial; the interpluvial corresponds to the major interglacial of the

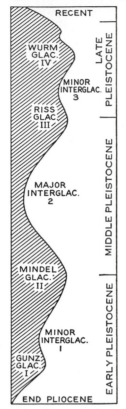

Middle Pleistocene, and the second pluvial, to the combined Riss and Würm stages.

Animal remains found with those of man are of importance. They will tell us whether the times were warm or cold—glacial or interglacial—and the presence or absence of certain forms may afford evidence that we are dealing with early or late Pleistocene times. Plants, too, furnish useful evidence, for they are even more restricted by differences in environment than are animals and hence can reflect the climatic fluctuations which occurred during the Pleistocene. An interesting recent development is due to the discovery that pollen grains may be preserved intact in peat beds for long periods of time and yield valuable information concerning the types of trees present when the peat bog was formed.

Man has long been a toolmaker and tool-user, and stone implements are found in many Pleistocene deposits. The general sequence of the human cultures to which these implements pertain is in many cases well established, and human fossils associated with stone tools can often be assigned to a definite cultural period. But this does not always assure an absolute date, for cultural stages may overlap. The reader of this book lives in an advanced stage of culture; the native Australian is still in the Stone Age. We have no guaranty that greater uniformity existed in the past.

In sum: We may attempt to date early human remains by the geologic conditions of the find, by the associated animal and plant remains, and by implements discovered with them. But this evidence is often difficult to

A diagram of Ice Age chronology. The distance from the bottom to the top represents the duration of Pleistocene and Recent times—perhaps a million years. The curved line represents successive advances (curvature to right) and retreats (curvature to left) of the glacial front in Europe and America. There were four such advances, named from Alpine localities and their intervening interglacial stages, the second one perhaps much longer than the others. For convenience the Pleistocene may be divided into Early, Middle, and Late portions, indicated here and as used in subsequent discussion in the text.

correlate with the glacial timetable, and this timetable, in turn, is still not too surely established. Hence dates in years assigned to earlier human finds must be regarded as only tentative in their nature. Helpful in age determinations is the recently developed fluorine test. Bone well buried in the earth is liable to be bathed, periodically or constantly, by ground water. This typically contains minute amounts of the chemical element fluorine, and some of the fluorine, in the course of time, comes to be deposited in the bone; the longer the bone lies in the ground, the more fluorine it comes to contain. Unfortunately, this test cannot be used to give an *absolute* determination of time, for the amount of ground-water fluorine to

Geological Time	Glacial Sequence	Rough Estimates of Time in Years since Beginning of Stage	Cultural Stages	Type of Culture		
Recent	Postglacial	1,400 B.C.	Iron Age	Use of metals		
		3,000 B.C.	Bronze Age			
		5,000 B.C.	Neolithic	Agriculture and animal domestication		
		13,000 B.C.	Mesolithic	Azilian, Tardenoisian, Maglemosian, etc.		
Upper Pleistocene	Late Würm glaciation (IV)	40,000	Upper Paleolithic	Aurignacian, Perigordian, Solutrian, Magdalenian in Europe; related cultures in other areas		
	Early Würm glaciation / Third interglacial	250,000	Middle Paleolithic	Mousterian culture of Europe and western Asia, Levalloisian (continued), etc.		
	Riss glaciation (III)			Hand-ax cultures of western and southern Europe, Africa, and east to India: Acheulean, preceded by Abbevillean	Flake and chopping-tool cultures of northern and central Europe: Clactonian, Levalloisian (part), etc.	Chopping-tool and flake cultures of southeastern Asia: Soan, etc.
Middle Pleistocene	Second interglacial / Mindel glaciation (II)	800,000	Lower Paleolithic			
Lower Pleistocene	First interglacial / Gunz glaciation (I)	1,000,000	Pre-Paleolithic	Crude beginnings of use of stone: "eoliths," etc.		

which a bone is exposed varies from place to place. It can, however, be highly useful in giving the *relative* age of bones supposedly from the same layer or deposit; it was, for example, useful in exposing the Piltdown hoax, discussed later.

But for giving absolute and fairly accurate dates for at least the tail end of the fossil story, two methods are available—one geological, one chemical. The abundant waters flowing off the margins of glaciers, particularly during the summer, carry with them mineral particles which are deposited as thin layers of clay, termed "varves," in nearby ponds or lakes. Summer and winter layers differ in color and thickness, and their thickness may vary from year to year, with variations in the weather. A vertical cut through a clay bank formed near the one-time margin of a glacier may reveal dozens or even hundreds of these annual varves. These can be measured for thickness and matched with overlapping varve series farther out from the glacial center or closer to it. As a result, in Europe and less perfectly in North America we can establish a fairly accurate timetable in years for the retreat of the last Würm glaciers and estimate rather accurately the age of any bones or implements that can be tied in to the varve sequence.

Unfortunately only a very small fraction of the material can be tied in thus. Much more important—in fact, causing a major revolution in work in human paleontology and archeology—has been the recent development of a chemical method using carbon-14 in age determination.

The element carbon is the basic element in all living matter. An atom of ordinary carbon weighs the equivalent of twelve atoms of the lightest element, hydrogen. But commonly present in the carbon dioxide in the air and hence taken up by plants is a small fraction of a different form of carbon—an "isotope" (as variant forms of an element are termed) with an atomic weight of 14 rather than 12. When wood and bones are first buried, they contain the normal fraction of carbon-14. But this isotope is unstable and gradually breaks down to become another common element, nitrogen. The rate at which the change occurs has been determined by the physicists; half of the amount present in an object, if buried, will disappear in somewhat over 5,500 years. In consequence, if a chunk of organic material, such as a piece of wood, is obtained from an ancient human site and chemically analyzed, the percentage of carbon-14 still present will give us a reasonably accurate determination, in years, of the age of the deposit. Use of this method in the last decade or so has shed much light on the later phases of human history. But there is an unfortunate limitation. The breakdown of carbon-14 is, from a geological point of view, a rapid one; after about 25,000 years, so little is left that accurate determination is impossible for sites of greater antiquity; even after 10,000 years or so, the dates obtained are none too precise.

HUMAN CULTURES

Although the skeletal remains of ancient man are our main concern in this chapter, cultural objects are generally found associated with such fossils, and we should therefore consider, if briefly, cultural as well as physical evolution.

Many of the works of man are made of perishable materials such as vegetable fibers, wood, skins, and hair, which disintegrate rapidly in the earth. For his primary tools and weapons, however, man has always used harder and more lasting materials—metals in more recent times, stone in earlier days.

Flint.—Among the types of stone available, man early discovered the value of flint as a material for the making of implements. Whereas many types of stone are composed of mixed materials which fracture irregularly and frequently abrade all too readily, flint is a homogeneous, siliceous material which is extremely hard and retains an edge well and yet can be readily chipped by an experienced hand into a variety of useful tools. Flint occurs as nodules in chalk rocks, common in many parts of the world, and hence was available for use by numerous ancient races of man. In the absence of flint other materials were used—quartzite, silicified tuff, hard shales, etc.—but these are generally inferior in their properties and make relatively crude implements.

We can distinguish two major types of stone tools used by man in his early history. In one, common in certain of the oldest recognized cultures, it is the nodule itself which forms the implement; flakes are struck from it until the remaining core has been given the desired shape. The second type, by far the more common, is that in which the flakes rather than the core are the implements.

In many cases the flakes appear to have been struck off the core more or less at random. Such flakes may be useful without further work being done upon them but frequently were retouched, a series of smaller chips being removed along the margins to give more perfect shape and greater usefulness. On the other hand, the shaping was sometimes done before the flake was struck off; the core was prepared by preliminary chipping, so that when the flake was dislodged by a skilfully directed blow, it was already of the desired contour and needed no further work done upon it.

As a result of these methods ancient man was able to manufacture a variety of implements. In some cases we can hardly guess as to the uses to which they were put, but they appear to include a variety of scrapers, knives, awls, chisels, etc., which experiment shows can do effective work upon such materials as wood and hides.

Below we will characterize briefly the sequence of cultures which the study of such remains shows to have occurred in the process of human development. These stages are termed "ages." Following a vague period in

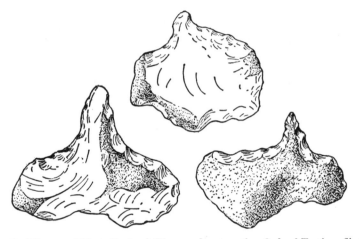

Eoliths from the Pliocene of Kent, England. These and many other finds of Tertiary flints supposed to be of human manufacture are, in many cases at least, probably of natural origin. (After Harrison.)

which man may have made the first steps in tool-making, they include: (1) the Paleolithic, or Old Stone Age, covering much of the Pleistocene and frequently divided into Lower, Middle, and Upper periods; (2) the Mesolithic Age of the early millenniums of geologically recent times; (3) the Neolithic, or New Stone Age, when man first began to lead a settled rather than a nomadic existence; and (4) the metal ages, Bronze and Iron, which carry the story down to historic times.

Eoliths.—In Middle and Upper Pleistocene deposits we find numerous stone tools which were definitely made by man and which fall into well-recognized categories in characteristic cultures. But before this level was reached, there must have been stages during which man made his first crude experiments as a toolmaker and tool-user, availing himself of such pieces of flint as came to hand or, at the most, knocking off a few chips to fashion the piece to his hand or to his desires. To flint finds believed to be of this nature, the term "eoliths," or "dawn stones," is often applied. But it is obvious that it is a difficult if not impossible task surely to tell such implements from stones chipped or worn through natural causes, and the status of many supposed discoveries of this sort is dubious. Numerous finds of flints from the early Pleistocene have been claimed as man's handiwork. This is, a priori, not unreasonable, and this period may perhaps be termed the Pre-Paleolithic; surely by the beginning of the Pleistocene man's ancestors must have advanced beyond the apes, and many of these flints give some appearance of having been worked. However, equally convincing eoliths have been described from the Miocene and Pliocene, when it is probable our ancestors were still on the ape level. These are open to considerable suspicion, and skepticism is even stronger concerning still older eoliths from the Oligocene and even the Eocene, when not even a monkey was in existence.

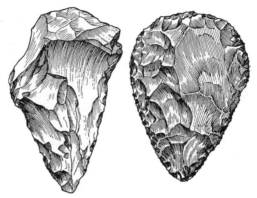

Examples of the hand ax, or *coup de poing*, from Abbevillean (Chellean) (*left*) and Acheulean (*right*) deposits. About ¼ natural size. The Abbevillean types are crudely made; those from the Acheulean are more firmly chipped and thinner and the edges straighter. (After Cole.)

Lower Paleolithic.—In contrast with this dubious situation is that which we encounter when we progress into the middle portion of the Pleistocene. Here man was a maker of stone tools which, although frequently crude, are readily identified and classified by the expert.

The classic Lower Paleolithic cultures are those identified nearly a century ago in the terraces along the Somme River in northern France at Abbeville, St. Acheul, and other towns in the Amiens district. In these cultures the characteristic tool was the hand ax, or *coup de poing* (although flakes were used as well). This was a large nodule of flint often weighing several pounds from which chips had been removed to give it an almond-like shape, the rounded end fitting the palm, and a cutting edge extending down either margin. The older phase of this type of culture, in which the hand ax was crudely prepared, is the Abbevillean;[1] in the succeeding and

[1] Formerly termed Chellean.

closely related Acheulean the implement was more carefully fashioned and the cutting edges straighter. Abbevillean and Acheulean types are widespread in Middle Pleistocene deposits in western and southern Europe; beyond this region they have been found as far east as India and south through Africa to the Cape.

But it has been found in recent years that there were other widespread cultural elements in the Lower Paleolithic. In Europe there existed, alongside the hand-ax cultures, others in which the implements were made exclusively from flakes, not cores. Their classification and relationships are by no means settled as yet. We may note the Clactonian culture, characterized by an abundance of small rough flakes, and the Levalloisian, in which the principal type was a large thin flake struck from a carefully prepared core.

A Levallois-type flake. Large flakes of this sort are characteristic of many horizons of the Lower and Middle Paleolithic. The flint nodule from which they were struck off was first prepared by chipping the surface to the desired contours. After the flake was detached by a sharp blow no further work was done upon it. In consequence one surface (the original inner one) is smooth, except for a swelling (termed the "bulb of percussion") near the point where the blow was struck. The specimen illustrated was nearly 5 inches long. (From De Mortillet.)

During the last two decades there has been emerging knowledge of still another culture complex characteristic of the Middle Pleistocene of southeastern Asia. In this we find, in addition to flakes, a core tool; this latter is not a hand ax but a crude, massive chopper, made by striking some flakes off a modest-sized cobblestone. Such cultures are present in India (the Soan culture), Burma, China, and Java and are of great interest because they appear to be associated with the early human types of that area—*Sinanthropus* and perhaps *Pithecanthropus* as well.

Apart from the bare knowledge of their tools, we can tell little of the life of men of the early Paleolithic except by inference. Occasional discoveries of hearths indicate that they were users of fire. Clothes, if present, can have been only in the shape of hides. But there is no indication of homes of any sort, probably nothing beyond the use of natural rock shelters and caves or of boughs arranged as a windbreak. They were without agriculture or domestic animals and hence must have been nomadic hunters and food-gatherers. All in all, their life, judged by modern standards, must have been on a low plane—and a most disagreeable one.

Middle Paleolithic.—Later in the Pleistocene, during much of the last (Würm) glaciation and the warm period preceding it, the characteristic European culture is the Mousterian, which derives its name from the rock shelter of Le Moustier in southern France. In the typical Mousterian the hand ax has disappeared, and we find instead a variety of flake tools which are carefully shaped by retouching. This suggests, as may well be the case, that the Mousterian has been derived from the flake cultures of the Lower Paleolithic. There is no sharp line of demarcation between Lower Paleolithic and this period, and earlier types of implements still persist in the Middle Paleolithic. Even the hand ax is found in some sites of this age (as at La Micoque in France), while the large flakes characteristic of the

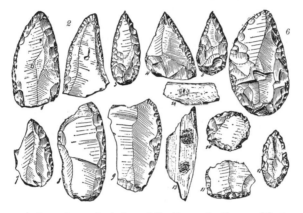

Mousterian implements from the rock shelter of La Ferrassie, France. Most are well-shaped flint flakes with the edges carefully retouched. (Bones *13, 14*) were used for flaking flints by pressure or as chopping blocks. (After Capitan and Peyrony.)

Levalloisian are frequently encountered with typical Mousterian implements. The Mousterian is common in many parts of Europe and on east into Palestine, west Turkestan, and Siberia, and closely related cultures, together with the Levallois flakes, are widespread in Africa. In many parts of Asia, however, the Mousterian is unknown. In the Middle Paleolithic, in contrast to earlier cultures, the makers of the tools are definitely known; in numerous instances the Mousterian sites have yielded remains of Neanderthal man.

Upper Paleolithic.—The last phases of the Würm glaciation (and hence the very end of the Pleistocene) are characterized by the Upper Paleolithic cultures. Here for the first time we are dealing with implements definitely associated with men of our own species, *Homo sapiens*. In Europe the cultures of this time are well known and have received names based on sites in central and southern France. At the beginning of the period, as much as 30,000 to 40,000 years ago, there appear to have been two somewhat distinct cultures, the Aurignacian and the Perigordian. Succeeding them, we find in many sites a Solutrian layer containing a culture of a different type, which may have represented a relatively brief invasion of western Europe from the east. This is in turn succeeded for a relatively short time by the Magdalenian and related cultures, representing a local evolution from the Aurignacian and Perigordian. The Solutrians appear to have been nomadic steppe hunters of wild horses; the "native" Upper Paleolithic peoples of Europe, although likewise hunters, of course, seemed to have been less nomadic, although probably migrating with the seasons between outdoor shelters and caves.

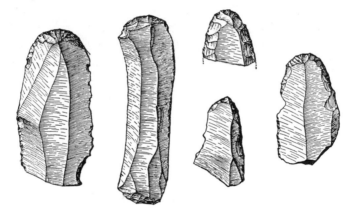

Upper Paleolithic scrapers from the cavern of Font-de-Gaume. Long blades of flint, simply but effectively worked, are characteristic of the Aurignacian and Magdalenian. (After Capitan, Breuil, and Peyrony.)

Flint work in the Upper Paleolithic cultures has as a major feature the production of a great variety of long slim blades, many suitable for knives, struck from a carefully prepared core and further "touched-up" afterward. A characteristic type is that known as a graver (or burin), in which the end of the blade was used as a knife-point. The Solutrian was characterized by large "laurel-leaf" and "willow-leaf" blades. These were carefully shaped by the removal of small flakes through pressure rather than hammer blows.

New cultural features made their appearance in the Upper Paleolithic. Bone, previously little used, is extensively employed, particularly by the Magdalenians. An interesting item is the discovery of bone needles. A nov-

Art of Upper Paleolithic man. Artistic impulses, unobserved in the cultural remains of earlier human types, were strong in the days of Cro-Magnon man in Europe. Antlers, bone, and soft-stone materials were carved or engraved, as in the examples seen at the lower left. Particularly striking are the finds of engravings and paintings on cavern walls in southern France and northern Spain. *Below, right,* a part of the cavern of Font-de-Gaume in the Dordogne region of France. A series of bisons were painted on the bosses of the rough rocky wall at the left of the photograph. One of these bisons is illustrated (*right*), as well as a pair of reindeer from another part of the same cavern. Many of the cave paintings are colored with local clays—dull reds and umber—and show artistic skill. (Mobile art objects courtesy American Museum of Natural History, New York; other figures from Capitan, Breuil, and Peyroney.)

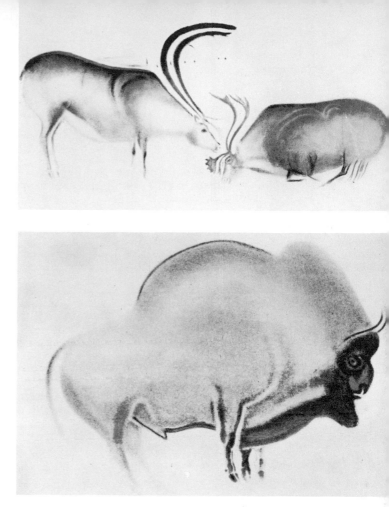

Cave paintings. *Left,* two of a magnificent series of murals discovered in the cave of Lascaux in the Dordogne region. (After the Abbé Breuil.) **Folsom points.** *Above,* flint points associated with an extinct bison at Folsom, New Mexico. This find, several decades ago, spurred the search for traces of early man in America. **Post-Paleolithic days.** After the close of the Ice Age, Europe was occupied by bands of hunters and fishers. *Below,* a camp scene of a band of these hunters. (Courtesy American Museum of Natural History, New York.)

el feature is a marked development of art, especially in the Magdalenian. Many engraved and carved bones and statuettes have been discovered, for example, and especially striking are the engravings and paintings which adorn the walls of a number of caverns in southern France and the Pyrenees. In a number of cases they are colored—for example, in red, yellow, and black, and most show a high degree of artistic skill. Human figures are rare; most are those of animals which were hunted, such as bison, reindeer, mammoths, and we can be confident that their purpose was a hunting magic, to give the hunters some sort of religious hold over their potential victims.

Although our present knowledge of them is limited, similar Upper Paleolithic cultures were widespread in Asia and Africa, in India and China to the east, and Kenya to the south. In some cases these appear to be closely comparable to the European Aurignacian; in others local terms have been given, as in North Africa, where Upper Paleolithic cultures termed the "Capsian" and "Oranian" are notable for the abundance of tiny flakes termed "microliths," which became widespread in the next cultural stage.

Mesolithic.—It was once believed that from the typical Old Stone Age man advanced at once into the settled, essentially "civilized" Neolithic stage. But in recent decades it has been increasingly realized that there existed transitional cultures, now classed as the Mesolithic, or Middle Stone Age. The Old Stone Age, for the most part, can be considered as having come to an end with the final stages of retreat of the glaciers, somewhere on the order of 13,000 years before Christ; the Mesolithic continued for a number of thousands of years beyond this time, terminating perhaps about 6000 B.C. in southwestern Asia (with an advance there to the Neolithic stage), but persisting a thousand years or two later in such outlying regions as Europe and continuing down to the present in a few areas.

Mesolithic man lived under climatic conditions which had changed greatly from those known by his predecessors. In much of Europe, with the retreat of the ice, there ensued a stage of damp climates with thick forestation, and many of the animals which man once hunted had vanished from the scene. Farther from the erstwhile glacial regions there also appear to have been climatic shifts, and great regions of the Sahara and central Asia, earlier well watered, appear to have become progressively drier and less adapted for human life.

Mesolithic man was still a wandering hunter and food-gatherer like his predecessors. He appears to have found the thick European forests almost impenetrable, so that a majority of our Mesolithic finds are from seashore camps or from sites well to the south, to which the forests had not spread. Although the dog appears as a camp-follower, there were no other domestic animals and no agriculture. Because of the difficulties of hunting in the forests, shellfish became a very important article of diet for European men

A Solutrian laurel-leaf blade. This cultural stage is characterized by finely worked flints. The desired form was attained by the removal by pressure flaking of small chips over the entire surface. (After Cole.)

of this age, and there are known many great shell mounds, the garbage dumps ("kitchen middens" is a more polite term) of Mesolithic men. Many of the flints found in Mesolithic camp sites are microliths—small chips in triangular and other geometric shapes. The invention of the bow, with stone-tipped arrows, appears to have taken place at about this stage, and other microliths appear to have been placed along the margins of spears or harpoons or in a row along a grooved stick to make a cutting tool. Fishhooks are known, and crude stone axes; the canoe makes its appearance among shore dwellers. In Europe, Azilian, Tardenoisian, and Maglemosian are among the more familiar names given to cultures of this age. Mesolithic cultures of various sorts appear to have been widespread also in both Asia and Africa, bridging the gap between the Old Stone Age and Neolithic times. Although most groups in the Old World have passed beyond this level, Mesolithic or even Upper Paleolithic conditions persisted in such outlying regions as Australia, and it seems certain that the early invaders of the Americas were in a Mesolithic (or even late Paleolithic) cultural stage.

Neolithic.—The initiation of the Neolithic marks the greatest revolution that has occurred in human cultural evolution. The age gains its name from the relatively unimportant fact that stone implements, although retained, were modified in type. Somewhat more important was the development of pottery, which improved facilities for food storage. Fragments of pots made of baked clay are practically indestructible; this has made them invaluable to the archeologist in identifying cultural groups through ceramic variations.

The major developments of the Neolithic, however, were in two other fields: the development of agriculture and the domestication of animals. Man thereby gained control over his food sources, could cease to be a wandering hunter, and could live in settled communities with all the implications for future social and economic advances that such a life implies. Among agricultural products, edible grasses, from which our modern grains have arisen, were prominent from the first. Of animals the pig, cow, sheep, and goat were early domesticated; the ass and, later, the horse were added as beasts of burden and transport. While there are many uncertainties, it appears that the wild progenitors of many of these plants and animals were present in a belt of country extending from Egypt east through Mesopotamia to northwestern India. The valleys of the great rivers of this area—the Nile, the Tigris, the Euphrates, and the Indus— were major centers of early Neolithic life. The time of origin cannot be definitely set. One site in Mesopotamia is at least as old as 5000 B.C., and since most of the typical Neolithic features were even then present, the beginning of this cultural phase must go back another thousand years or so. Neolithic culture, with established villages of mud-walled or brick houses, was widespread in the Near East by 4000 B.C., and from that time on this new culture appears to have spread rapidly over vast areas of the Old World. Europe was a bit backward, for it was not until later than 3000 B.C. that agricultural villagers began to appear in numbers to replace migratory hunters over most of that continent.

Introduction of metals.—Having exploited the animal and vegetable resources about him, Neolithic man was not slow to explore the possibilities of the mineral world. Copper objects soon appear in Neolithic materials. This metal by itself is too weak to form good tools, but it was soon found that alloys with other metals gave a much harder product. Particularly useful was bronze, a copper-tin alloy, and by about 3000 B.C. a Bronze Age had begun in the Near East, with this metal supplanting stone in the more important tools and weapons. From this center bronze spread widely over Europe and Asia so that this metal was common in most of Europe by 2000 B.C. and had extended east to China by about 1400 B.C.

Later—again apparently in the Near East—man mastered the methods of preparation of a second, more difficult but still more useful, metal— iron. Beginning about 1400 B.C., the use of iron spread rapidly, and within a relatively few centuries iron had supplanted bronze in great areas of Eurasia and is today the dominant toolmaking material of the great majority of Old World peoples.

In Africa, however, bronze did not penetrate far south, and iron was introduced but slowly and to a limited degree. In consequence, many peoples of that continent are still in an essentially Neolithic stage, and the same is true of the more backward areas of Asia. America was still less progressive, for the Indians had barely begun the use of copper at the time of discovery and were thus far behind the Old World.

HUMAN SYSTEMATICS

Generic names given by their describers to early human remains, and used in the pages that follow, lead us at this point to a brief consideration of the classification of human types. We have discussed in the opening chapter the general nature of animal systematics. The term "species," we have seen, is usually given without question to the whole of a large but interbreeding population of animals of any sort, even though there may be considerable minor variations within this population. And genera are also usually conceived of broadly, so that, for example, most members of the cat tribe, from lions and tigers to the domestic pussy-cat, are included by most (although not all) workers in the single genus *Felis*.

What about man? Here many authors have tended to adopt a different policy. Just as, we have seen, we tend to look into our own human pedigree more minutely than we do in the case of other phylogenies, so many writers have tended to split up human and near-human types into genera and species with abandon, utilizing for purposes of definition differences which would be held of little account in other groups.

Modern man is the type form of the genus *Homo*. Students of fossil types have tended to place such forms in distinct genera on such grounds as shape of the forehead, presence or absence of brow ridges, degree of development of a bony chin. But these differences are certainly no greater than those used for merely specific distinctions in other cases. In consequence modern students with a broad view of the subject advocate placing all human remains in the single genus *Homo*. We shall adopt this point of view here but will nevertheless use such "generic" terms as *Pithecanthropus* as convenient handles for referring to various fossil human finds.

What about species within the genus? In general even though its members differ considerably from one another, a large population of similar animals is considered to form a single species if its members or "races" can breed freely with one another. Modern man is by most considered as constituting a single species, *Homo sapiens* (even though the choice of a specific designation seems a bit conceited). A few anthropologists have advocated "splitting up" modern man into several species—black, white, and yellow, for example. But this is, when one looks the situation over, absurd. Any two modern human types *can* interbreed and *do* quite commonly interbreed. And although intermarriages between tribes or racial groups were surely less common in earlier days, one can readily believe it occurred to some extent, at least, far back of historic times. Modern man is surely one species.

What of the geologic past? Were ancient centers of population of archaic human types so isolated in remote stages of the Pleistocene that there existed, side by side, distinct human species in, say, eastern Asia, Europe, and Africa? Or was there, even then, some intermingling of blood between the various Old World areas, despite a presumed sparseness of the human population? We do not know. And again, if we were to claim that a certain fossil human type is specifically distinct from modern man but gave rise to him by gradual change, at what stage of his evolution shall we arbitrarily say that he has shifted from the one species to its successor? On the whole, the problem of human species is one that we may well avoid for present purposes, although I shall use in the account which follows some of the specific names that have been given to human fossils.

PRIMITIVE MAN

For the late phases of the human evolutionary story, much of the evidence is derived from finds in Europe and the general Mediterranean region. For the earlier parts of the story, however, the European data is, as yet, unsatisfactory. We have already seen, in the last chapter, that it is Africa which furnishes the major data as to forms structurally intermediate between apes and men. And it is from the Middle Pleistocene deposits of Asia that there have come the two famous and closely related types, *Pithecanthropus* and *Sinanthropus*, which furnish valuable knowledge regarding the most primitive of definitely human stages of evolution.

The Java "ape man."—In the late eighties Dr. Eugene Dubois, of the Dutch army medical service, went to the East Indies with the announced intention of finding primitive man. Many have made such announcements; the surprising thing is that Dubois did actually find what he was looking for. Remains of extinct animals had been discovered in deposits of volcanic ash on the banks of the Solo River in eastern Java. Dubois set to work, in 1892, to excavate such a bone bed near the village of Trinil. Soon, among the mammal remains, he came upon a tooth of human appearance. Close by was found a skullcap and later, and some distance

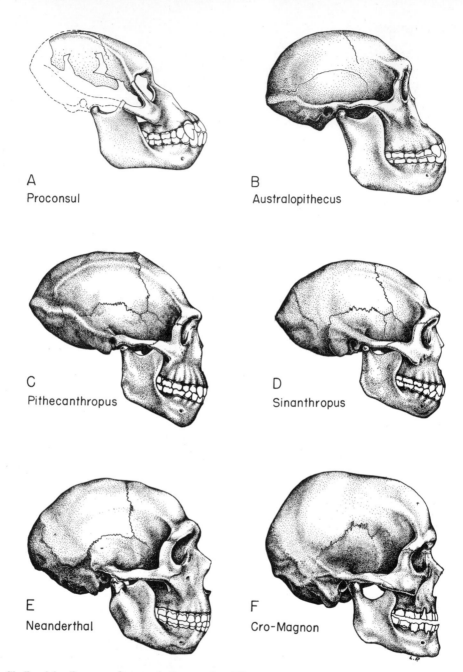

A
Proconsul

B
Australopithecus

C
Pithecanthropus

D
Sinanthropus

E
Neanderthal

F
Cro-Magnon

Skulls of fossil apes and men. *A, Proconsul,* a Miocene great ape; *B, Australopithecus,* the South African man-ape; *C, D,* the primitive human types from Java and China; *E,* a typical Neanderthaloid; *F,* Cro-Magnon, an Upper Paleolithic man of modern type. (*A, B,* after LeGros Clark; *C, E, F,* after McGregor; *D,* after Weidenreich.)

away, in the same excavation, a thigh bone (femur) and two other teeth. Dubois returned to Europe, gave a scientific description of the fragments, which he named *Pithecanthropus erectus*, the "erect ape man," and turned the remains over to a Dutch museum. With this, additions to our knowledge of this Java man ceased for three decades. A German scientific expedition made vast excavations at the Trinil site, found numerous animal bones, but failed to find the slightest further trace of *Pithecanthropus*. And, for some reason difficult to understand, even the original find itself was during all this time kept almost completely hidden from both the public and the scientists.

The nature of the skullcap, as described by Dubois, was so unusual that much debate and discussion ensued. It was obviously far more primitive and apelike than anything previously regarded as human. The bones are exceedingly thick. There were heavy, apelike brow ridges above the eye sockets, and the vault of the skull was extremely low, indicating a brain far smaller than in modern man. But brow ridges are retained in some definitely human types (the later Neanderthal type, for example), and the skull lacks the median longitudinal crest (sagittal crest) found in the larger apes. (The skullcap has a slight ridge in the center, but this is due to the manner in which the paired bones of the forehead became fused, not to the attachment of heavy jaw muscles, as in the ape crest.)

Nothing could be said, from the first remains, about the face and jaws, and the teeth were inconclusive (and in fact two may belong to a contemporary orangutan). The femur was definitely human (a peculiar bony outgrowth near its upper end is due to a diseased condition frequent in man and not of particular interest). The position of the articular surfaces at the ends of the bone and the straightness of the shaft prove conclusively that its owner walked erect. However, since this bone was found some distance away from the skull, there is no definite proof that it came from the same individual, and some writers are inclined to think that the femur belonged to some other form of man.

In summing up this evidence, Dubois himself was at first inclined to consider that the creature was halfway between man and ape—the "missing link" of popular terminology—a form with many apelike features but walking erect in human fashion. From this point of view the scientific name is an appropriate one. As to its age, its discoverer believed it to be Pliocene and hence of remarkable antiquity.

There were, however, various dissenting opinions. As to its geological position, it soon became apparent that it was rather later than at first appeared; that it was Pleistocene, not Pliocene, and not even early Pleistocene at that. It comes from beds at about the middle of the Pleistocene—probably at about the time of the second glaciation farther north, a period thought to have been about 500,000 years ago according to some current estimates. As to its evolutionary status, a majority of students were inclined to believe that, although unquestionably primitive, *Pithecanthropus* was man rather than ape. But, on the other hand, one authority claimed that it was nothing but a gigantic gibbon, for the contours of the gibbon skull are similar (although on a much smaller scale), and the gibbon, too, can walk erect.

Recent decades have seen a great increase in our knowledge of *Pithecanthropus*. In 1923 the original remains were made available for study, and it was disclosed for the first time that the interior of the skull had been excavated, thus affording valuable data on the brain. Dubois further revealed that he had other materials which were from the same beds (although from other localities) and hence may well have belonged to *Pithecanthropus*. These materials included four other femora and a partial lower jaw. More recently, paleontological work in Java by von Koenigswald has unearthed a number of new finds of this apeman. All are from the valley of the same Solo River, but from other localities, and most from Sangiran, farther upstream. At this last place were found remains of three more skullcaps, one very well preserved, a lower jaw, and an upper jaw which gives much information, previously lacking, concerning the face. One skull is of a very large and massive type and so distinctive that its finder considered it as representing a new form. From Modjokerto was obtained a child's skull in somewhat older beds; this was at first thought to be of a different type but is now recognized to be a young *Pithecanthropus*.

If we check over this list of material now available, it is seen that we know little except the skull and jaws. These, however, are fairly adequately represented, and from these structures the expert can extract much valuable data as to the nature and relationships of Java man.

The mouth region is protruding because of the large size of the teeth, much as in the South African man-apes but to a lesser degree. As has been said, the braincase was extremely low and the brow ridges enormous. It is, further, rather narrow compared to its length, a feature common to most early human forms. The breadth-length ratio is one universally used in the study of human skulls. Those in which the breadth is less than 75 per cent of the length are termed dolichocephalic—long-headed; if the percentage is above 80, brachycephalic, or round-headed; intermediate figures indicate a mesocephalic condition. The average index of the three skull-caps measured is 75.

From the contours of the interior of the braincase there can be made an endocranial cast. This mirrors the folds of the tissues wrapping the brain and hence gives valuable clues as to brain shape and size. A study of such a cast of the type skull shows that the brain cavity had a capacity of about 940 cubic centimeters; in the case of the other good skullcap, the estimate is lower, 835 cubic centimeters. The endocranial capacity of a modern male European is about 1,500 cubic centimeters; that of some primitive existing races averages about 1,200; few great ape brains exceed 600. *Pithecanthropus*, from these figures, was thus about midway between the great apes and living men in the size of his brain. These figures of brain size are rather smaller than is the case in any other group of fossil men. However, early human types tend to show a considerable difference between sexes in brain capacity, the female being the smaller, and a study of the materials suggests that the two good skullcaps belonged to early Javan ladies. In brain, *Pithecanthropus* may be rather on the human side of the apeman "divide." That this is the case is shown by the brain contours seen in the cast. These indicate an arrangement of the "gray matter" in an essentially human rather than a simian pattern. The development of areas related to speech and its reception suggests that *Pithecanthropus* might have had at least the rudiments of language.

The upper-jaw specimen shows that *Pithecanthropus*, like certain other related types, differed markedly from modern man in a more projecting and broader face, with a large and probably flattened nose. The teeth are rather large and show some apelike features of the molar pattern but, on the whole, are human rather than apelike. The "wisdom tooth," small in many later fossil and living men, is not yet reduced. In apes a gap exists in the upper tooth row into which fits the projecting canine of the lower jaw. In the only upper-jaw specimen of *Pithecanthropus* this gap is present. Apparently this is a male; the female presumably had a smaller and more normal human canine, and this simian feature has not been reported in any other human type.

If the evidence now at hand concerning *Pithecanthropus* be summed up, it is clear that this apeman was a man and not an ape. Nevertheless, he was a man of a very low and seemingly primitive sort and had retained many archaic features absent or reduced in most other known fossil men. We may postpone further discussion of his position in human evolution until we have examined the closely related *Sinanthropus*.

Peking man.—In the early 1920's a Swedish scientific mission in China investigated a series of caves in the limestone hills near Choukoutien, some thirty-seven miles southwest of Peking, China. These contained an enormous mass of deposits which included bones of a variety of animals, many of which are extinct. This fauna is now known to indicate definitely a Middle Pleistocene age for the associated materials—probably the second interglacial stage.

Included in the finds were two teeth which were either simian or human and attracted the interest of Dr. Davidson Black, anatomist at the medical school in Peking. Study of a third tooth, found in 1927, convinced him that it was human but of a primitive sort. On this slender basis he established a new genus and species of man, *Sinanthropus pekingensis*, and undertook further investigation of the site.

His seemingly rash conclusions were, happily, verified. In 1929 a well-preserved skullcap of a very primitive sort was unearthed. Adequate funds were procured for extensive excavations, and, in the course of a decade, under Dr. Black and, after his untimely death, Dr. Franz Weidenreich, there came to light a whole series of remains belonging to about three dozen individuals including both sexes and both young and adults. There was little postcranial material, although there were specimens of humerus and femur and a few other fragments, and hence we have very little idea of the general bodily build. No skull was complete, but between the various specimens nearly the entire cranial structure is known.

The braincase is quite similar in its proportions to that of *Pithecanthropus*, with large continuous supraorbital ridges and a very low retreating forehead. The skull is long and narrow posteriorly, the bones much thickened. The face, too, is quite similar to that of *Pithecanthropus*, and the jaws show a similar chinless, primitive, and yet essentially human structure. The brain size is, however, rather greater than in the Java man, for the endocranial capacities reported are 915, 1,050, 1,100, and 1,200 cubic centimeters.

Several points of interest can be deduced regarding the life of Peking man. He was already a tool-user, for a number of implements have been found in the cave. These are of the crude chopper and flake types, which we have noted earlier to be characteristic of the Lower Paleolithic period in eastern Asia. He was, further, a fire-user, for a number of hearths were found.

Still another characteristic—not so pleasant—is suggested by the nature of the fossil finds themselves. The remains consisted, we have noted, al-

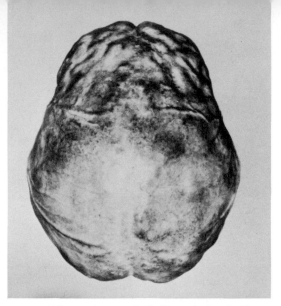

Recent Pithecanthropus finds. New finds of this ancient type have been made in recent years. *Left*, side and top views of one of the new, more complete, braincases (from von Koenigswald). *Above*, a cast of the upper surface of the brain cavity. The front end (*top*) shows clearly the impressions of the convolutions of gray matter. From such casts many deductions may be made concerning the topography of the brain. (From Dubois.)

inanthropus. *Below*, a restoration of Peking man and the restored skull upon which it was based. Other figures of this fossil man are shown overpage.

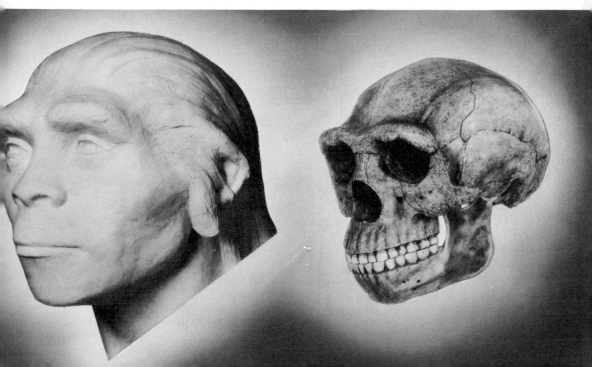

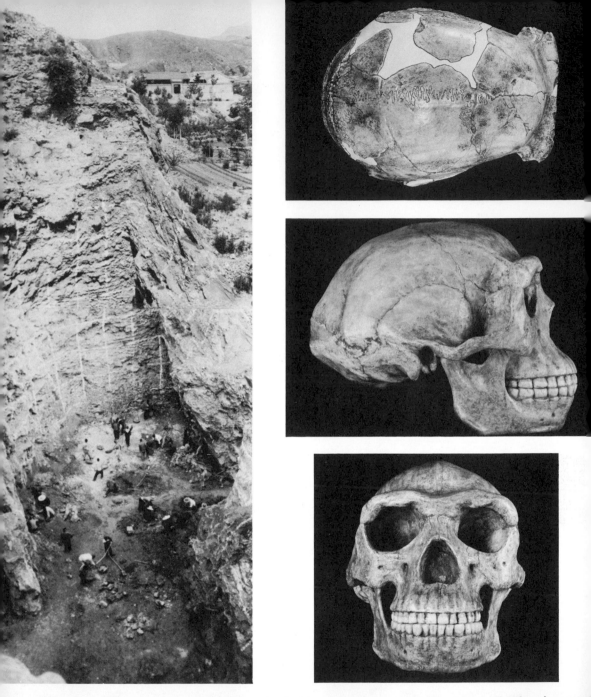

Peking man. The remains of this early and primitive man have been obtained from Choukoutien, near Peking, where former caves in a limestone hill have been filled with hard materials containing remains of man and many other Pleistocene mammals. *Left*, part of the vast excavations that have been made there in the search for this fossil man. The white lines mark off numbered sections of the quarry so that finds can be accurately located. *Above, right*, is a view of one of the skullcaps seen from above (compare with *Pithecanthropus*). *Below this* are side and front views of a skull reconstructed from numerous specimens. (All *Sinanthropus* photographs by Dr. Franz Weidenreich, Cenozoic Research Laboratory, Peking, courtesy American Museum of Natural History, New York.)

most entirely of skulls. These were not burials, as far as can be determined from the conditions under which they were found, and in every instance the braincase had been broken open from below. It seems highly probable that the brains, and probably the missing bodies, were used as food. By whom? There are no traces there of any other creature capable of doing this except Peking man himself. He may thus be reasonably accused of cannibalism.

The story of the *Sinanthropus* finds had a sad ending. The materials were assembled and studied in a laboratory in Peking. At the outbreak of World War II, in 1939, it was decided to take the specimens to the coast and place them, for safekeeping, on an American warship. En route to the coast, the train carrying them was seized by Japanese troops, and the fate of the fossils is unknown to this day. Were they, in ignorance of their scientific value, thrown into a train-side ditch? Or were they, like many fossil bones in China, ground to powder and used as medicine? We may never know.

Relationships of *Pithecanthropus* and *Sinanthropus*.—When the first braincase of Peking man was discovered, it seemed fairly sure that this type and *Pithecanthropus* were related, and this impression has grown continually stronger as more material of both forms has become available. The major workers on both forms have agreed that the two are exceedingly close to each other and represent little more than racial variants of a single human type. There are a few differences. The figures on brain capacities run higher for *Sinanthropus* than for *Pithecanthropus*, but this may be due to sexual variation; the Javan skulls measured appear to be those of females, while a larger proportion of the Peking skulls may be those of males. In *Pithecanthropus* the supraorbital ridges are rather more prominent, and *Sinanthropus* appears to lack the gap in the tooth row for the canine, noted in the male *Pithecanthropus*. The skull bones are somewhat thinner, the teeth somewhat smaller, and hence the mouth less protruding. But most differences are of a minor nature. The two are so close that they certainly belong to the same genus and quite probably are only variants of a single species. We shall in later sections of this chapter use the term *Pithecanthropus* to refer to both of these eastern apemen.

That these early Asiatic finds belong to a very primitive human type is obvious. We no longer believe, as Dubois did originally, that the Java apeman is halfway between ape and man. He is definitely on the human side of the boundary; but we are certainly dealing with a lower type than any other adequately known human fossil. We can say with confidence that *Pithecanthropus* represents a structural stage through which the ancestors of later men may well have passed on their way upward from the ape.

In contrast to the simple and seemingly clear-cut story of earliest man in eastern Asia is the situation regarding men of the early and middle Pleistocene in other regions of the world. There are no finds of like antiquity in central or western Asia nor in most parts of Africa, although some skull fragments from Tanganyika have been thought to be of this type. Nor are we much better off when we come to consider Europe and the Mediterranean region, although these regions have been much better explored for fossil material than any other part of the Old World. We have abundant specimens from Europe and Africa of Lower Paleolithic cultures. Man was surely present and reasonably abundant here at the time; but of the makers of these flint tools we have only a few specimens of lower jaws.

The Heidelberg jaw.—Near the little village of Mauer, not far from Heidelberg, is a large sand deposit formed in an old oxbow of the Neckar River at a rather early stage of the Pleistocene—possibly the first interglacial stage. This has been intensively worked for commercial purposes and was found to contain numerous animal bones, including, for example, ancient elephants, rhinoceroses, and lions. In consequence, a careful watch has been kept on the pit for decades. In all this time only one human bone has been found; but this one is a specimen of great interest, the so-called Heidelberg jaw, excavated in 1907 at a point about 80 feet below the surface. The most striking impression given by this specimen is its large size and heavy build. There is no chin prominence, and hence at first sight the jaw appears somewhat apelike; but the teeth and dental arch are definitely human, and the tooth row itself is no larger than in some modern types—this in contrast with the Asiatic forms just discussed. The canine does not project at all above the other teeth, and the wear on the molars shows that the chewing motion was of a human rather than apelike sort. An unusual feature is the great width of the ascending ramus (the upper part of the jaw behind the tooth row) and the shallowness of the notch at its top.

The essentially human character of this ancient type finds expression in the fact that even those "splitters" who tend to separate human types into distinct genera and species commonly assign this jaw to our own genus (although to a different species) as *Homo heidelbergensis*. No implements have been found in the deposit, but the date and locality of the find suggest that this man may have been associated with the rather crude flake industries of the Lower Paleolithic, from which the Mousterian later arose.

It has long been recognized that, except for the broad ascending ramus and its rather unusual size, the jaw can be compared in many respects with that of Neanderthal man, who later occupied this territory. On the other hand, except for its rather heavier build and relatively smaller teeth, the jaw is not markedly different from that characteristic of the men of the *Pithecanthropus* group, who were roughly contemporaneous.

The Ternifine jaws.—Nearly a century ago an excellent fossil deposit was discovered near Ternifine in the Oran region of Algeria. There were nu-

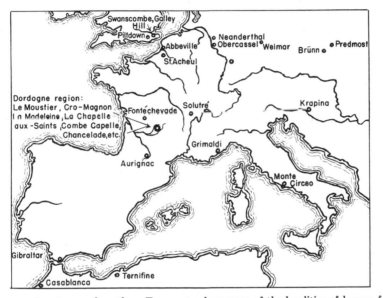

A sketch map of western and southern Europe, to show some of the localities of human fossil and cultural finds.

merous animal remains and flints of a primitive Lower Paleolithic type, dating the deposit as from the beginning of the Middle Pleistocene times, but it was not until 1952 that the French paleontologist Arambourg discovered here two human lower jaws, one nearly complete, and later a third jaw and a skull fragment. Further work in Barbary soon after yielded a fragment of another and almost identical jaw of somewhat later date near Casablanca, Morocco. As in the case of the Heidelberg specimen we are dealing with a type that certainly is primitive; but, in contrast with that type, the Ternifine jaws are clearly and closely comparable to those of the Asiatic apemen. Quite possibly a good skull, when found, might—and very probably would—show some minor differences from *Sinanthropus* or

Pithecanthropus. But we can now reasonably assume that in Middle Pleistocene days a single general primitive human type was spread broadly over the Old World. Cultural differences existed between one area and another and presumably between one race and another, just as differences exist today; and also as today, there were surely racial differences, some of which we have seen in comparing *Sinanthropus* and *Pithecanthropus* and in comparing the Heidelberg jaw with either. But it is logical, as far as our evidence goes, that mankind was then a single species, well advanced over the australopithecine level but very far below modern standards in skull and jaws, in brain development, and in habits and culture.

The Piltdown hoax.—We cannot refrain here from discussing the famous (or rather notorious) Piltdown finds which long perplexed the scientific world. In 1912 an amateur archeologist, William Dawson, turned over to the distinguished paleontologist Sir Arthur Smith Woodward fragments constituting most of a human braincase which he said had been found in gravels near the manor of Piltdown in Sussex, England. This gravel deposit is one of considerable antiquity, not improbably dating from the first interglacial stage. A number of fossil mammal fragments were reported from the gravels. Some appeared to be as old as the Pliocene, but their presence here, it was suggested, was due to redeposition. In addition, there were a few flints of archaic type and a crude tool fashioned from an elephant limb bone.

Although the skull was incomplete and fragmentary, a reasonable reconstruction could be made. Except that the bones are a bit thick, the skull is essentially modern in build, lacking the strong, bony, brow ridges expected in early human types, and with a high forehead. This was surprising for a specimen of such early date. What sort of a jaw would the Piltdown man have had? The answer was soon forthcoming. Further digging was done, under Dawson's supervision, in the gravels, and a large part of a jaw was unearthed, as was a large canine tooth and nasal bones. The jaw proved even more of a surprise than the skull. For it was more primitive than even those of the archaic men of Java and China and, in fact, was identical in structure with that of such a great ape as the orang! The Piltdown man, very modern in his skull, very primitive in jaw pattern, did not fit at all in the picture of gradual simultaneous evolution of both structures suggested by the Java and Chinese finds, and one school of thought advocated the idea that skull and jaw did not really belong together, that while the skull was human, the jaw was that of an ape. But fossil Pleistocene apes are otherwise unknown in England; skull and jaw

The Piltdown hoax. Probably the greatest hoax ever perpetrated on a group of scientists was that concerned with "Piltdown man." About 50 years ago Dawson, an amateur archeologist and paleontologist, reported to Sir Arthur Smith Woodward, a distinguished English paleontologist, the finding of part of a fossil human skull in very old Pleistocene gravels near Piltdown Manor, Sussex. Further excavation of the gravels (under Dawson's direction!) brought forth a very apelike jaw and canine tooth. *At the right* are shown Dawson (with the bald head), Smith Woodward, and a laborer, digging and sifting gravels in the Piltdown lane; *next below*, the skull fragments, jaw, canine tooth, and a flint implement found there as well. The human skull and ape jaw formed an incredible combination despite attempts (as in the bottom figures) to restore the skull and facial appearance of this supposed ancient man. Modern methods plus careful study of the materials have recently proved that the jaw is that of a modern ape and that this and other finds were "planted" in the gravels by someone with the deliberate intent of deceiving the scientific world. (Photographs courtesy American Museum of Natural History.)

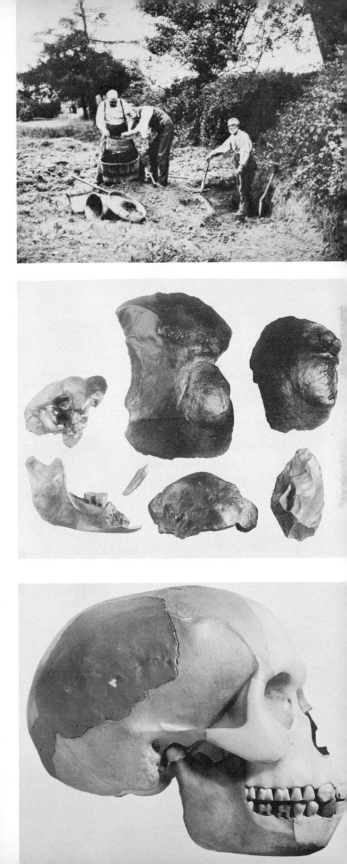

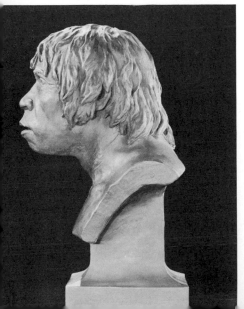

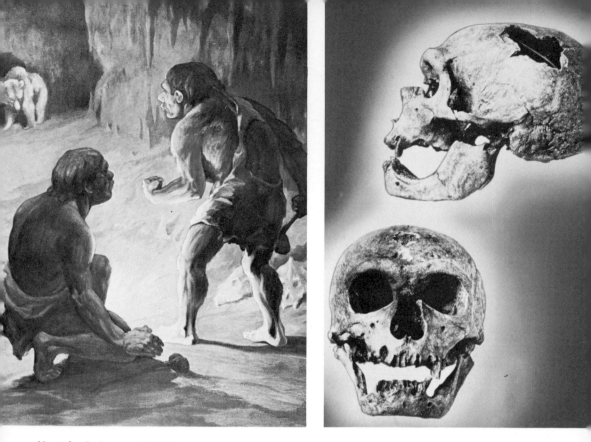

Neanderthal man. This inhabitant of Europe in the earlier part of the last glaciation is known from numerous skulls and considerable parts of the skeleton, so that attempts at complete bodily restoration (as in the scene, *upper left*) are warranted. The most adequately known specimen is the skeleton of an old man from the cave of La Chapelle aux Saints, described by Marcellin Boule. This skull is pictured above (*right*). *Lower left*, the same skull with teeth and other missing parts restored by Dr. J. H. McGregor. *Lower right*, the skull of a female from Gibraltar; the contours are more rounded, brow ridges less exaggerated, etc. (Restoration at upper left by F. L. Jaques, under the direction of W. K. Gregory; photograph courtesy American Museum of Natural History, New York; La Chapelle skull unrestored from Boule; lower figures courtesy Dr. J. H. McGregor.)

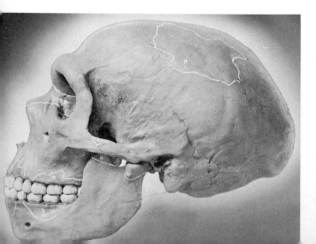

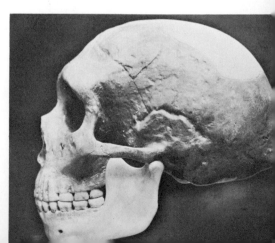

appeared to have been found close together; and as further evidence for association of the two was the report by Dawson that he had found in a neighboring locality a scrap of a second skull similar to the first, together with teeth like those in the original jaw.

There the matter rested, a seemingly insoluble enigma, for several decades. No one doubted the authenticity of the finds; and whether or not skull and jaw were associated, Piltdown man presented an ever present headache to one attempting to work out a reasonable scheme of human evolution.

A few years ago an Oxford anatomist, while making a detailed study of the jaw, noticed that the wear on its molar teeth looked unnatural, as if made by some abrasive instrument rather than by normal chewing action. Following up this clue, the whole set of Piltdown material was gone over carefully by a series of competent scientists; and in 1954 the announcement was made that "Piltdown man" was a gigantic fraud! The skull pieces may be of some antiquity, but have been "doctored." The jaw is that of a modern ape, with diagnostic regions broken off, the teeth filed to obscure the pattern, and the bone stained to give it an antique appearance. Even the other fossil teeth and bones had been "planted" and hail from a variety of other places—two from as far away as Malta and Tunisia.

Who perpetrated the hoax, and why it was done, cannot be said positively, although suspicion points strongly to Dawson (who died a few years after the "discoveries" were made). Scientists may differ widely (and often violently) as to the conclusions to be drawn from their work, but the basic principle of the scientist's code is absolute honesty in presentation of their facts. And because of their own high standards of truth, the scientist is liable to be taken in when he comes in contact with an amateur lacking the same code of ethics.

LARGE-BRAINED MEN; THE NEANDERTHAL PHASE

To evolve modern man from the primitive apemen of early Paleolithic times requires considerable progress and refinement in the structure of skull and jaws and, one may suspect, other parts of the skeleton. Most especially, however, there has to occur a major increase in intelligence and cultural potentialities. Men with large brains were present by Middle Paleolithic times, probably a good quarter of a million years ago and perhaps even earlier. But historically our study of early large-brained men has been given a peculiar twist by the fact that in Europe the common inhabitant in the Middle Paleolithic and the maker there of the characteristic Mousterian stone industries was a very specialized form, Neanderthal man.

This race, usually called *Homo neanderthalensis*, was the first fossil human type to be discovered and recognized and is still the best known of early races. The first specimen to gain wide attention and from which the

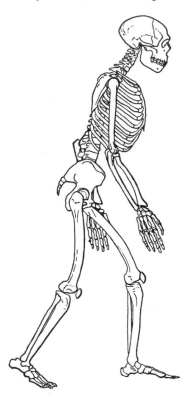

race takes its name was found in a cave deposit in the Neander Valley near the Ruhr district in western Germany in 1856. Although a complete skeleton may have been present originally, little was preserved except the skullcap. This immediately excited interest because of the low forehead and prominent brow ridges, but, in the absence of anything else comparable, its status was debated. Most rightly recognized it as a primitive human type, but others claimed that it was merely a congenital idiot of our own species or (according to one eminent scientist who appears to have been a Russophobe) the remains of a Russian soldier killed in the Napoleonic wars!

Neanderthal distribution.—Later, other and more complete remains of this type of man were found in many localities. Numerous finds have been made in western Europe, particularly in France. To the southwest, two skulls of this species have been found at Gibraltar. The Neanderthal type was long unreported from Italy, but in the 1930's three characteristic skulls

The skeleton of Neanderthal man. (After Boule and Weinert.)

were found in the general Rome district. Farther east the rock shelter of Krapina in Croatia has yielded skeletal remains of a considerable number of Neanderthal individuals (there are suggestions of cannibalism at this locality). To the southeast recent discoveries have shown Neanderthal man (including remarkable variants from the general type) to have inhabited Palestine and Mesopotamia. Farther to the northeast this race was present in the Crimea and northern Iran, and a Neanderthal child's skull has been discovered in a cave in western Turkestan once occupied by a group of Mousterian goat hunters. Neanderthal men ranged widely over northern Africa, at least, for specimens are known from Morocco, Libya, and Ethiopia.

Neanderthal man thus occupied a considerable territory. As yet, however, he is not known from any part of Africa except its northern coast, nor from central or eastern Asia. His absence may, of course, be only apparent, and new finds may presently extend this range. We have noted, however, that he is, where known, associated with Mousterian culture. The Mousterian appears to be absent in much of Asia and Africa, although other equivalent Middle Paleolithic cultures may be present. We have no strong reason to believe that the other cultures were made by Neanderthal man, and it may well be that known finds do actually outline the territory he occupied.

Typical Neanderthal structure.—Most of the specimens of this race conform closely to a single type, best described in a classic monograph by the French paleontologist, Marcellin Boule, based on a well-preserved skeleton from the rock shelter of La-Chapelle-aux-Saints in southern France. The data gained from this specimen can be confirmed from many others, for there are a number of intact skulls and skeletons. This statement, by the way, reveals a significant point as to the *mores* of Neanderthal man. We tend to think of him as a rather brutal creature; but there is here evidence that, at any rate, he differed from his predecessors in not being a cannibal, and in at least one case there is evidence of a ceremonial burial of the dead. At first glance one is impressed by the somewhat apelike appearance of the skull. The head is massive, with heavy supraorbital ridges, a receding forehead, and a low vault to the braincase. This, however, does not indicate that the brain was small, for the braincase juts out strongly posteriorly, with a marked projection which contrasts sharply with the rounded contours of modern man in this region. The brain, thus, is large, with an average capacity in typical males of about 1,550 cubic centimeters, a figure as large, or larger, than that found in the highest of living races. Sexual differences are marked, for females average about 200 cubic centimeters less. Although this brain is of modern size, its proportions are, as would be expected, quite different from that of *Homo sapiens*. The frontal region of the brain hemispheres, which is thought to be important in the development of the highest mental faculties, is relatively small.

Details of the ear region and the heavy arch at the side of the skull are also quite different from those of neanthropic man. The face is very long and large in proportion to the braincase, the jaws projecting, the eye sockets large and rounded in form, the nose opening broad. The cheek bones of modern man have a characteristic hollow (fossa) excavated above the canine teeth; this is lacking in the Neanderthal type. The size of the face is related to the fact that the teeth are in general large and powerful, the dental arch being much larger than in typical modern men. The patterns of the molar teeth retain some apelike features, but the dentition is on the whole essentially human in character, and there is none of the ape projection of the canines. The jaw, as one would expect, is also massively built but essentially human. There is, however, no chin projection in typical members of this group, the front edge of the jaw falling away vertically below the anterior teeth.

In contrast with other primitive types of man, the skeleton of Neanderthal man is nearly completely known. This was a race of short stature; the average height of males seems to have been about five feet, four inches; the females may have been about six inches shorter. Although short, the body was powerfully built, the chest barrel-shaped. The forearm and lower leg were proportionately much shorter than in modern man. The carriage is thought by some authors to have been somewhat stooped, the head inclined somewhat forward on the short neck, the knees habitually slightly bent. The ankle joints indicate that the squatting pose was frequently assumed.

Early Neanderthalers.—These typical Neanderthal men are known mainly from deposits laid down in the earlier part of the Würm glaciation. A few specimens are considerably earlier—from the preceding Riss-Würm interglacial. Some of these appear to be typical members of the group, but others appear to depart from the general structural pattern in various features. The specimens from Italy are found with a fauna characteristic of a warm climate and hence may come from this interglacial stage. The Krapina finds mentioned above may also be of this age. Although the Krapina remains are fragmentary, they appear to represent a subrace characterized by a more slender build and smaller brains than the typical Neanderthal man. Specimens from the Ilm River Valley near Weimar in Germany include two typical jaws and a skull which appears to have a little higher vault to the braincase than is generally true of this race.

A still earlier Neanderthaloid specimen is that found at Steinheim, Germany, which may date from the time of the third glaciation. This, a female skull, has a brain capacity of only 1,100 cubic centimeters, a remarkably low figure even for a female and well down toward the *Pithecanthro-*

pus level. On the other hand, the face has less of a snoutlike appearance than most later Neanderthals, and the back of the braincase has more rounded, "neanthropic" (i.e. modernized) contours than is usually the case.

The Mount Carmel race.—In Palestine, following the finding of a Neanderthal skull near Galilee, excavations made during 1929–35 in caves on Mount Carmel revealed a series of twelve or more skeletons of people who dwelt there in Upper Pleistocene times—probably third interglacial—and are associated with a typical Middle Paleolithic culture of flints of Mousterian and Levallois types. These remains have been carefully studied by McCown, one of their discoverers, and Sir Arthur Keith and show a most interesting series of variations in structure. Their describers agree that all are to be classed with the Neanderthal group. But while a few individuals, including a woman from a cave called Mugharet et-Tabun, are typical members of that group, the others, mostly from a second cave, Mugharet es-Skhul, show to a variable degree numerous neanthropic features which are suggestive of *Homo sapiens* rather than *H. neanderthalensis*. Typical Neanderthalers are short; at Mount Carmel the men tend to be tall, for they have heights of 5 feet 7 inches to 5 feet 10 inches, and even the females are of moderate height. The face is not as large as in Neanderthals; the skull vault is not as flattened; the supraorbital ridges, while fairly prominent, are not as continuous; and the forehead is moderately full rather than receding. In Neanderthal man, we have noted, the occiput projects backward with a rather vertically compressed shape; here the contours tend to be more rounded, halfway toward modern conditions. The face is not as long as in typical Neanderthals, and there is little of the prognathism, the forward projection of the front teeth, seen in that form. In Neanderthal man the eye sockets are very large and round; here they are wide but not high. The nasal opening, primitively broad, is rather variable in these finds but sometimes approaches the narrower type of modern man. The teeth, rather large in typical Neanderthal men, are variable in size, although the pattern of the molars remains primitive. In the jaw the angular region at the lower back corner is rather more developed than in the ordinary members of this race, while an obvious and striking feature is the fact that there tends to be more or less of a projecting chin as in modern man. In the postcranial skeleton the posture is rather closer to that of modern man than that of the typical Neanderthal, the distal segments of the legs are long rather than short, and the structure of the backbone is intermediate between Neanderthal and modern types. The brain capacity of the typical Neanderthal woman of Tabun is 1,271 cubic centimeters; of the women of the Skhul group, 1,300 and 1,350 cubic centimeters; of the men, 1,518–87 cubic centimeters. While these figures are such as might be found in either Neanderthal individuals or large-brained modern men, the brain shape, including a better development of the frontal region, is closer to the modern type.

We have, thus, at Mount Carmel a group of people who exhibit, to a variable degree, a mixture of features characteristic of Neanderthal man and of *Homo sapiens*, particularly of the general Cro-Magnon type which was soon to appear in Europe. In summarizing the evidence, McCown and Keith pointed out that of twenty-five important features which they list, the Mount Carmel individuals are Neanderthaloid in but three, intermediate in fourteen, modern in eight. In a long series of features of lesser importance, the figures again favor modern man by a "vote" of 46 to 16. They point out, however, that too much reliance should not be based on mere statistics and that in really basic body pattern Mount Carmel man is still fundamentally a Neanderthal type, although with many neanthropic features.

Neanderthal relationships.—Before discussing the possible implications of these recent Palestinian finds, let us consider the general evolutionary position of the Neanderthal race as found in the European region. As to its origin, there is general agreement that Neanderthal man is a descendant of the apemen of the general *Pithecanthropus* type. But is he an immediate, direct descendant? Neanderthal man and modern man both have large brains. Did they evolve this character independently, or did they have a common ancestor with a large brain above the *Pithecanthropus* level? In Europe, Neanderthal man disappears, apparently rather suddenly, toward the end of the Pleistocene and is succeeded by Cro-Magnon and related forms of our own species. What were the relationships between these successive groups?

A number of alternative answers to these questions are possible and have been advocated. As to the relations of the typical Neanderthal man in Europe to his "modernized" successor, some have maintained that typical modern man evolved rapidly in Europe from his Neanderthal predecessor there. To support this it is pointed out that various Upper Paleolithic European specimens (Predmost, etc.) show certain features suggestive of the Neanderthal type, such as rather massively built skulls and brow ridges more pronounced than is usual in *Homo sapiens*. But this appears none too probable. It seems clear that the seeming Neanderthaloid

features of these Upper Paleolithic men are but minor elements in their
makeup; these individuals are very definitely neanthropic men who have
a few archaic features; there is no European transitional type. It seems
more likely, that as a majority of workers would advocate today, the European Neanderthal was harried and wiped out (with perhaps a bit of interbreeding in the process) by invading men of modern type—men who had
evolved outside the region occupied by Neanderthal man and his typical
Mousterian culture.

But does this mean that Neanderthal man and modern man both trace
back separately to the archaic apemen of the *Pithecanthropus* type? Not
at all necessarily. Quite possibly there existed at an earlier time a man of
large brain size, with a rather primitive build of skull but lacking the extreme specialization of the typical Neanderthal man. From such a form
might have come, in one line, the later and more specialized Neanderthal
men of Europe, with their flattened skull shape and other exaggerated
features, and on the other hand, man of modern type with emphasis on
such progressive features as a high skull vault and developed chin.

That this hypothesis is the true one cannot be definitely proved in the
present state of knowledge, but there is an ever growing body of facts to
support it. We have noted above that the older specimens assigned to
the Neanderthal group are less specialized and have more highly vaulted
skulls than do later members of the race, suggesting an approach to the
common ancestral type.

The Mount Carmel skeletons can be interpreted as also favoring this
hypothesis. As we have said, they show a "mixture" of Neanderthal and
neanthropic characters. Possibly, as some have suggested, that may be
due to interbreeding of two races. However, the describers of the material
favor an interpretation in harmony with the thesis presented here. They
suggest that after Neanderthal man had entered Europe, the unknown
Asiatic or African large-brained stock from which he came progressed
onward toward a neanthropic condition. The Mount Carmelites are examples of these transitional forms which had wandered from their unknown homeland into Palestine. Later, when the evolution to the modern
type had been completed, the end products of this series migrated, they
believe, into Europe as definitive Upper Paleolithic members of *Homo
sapiens*.

Early evidences of large-brained men.—The evidence thus suggests
that Neanderthal man is merely one side line of a stock of men with primitive features but large brains, and that neanthropic man is the end prod-

uct of another line developing from this stock parallel to him and finally becoming dominant. Further evidence to the same effect is presented by a number of other finds not considered as Neanderthal which exhibit large brains or other neanthropic features but are definitely or probably of considerable antiquity. And some are very ancient indeed.

The Swanscombe fragment.—The occipital bone of a human skull, perfectly preserved, was found in 1936 in one of the terraces of the lower Thames Valley at Swanscombe (Kent). The following year renewed search brought to light an equally well-preserved parietal. Nothing further has been found of this individual. The 100-foot terrace deposit in which it was found was formed during the great interglacial stage of the Middle Pleistocene; and in the undisturbed layer with the human remains were typical Acheulean implements.

But while we have here a very ancient European, our knowledge of him is distressingly limited. The occipital is a bone which includes the posterior part of the braincase; the parietal covers much of the roof. We have then considerable data as to the back part of the skull. The bones are rather thick but in other respects are modern in type, with rounded contours contrasting strongly with those seen in paleoanthropic forms. We have even sufficient data to make a reasonable estimate of the brain capacity (about 1,350 cubic centimeters). He was thus a large-brained early man and lacking the peculiar Neanderthal skull contours in the occiput. Thus as far as the evidence goes, he may represent the common ancestor we are seeking. It is only too regrettable that we have no further knowledge of him.

Reinforcing the evidence of the Swanscombe specimens is a recent find of two fragmentary skulls from Fontéchevade (Charente), France. The time of burial appears to have been the third interglacial interval; a pre-Mousterian culture accompanied the remains, and over them is a Neanderthal layer. One specimen is a skullcap much like that of a modern man, with evidence of a vertical rather than a sloping forehead. The second consists only of a forehead—again vertical and with no more signs of brow ridges than in the most "advanced" of living types. We are, of course, in the dark as regards the rest of the skull, jaws, and skeleton, but here is a concrete piece of evidence showing that at least a modern type of forehead had evolved-long before modern man became common.

A number of other finds from various areas of Europe, and from Africa as well, have been reported from time to time which have tended to confirm this thesis that men with large brains, but not Neanderthalers, and

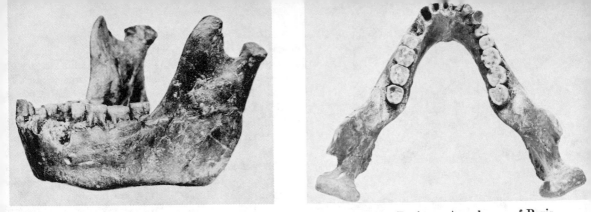

Early man from North Africa. Important recent finds have been made by Professor Arambourg of Paris in the fossil deposit of Palikao (Ternifine) in the Oran region of Algeria. These include three jaws, one of which is shown in the two figures above, and a skull fragment. Apart from the two specimens illustrated below, these are the only finds, west of Java and China, which are as old as Lower Paleolithic times.

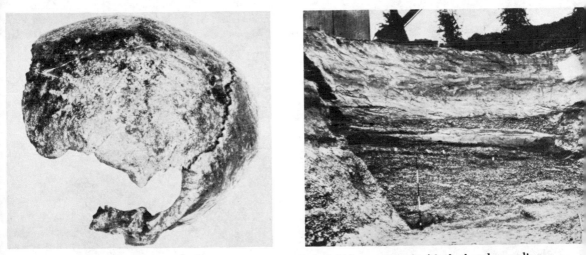

The Swanscombe skull fragment (*above*) is a human fossil, definitely associated with the hand-ax cultures of the Lower Paleolithic. It includes only the large parietal bone, covering much of the skull roof, and the occipital bone, forming the back of the skull. These are shown articulated, from the left side. The rounded skull contours are more similar to those found in our own species than in Neanderthal. *Right, above*, Acheulean deposits, in which the specimen was recovered; white spots above the level of the rod indicate the exact layer. (Photographs from Oakley and Morant.)

The Heidelberg jaw. This massive specimen was found deep in a commercial sand pit at Mauer near Heidelberg. Despite close watch for half a century, no further human remains have been found in this deposit, which was certainly formed well back in the Pleistocene and possibly during the first interglacial stage. (After Schoetensack.)

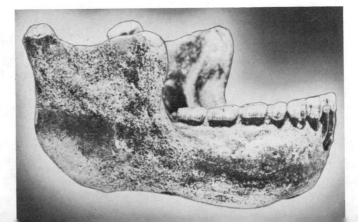

Neanderthal variants. While typical Neanderthal remains ₐ those from Europe during the Würm glaciation, other spe mens are found which are earlier in time or from western A₅ rather than Europe; these may vary considerably from t typical Neanderthal structure. *Upper left*, a skull from Mou Circeo on the Italian coast near the Pontine marshes, which from the last interglacial stage but is structurally similar later finds. *Next below*, two views of another skull from t last interglacial (from Steinheim in southern Germany), whi is less typically neanderthaloid in its skull contours and clo₅ to modern man. Of great interest are the finds from Mou Carmel in Palestine, which show many structural intergrad between Neanderthal man and our own species. One of t Mount Carmel skulls is shown below. *Below, left*, one of t Mount Carmel skeletons in process of excavation. *Below, rig* the bluff containing the Mount Carmel caves. (Mount Circ skull after a photograph by Dr. A. C. Blanc of Pisa, its di coverer; Steinheim skull from Weinert; Mount Carmel sk from McCown and Keith; Mount Carmel scenes from *Bullet American School of Prehistoric Research*, courtesy Dr. G. MacCurdy.)

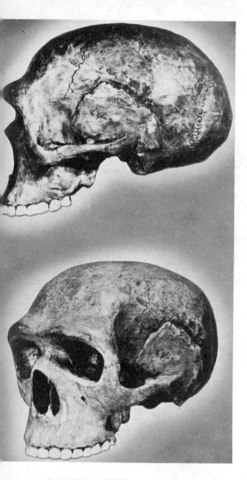

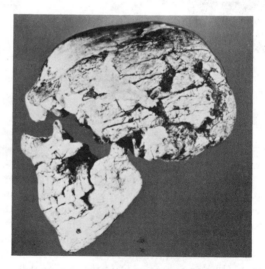

with vaulted skulls were in existence in Middle Pleistocene days. On the whole, however, no great weight has been placed on such reports by most authorities, because the skeletal remains concerned have been either fragmentary or of dubious age or both. Of these perplexing finds we will cite as typical the case of the Galley Hill skeleton.

Galley Hill is near Swanscombe, mentioned above, and the site is the same 100-foot terrace containing Acheulean deposits. In gravels of this age, in 1888, a workman and local amateur scientists excavated a nearly complete human skeleton, including most of the skull and jaws, except for the bones of the face. The whole structure is typically neanthropic, including a jaw of modern type; in fact there is nothing to distinguish this Galley Hill man from *Homo sapiens*. If this skeleton had been found in a later deposit, its authenticity would not have been doubted. But when claimed to come from a horizon this early, it ran counter to the evolutionary beliefs of many workers, and hence a loophole was sought. This was found in the fact that the disinterment was done by amateurs. Several scientists who made careful inquiry later were satisfied as to the authenticity of the find. But, nevertheless, doubts may legitimately remain; possibly some mistake may have occurred, and the skeleton may have been an intrusive burial rather than "native" to these old gravels.

Were it not for a dispute over the dates of the specimens, finds from two sites in the Victoria Nyanza region of Kenya, East Africa, would be regarded as of the highest importance in the history of modern man. Here in the 1930's the anthropologist Leakey discovered a jaw fragment with a modern type of projecting chin. It came from a locality (Kanam) which appears to be of very early Pleistocene date; at nearby Kanjera, in a Middle Pleistocene site, were found, together with two limb bones, fragments of four high-vaulted skulls of neanthropic type.

Solo man.—We may conclude this section with a description of two other types of men which are large brained but are neither neanthropic in appearance nor Neanderthals in the proper sense of the term; they may represent stages in the evolution of modern types from archaic ancestors in Asia and Africa.

From the banks of the Solo River in Java, where *Pithecanthropus* was discovered, have come the remains of a second human type, described as *Homo soloensis*. The materials representing this form come from the neighborhood of Ngandong, only a few miles from Trinil. Despite the geographical nearness of the two sites, the two human types are clearly separated chronologically, for the Ngandong deposits are apparently from a later period of the Pleistocene, certainly no earlier than the third glaciation.

Of Solo man there have been found eleven partial skulls, as well as two thigh bones. In all of the skulls the base had been removed, suggesting, as in other cases noted, a cannibalistic fondness for brain food; and face, jaws, and teeth are as yet unknown. The brain was rather small, for a male skull has a capacity of only 1,316 cubic centimeters, and several females average but 1,175 cubic centimeters. The forehead was low, the brow ridges heavy—archaic, paleoanthropic features suggestive of Neanderthal relationships or of a descent (as in the case of that form) from the much earlier *Pithecanthropus* type. But the contours of the rest of the braincase and a number of diagnostic technical features are not at all those found in these more primitive human types but are, on the contrary, much more like modern man; so much so that it is more probable that Solo man should be considered not as a separate species but as a primitive type of *Homo sapiens.* If so, how are we to account for the suggestively massive brow ridges and low forehead? It may be that Solo man represents an evolutionary progression upward in eastern Asia from a primitive large-brained ancestor to such modern forms as the living Australoids. On the other hand, we may have a blended type in which, as some believe to be true in Palestine, there is a mixture of races, an infusion of paleoanthropic blood into an early neanthropic race.

Rhodesian man.—It is generally agreed that Solo man presents close similarities to the Rhodesian skull, which may be considered at this point, although its geologic age is uncertain.

At Broken Hill, Rhodesia, valuable ore deposits were present in a hill which has, in consequence, gradually disappeared in the course of mining operations. A long cave passing through much of the hill contained many bones of animals so heavily impregnated with minerals that they were fed to the smelters. Fortunately a human skull was rescued, together with a few other bones of doubtful association including much of a leg. The skull is uncrushed and almost perfect and represents an unusual human type, with a number of primitive and specialized features. At first sight one tends to compare this bestial form with Neanderthal man because of such features as the exceedingly heavy brow ridges and the low forehead. The face is enormously developed and the palate as well, features which are specializations not known to be developed to such a degree in other known fossil types. But the contours of the braincase and other details of the skull show none of the diagnostic characters of Neanderthal man. The

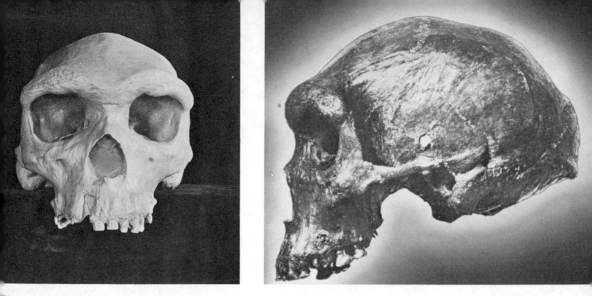

The Rhodesian skull comes from a cave deposit at Broken Hill, northern Rhodesia, which is of unknown date, although not improbably late Pleistocene. Front and side views of this massive skull are shown above (from Pycraft). Although the brow ridges suggest comparison with Neanderthal man, other features are rather more "modern," and close comparison is possible with the braincase of Solo man, seen below at the left.

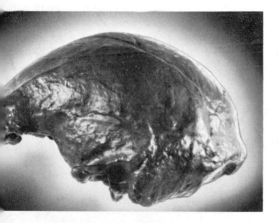

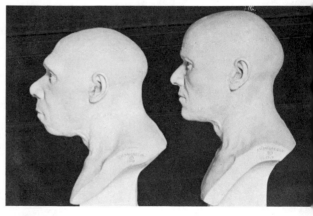

Solo man (*Homo soloensis*) from the late Pleistocene of Java is known from a number of skullcaps; the most complete is shown (*above, left*). (Originals described by Oppenoorth and von Koenigswald.)

Restorations of Neanderthal and Cro-Magnon types by Dr. J. H. McGregor. *Above*, the soft anatomy of the head has been carefully restored. *Left*, the hair has been added (although of course we have no certain knowledge of this) to give a more "natural" appearance.

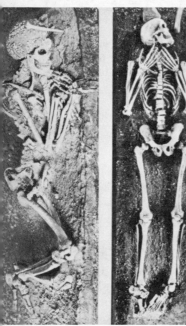

Cro-Magnon man. The Upper Paleolithic men of western Europe, oldest known representatives of our species in that region, are generally termed the Cro-Magnon race. The name is derived from the French rock shelter of that name (seen on a later plate), whence a typical example of the race was derived. Although variations occur, most of the skeletons from the Upper Paleolithic of Europe vary little from the typical Cro-Magnon type. The men were tall, ruggedly built, and large brained, with high foreheads and long heads. *Left*, two typical skeletons of tall individuals from the Grimaldi caves of the Riviera. *Below*, two skulls. *Left*, the type of the race, an old man from the Cro-Magnon shelter (teeth restored); *right*, a closely comparable skull from Grimaldi. Cro-Magnon man is of interest for his artistic productions (examples are shown on other plates); *above*, a restoration by Charles R. Knight of a group of Upper Paleolithic artists. (Upper figure courtesy American Museum of Natural History, New York; old man of Cro-Magnon, restoration by Dr. J. H. McGregor; other figures after Verneau.)

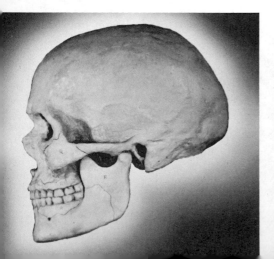

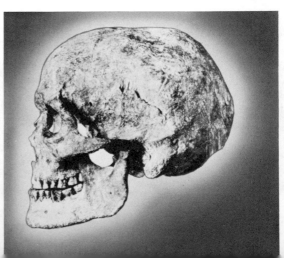

brain was small, with a capacity of only 1,300 cubic centimeters, strikingly low for such a large skull. If the limb bones, which are modern in type, belong to the skull, they indicate a man of average modern proportions. We have no data from which to estimate the age of the skull, but the similarity of its braincase to that of Solo man suggests that it may well be late Pleistocene. A comparative braincase found a few years ago at Saldanha, north of Cape Town, seems, from associated bones and implements, to date from about the third interglacial stage. The problem of the relationships of Rhodesian man is the same as in the case of Solo man. There is a fairly general belief that Rhodesian man may be related, through other extinct types considered later, to the ancestors of the living Bushmen.

MODERN MAN

Cro-Magnon man.—Toward the close of the glacial period, during the latter part of the Würm glaciation and the phases of the final retreat of ice, men of our own species entered Europe. Succeeding the Mousterian implements, we find in numerous deposits a layer containing new industries, those of the Upper Paleolithic, with flint blades and numerous bone implements replacing the flake tools of Neanderthal man and with evidence of contemporary activity in art and other fields, which indicates that the newcomer was a man of wider intellectual horizons and activities than his predecessor. This new European, we know from numerous associated skeletal finds, was, as Cro-Magnon man and related types, a representative of our own species, *Homo sapiens*.

Whence he came is a question to which we have as yet no satisfactory answer. We have seen that there is considerable evidence suggesting that such a modern type of man may have been evolving much earlier—in Middle Paleolithic days and perhaps even earlier. As to the place in which he arose, the Mousterian culture generally associated with Neanderthal man does not spread far into Asia, and the modern type of man may have been present in that continent while the typical Neanderthals still flourished in Europe.

The Upper Paleolithic stage in Europe may have begun as early perhaps as 30,000 B.C. and persisted until approximately 13,000 B.C. It is customary to refer the human finds of this age to the Cro-Magnon race, which takes its name from a French rock shelter where characteristic skeletons of this type were discovered. Remains of this sort are particularly abundant in the caves of France but are found widely distributed in various regions of western and central Europe; about one hundred individuals are known.

The head is rather large and massively built, features which some suggest are due to a dash of Neanderthal blood in the Cro-Magnon race. In general, however, one can state with confidence that paleoanthropic features are absent. The great brow ridges are lacking, the forehead and skull vault are high, the occiput rounded, the face reduced to modern proportions, the nose narrow and prominent, the chin highly developed, and in every respect Upper Paleolithic men were on a high plane of development. The skulls are almost always dolichocephalic, with cranial indices generally ranging from 65 to 76 (an exception, however, is one brachycephalic skull). These large heads contained large brains, averaging in males 1,700 or 1,800 cubic centimeters; again, however, sexual differences are marked, and the females were smaller brained. As racial peculiarities we find that the face was short but broad and the eye sockets correspondingly wide but low.

This was a tall race, on the whole, many males being six feet or so in height. In the body as in the skull, build and proportions are modern, without any trace of Neanderthaloid characters.

Some degree of variation existed, of course, among these ancient men. The Predmost type, for example, is that of a group of mammoth-hunters of ancient Moravia, in which the face was rather long, the chin rather less prominent, the supraorbital ridges more than normally developed—features which give a slight Neanderthaloid aspect to their skulls. A skeleton from Combe Capelle in France is that of an individual with both skull and body rather smaller than usual in the Cro-Magnon type; the jaw and brows were rather like those of the Predmost specimens. Skeletons from Chancelade in France and Obercassel in Germany, which date from a rather late stage in the Upper Paleolithic, show more rounded orbits, longer faces and shorter bodies, and a general appearance which has suggested to some workers a comparison with the Eskimos or at least with Mongoloids of some sort. A burial in the caves of Grimaldi on the Riviera contains two skeletons, probably a mother and her adolescent son. Here, as in the last instance, body and head are relatively small. The general proportions, projecting teeth, and rather broad noses have suggested to some writers Negroid affinities, although this belief is disputed.

Variations do exist, but, on the whole, they are not marked. A recent objective metrical study of the entire series of known Upper Paleolithic residents of Europe shows that, all in all, there are no greater variations than one finds today in a relatively pure racial group.

With the withdrawal of the ice and related changes in climates, plants,

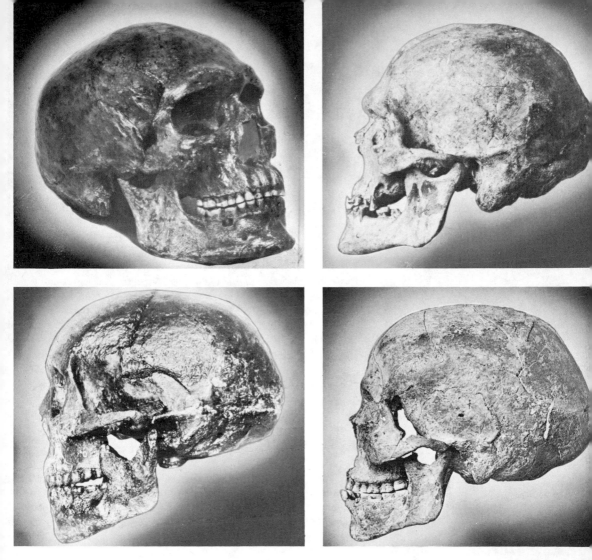

Upper Paleolithic men of Asia and Africa. Although most of our knowledge of ancient races of our own species has been obtained from Europe, the last two decades have seen the extension of our knowledge to other Old World areas. The Choukoutien cave in China has yielded from an upper level several skulls (*upper left*) which show few if any of the specialized features of modern Mongoloids.

North Africa has yielded a number of Upper Paleolithic skulls. *Upper right*, a male from the cave of Beni Segoual in Algeria, very similar to Cro-Magnon man. *Lower left*, one of several skulls from Gamble's Cave in Kenya Colony, East Africa, where an Upper Paleolithic culture was present; these men of Kenya show little trace of Negroid affinities. *Lower right*, a skull from Asselar in the southern Sahara; the skeleton is shown at the right. The specimen appears to be of considerable antiquity and does show Negroid features. (Choukoutien from a cast, courtesy Dr. Franz Weidenreich, Cenozoic Research Laboratory, Peking; Beni Segoual from Boule, Vallois, and Verneau; Gamble's Cave from Leakey; Asselar, from Boule and Vallois.)

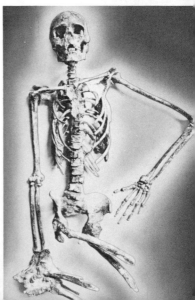

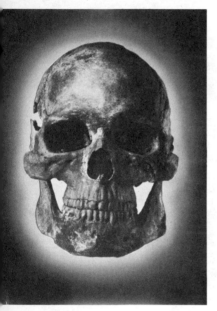

Two famous Upper Paleolithic sites. *Above, left,* the village of Les Eyzies in the Dordogne region of southern France. Behind the houses at the base of the cliff was the Cro-Magnon rock shelter. (Photograph courtesy American Museum of Natural History, New York.) *Upper right,* the grottos of Grimaldi, on the Riviera. Several caves with Upper Paleolithic implements, human skeletons, fossil animals are seen at the left as crevices in the cliff. (After Verneau.)

The Chancelade-Obercassel race. A Cro-Magnon variation of Magdalenian age is represented by skulls from Chancelade in France and Obercassel in Germany. *Left,* a figure of a male from Obercassel (the teeth restored). Some writers believe these specimens to show Eskimo-like features. (After Bonnet.)

The Grimaldi race. At Grimaldi, besides typical Cro-Magnon finds, a lower Aurignacian level revealed the skeletons of two distinctive individuals, with projecting teeth and other features suggestive of Negroid affinities. These skeletons and one of the skulls are shown below. They represent a middle-aged woman and a youth. (After Verneau.)

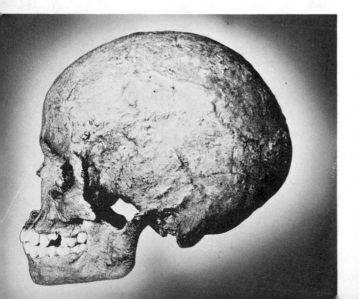

and animals, the Upper Paleolithic culture came to an end, and presently
new racial types appeared in Europe. What was the fate of the Cro-
Magnon race? The details are obscure, but, as suggested in the next chapter, their blood may still flow in the veins of many modern inhabitants of
western Europe.

Africa.—We have thus far concentrated our attention on the appearance of *Homo sapiens* in Europe and have done so because our knowledge
of his early history there is much more adequate than it is in other areas.
In recent decades, however, many new finds of ancient representatives of
our species have come to light in various parts of Asia and Africa. While
the story is none too clear as yet, we may briefly mention some of the
more interesting discoveries of this nature.

In northern Africa, finds in Algeria indicate that Upper Paleolithic
races of that region were fairly similar to their Cro-Magnon contemporaries across the Mediterranean. Not too dissimilar, also, are certain
Upper Paleolithic finds from East Africa, including a series of skeletons
from Gamble's Cave in Kenya and another individual from Oldoway in
Tanganyika (the latter was once believed to be still older). In many regards these men are comparable to the Cro-Magnon race, but they are
rather more slenderly built, with longer and narrower faces and rounder
eye sockets. They appear to be rather similar in type to the basic stock of
brunet "whites" which form the modern Mediterranean race (in a broad
sense of that term).

Africa today is the great center of the Negro race, and there are some
evidences of ancient members of that stock. A skull from Asselar, in the
southern Sahara, which cannot be accurately dated geologically but appears to be of respectable antiquity, shows Negroid features, and the same
is said to be true of other skulls from East Africa.

Workers in South Africa have accumulated considerable fossil evidence
which suggests that the Bushman is a descendant of ancient inhabitants
of that region. Among others a skull from Boskop in the Transvaal appears to represent an ancestor of this interesting race, although one with a
larger body and larger brain than in his supposed descendants—perhaps
physically degenerate. A still more archaic-looking skull from Florisbad in
the Transvaal is thought to be a connecting link between the Rhodesian
type and these Bushman ancestors.

Asia.—Upper Paleolithic cultures, rather comparable to those of Europe, stretch away eastward through northern Asia, and their makers
may have been not too dissimilar to the Upper Paleolithic men of Europe.

In China the very caves of Choukoutien have in their upper and later layers skull remains of this age which, while varied in details, are of a "high" human type.

We have no evidence of any sort regarding the early history of the Mongoloid peoples, now so important in Asia; possibly they may have developed at a relatively late date. Deeply regrettable is the lack of knowledge of Upper Paleolithic times in central Asia and in Iran and India. These areas have probably been of the greatest importance in the early history of our species, but we do not have the slightest scrap of a bone from these regions.

The native Australian is rather generally considered to be an archaic type of our species which has reached his present home by a migration down the East Indian island chain. Several fossil "documents" appear to record this migration. In Java still another type of some interest is Wadjak man, represented by two skulls. The age is uncertain but appears to be close to the end of the Pleistocene. This man definitely pertains to the modern species but is rather primitive, with somewhat prominent supraorbital ridges and a massive skull build; the brain was large. Wadjak man may have descended from the Solo man mentioned in the last section; on the other hand, it is generally agreed that he may be a primitive Australoid. Other early Australians are found in Australia itself; skulls excavated at Talgai (Queensland) and Cohuna (Victoria) have unusually massive faces, enormous palates, and large teeth but are in most respects very similar to the living Australians.

America.—Despite a considerable amount of investigation, we have failed to find in America any story of early human evolution comparable to that of the Old World. Apparently primates had become extinct in North America early in the Tertiary, before they had even attained the monkey stage, and manlike apes are unknown in the Western Hemisphere. Some years ago teeth were found in the late Tertiary of western Nebraska which appeared to be rather manlike and were given the name of *Hesperopithecus*, the "western ape." But further research disclosed that they were merely rather aberrant teeth of a fossil peccary, a member of the pig family; teeth are one of the features in which man and swine show embarrassing similarities. In South America a patriotic paleontologist, ambitious to show the evolutionary importance of his country, once described a series of "protohomos" supposed to show the evolution of man from ape in Argentina. But these remains, on critical examination, appear to be far from convincing evidence. They include, among other specimens, part of the skeleton of an ordinary South American monkey and a bone dredged out of the Buenos Aires harbor, which we may suspect to be a souvenir of a sailor fallen overboard after a large evening in that gay city. The earlier stages in human history, it would seem, were already past before man reached the western world.

The time of his arrival, however, is a problem of considerable interest. There is an increasing body of evidence that when Columbus and later voyagers arrived to discover the Indians in possession, man was far from being a newcomer and may have been here for many thousands of years. As we have noted, the Ice Age fauna of the Americas was a far different and richer assemblage of animals than those now living. For many decades there have been repeated finds which, when summed up, definitely show that man had entered this hemisphere before the Pleistocene fauna had become extinct and was a contemporary here of native American camels, ground sloths, mastodon and mammoth, and fossil species of bison.

But even so this does not necessarily mean any remote antiquity. At one time it was thought that many of these extinct animals had died out rather early in the Ice Age. This is no longer believed to be the case, and it seems certain that many elements of the older fauna survived to a late date. Camels, proboscidians, and sloths lingered on in North America well after the retreat of the last glaciation and may have been present in this country until a few thousand years ago.

If man had been an inhabitant of America during any great extent of the Ice Age, we would expect to find flint implements comparable to those of the Lower or Middle Paleolithic of the Old World. This is not the case. The oldest types of flint implements known at present are fluted dart points, termed the "Clovis" and "Folsom" types after the New Mexican localities where they were first discovered. This type of implement differs from any used by recent Indians in the West. But this is not in itself any proof of great antiquity and such a flint might have been produced by a people in the Mesolithic cultural stage or, at the most, the late Paleolithic.

Few skeletal remains are known in North America which can lay claim to any great antiquity, and none appears to be in any way primitive. The "Minnesota lady," for example, is the skeleton of a maiden found in a supposed Ice Age deposit in that state. The skeleton indicates that she had a powerful jaw (a feature upon which we shall not comment further) but in all other regards appears to be similar to members of recent Indian tribes of that region. A skeleton found in an old lake bed near Mexico City, dating from about 10,000 B.C. and associated with mammoth remains, likewise fails to show any major differences from later Indian types.

All the evidence so far cited is thus inconclusive in nature. Careful geological work has aided in dating the arrival of man. The "Minnesota lady" is claimed to have come from a deposit laid down in a Pleistocene lake, about 20,000 B.C., but the skeleton was excavated during a road-repair job, and we cannot be sure that she actually came from these lake beds. Geological investigations of sites where Folsom implements are present suggest very late glacial to early post-glacial times.

Current work in carbon-14 dating is aiding greatly in the solution of the problem of when man reached the Western Hemisphere. The lowest cultural level in the Sandia cave in New Mexico—well below a level with Folsom points—gives a date of approximately 20,000 years, indicating that man arrived here well before the close of the Paleolithic stage in the Old World. A number of quite reliable determinations are in excess of 10,000 years. Makers of the Clovis and Folsom points appear to have hunted the western plains 10,000 to 13,000 years ago. Bones from a cave in Patagonia show that man had penetrated south to that remote region, where he was a contemporary of ground sloths and native horses, by about 7000 B.C.

How did man reach America? Probably not by any lengthy sea voyage, for although a stray canoe or so might have reached America by sea, and there is some slight suggestion of cultural interchange across the Pacific, there is not the slightest trace of racial similarity between the Indians and the inhabitants of the Pacific islands to the west or Africa to the east. Presumably migrations were mainly by land, via the Bering Strait route from Asia. The crossing there today is a short one, a trip which would not be too difficult for any coast-dwelling primitive people which could make a crude craft of some sort. Further, geologists have pointed out that sea levels fluctuated greatly during the Pleistocene; a drop of little over 100 feet would effect a land connection, and the first comers may have entered America dryshod.

Once across, penetration of the continent would have been accomplished without too much difficulty. Even during the peak of the glaciation it seems that considerable portions of Alaska and British Columbia were free of ice; there existed down the west flank of the Rockies a corridor through which a hardy folk could have passed to reach the more hospitable lands to the south. We can picture the peopling of America as occurring by the passage into Alaska of successive waves of Asiatics who drifted slowly southward and spread out to populate the temperate and tropical regions.

CHAPTER 18

Human Races

In our last chapter we discussed the position of our own species, *Homo sapiens*, in the human family; here we shall make some attempt to differentiate between the various racial types comprising this species.

RACIAL DETERMINATION

Possible nonphysical criteria.—Upon what basis can we discriminate between one race and another? One obvious suggestion would be to use nationality as a basis. An Englishman, a Frenchman, a German, are all definite types to us, and we associate these types with definite governmental units. But we have only to look to Europe to see that national and racial boundaries need not coincide; many of the conflicts of that troubled continent have been, and still are, due to the inclusion of alien populations within national boundaries.

Language is another possibility. Man tends to retain the speech of his fathers even in a foreign land; various types of language show clear evidences of relationship and common descent. Cannot the evolution of languages be correlated with the history of the peoples who speak them?

Language may afford a clue as to racial affinities but often a treacherous one. Chief Two-Guns-White-Calf and the Pullman porter may speak English, but they are obviously unrelated racially to the English people among whom this language originated. In modern Europe there are numerous areas where a change in language has been made or attempted without any change in the racial makeup of the population, and there are many instances where the conquest of a country has resulted in the replacement of the native language by that of numerically insignificant conquerors. A century ago it was discovered that almost all the languages of Europe and those of Persia and much of India were descendants of a common tongue, of which Sanskrit is a little-changed relict. On the basis of this discovery was erected a theory of common descent of the peoples who speak Indo-European languages. But we now realize that this language stock is native to only a small proportion, at the most, of those who now utilize it; it may have been carried by migrations and conquests, but the Indo-European languages have spread far more vigorously than the peoples who have carried them.

Can culture be used as a basis for studying racial history? We tend as a rule to be conservative in our customs, our beliefs, the types of material objects which we fashion for ourselves. But here again we must be wary, for cultural traits often spread widely by adoption or are superimposed by conquest. The American Indian of pilgrim days traveled on foot and fought with bow and arrows, but the horse and rifle, both taken over from Europeans, were major elements in the life of the Indian of the covered-wagon era. The Mexican peon is a Christian, but the blood in his veins is almost identical with that of his predecessors who worshiped Quetzalcoatl.

Nationality, language, cultural traits, and cultural objects—all these may afford suggestions as to racial affinities or origins; but they are to be used with caution. The only definite and fairly positive knowledge which we can obtain regarding the pedigree of an individual or a group is from a study of its physical characteristics.

Racial mixtures.—When we discuss the relationships of other animals, we do so in terms of species or larger units. These groups are generally clear cut; species seldom breed with one another and are usually separated by clearly defined structural differences.

But in the attempt to study human races we are dealing with quite another type of problem. All living men are seemingly members of a single species. Members of all human groups can breed with one another and frequently do so; in the Hawaiian Islands, for example, may be found almost every conceivable type of cross between the major racial groups of the Orient and Occident.

Such wide racial intermixtures are facilitated, it is true, by the development of modern communication methods, but crossing between races must obviously have been a common occurrence throughout human history. Intermediate border groups have probably been common at all times, and every war or conquest has naturally resulted in a mixing of the strains of conqueror and conquered. We would think, for example, of the offspring of an Englishman and a French woman as products of a mixture of two races. But what is the English race? Within historic times we know of three invasions which have added new strains to the pre-existing population, and there is good evidence that a number of similar invasions have taken place in late prehistoric times. The history of the French "race" is a similarly complex story.

Probably there is not, and never has been, a really pure race. When we attempt to establish a racial type, we attempt to set up a series of characteristics commonly found in the people of a certain area or of a certain caste; few will exactly meet our specifications in all respects. A pure racial type is a man-made ideal, seldom perfectly realized.

Migrations.—It would be difficult enough to deal with racial problems were peoples stationary. But our problems are further confused by the fact that movements of peoples appear to have been common throughout human history.

The overseas movements of the European peoples are only the most recent and striking of human migrations. Throughout the historic period, wave after wave of peoples have swept into western Europe from the east—Celts, Germans, Huns and Avars, Turks—and archeology reveals similar prehistoric events. A great southern migration of Negroes into South Africa was in progress when England took over that region. The Chinese people have spread southward over a vast area during the time covered by their written history. Many migrations of American Indian tribes can be definitely plotted.

These are but a few of the more striking of known migrations of peoples. Mankind has been, and still is, in motion. Groups once strangers have come into juxtaposition or have fused; related types have been isolated from one another and their affinities obscured or rendered doubtful.

Can we trace any general pattern in these migrations? The movements of a people may be due to a variety of causes; they are, in physical terms, a "resultant of forces." Better lands ahead to hunt in, to till, or to plunder; powerful enemies behind; a need for more territory for increased numbers; these and many other factors may make for migration.

In the evolutionary history of mammals we find that the main, early center of group dispersal appears to have lain in the northern land areas and particularly in Asia (although other continents, particularly North America, have acted as minor centers). From that continent appear to have radiated many mammalian groups to migrate eastward via Alaska to the Americas, southward to Australia or Africa, westward into Europe. In agreement with these findings is the fact that the most primitive of living mammals are found in comparatively isolated areas, farthest from their original Asiatic homes, whence they have been pushed by more advanced types. Australia is the home of the most primitive of existing mammal faunas; archaic ungulate types persisted in South America long after their extinction elsewhere; the modern African fauna is a survival of that of Eurasia two epochs ago.

The distribution of modern human races and the known history of human migrations suggest that the geographical story of man has been similar to that of many other mammalian types. A majority of known migrations have emanated from an Asiatic center; and some of the seemingly most primitive of human races are found in the Australian region and Africa. Whether or not Asia is the original home of man cannot be said with certainty; but Asia has surely been a major center of human dispersal.

Physical criteria of race.—Attempts to characterize racial groups should not be made merely upon casual and random observations but upon careful study of numerous individuals, such studies being of a quantitative nature as far as possible. Many features may be observed or measured on the living; numerous standard types of measurement may be made upon the skeletons. Naturally, skeletal measurements alone are available for prehistoric peoples, and even for well-known historic races accurate data as to the soft parts, such as hair, skin, pigment, etc., is usually woefully inadequate.

We must, of course, use some caution in drawing conclusions regarding racial evolution from physical data. Environmental factors, particularly, may cause changes in the development of an individual to such an extent that we might incline to place him in a racial category rather different from that to which he belongs. We are all familiar, for example, with the fact that hardships in youth, particularly poor diet, may make for restricted and abnormal growth. On the other hand, studies of American college fathers and their sons indicate that the younger generation is distinctly taller—by an inch or so—on the average. This does not indicate that there is any change in inherited factors but is due, we fondly believe, to more intelligent upbringing and better diets, which have given an environment more favorable for growth. Measurements of the children of Italian immigrants indicate that their head shapes are rather different from those of their parents. It has been suggested that in such a case the new environment has resulted in a permanent hereditary change in this character. There is, however, a reasonable alternative explanation. In their native land the inhabitants of each region mate only with their near neighbors, and stable local types develop. In the New World, however, Calabrian may wed Milanese, Sicilian may wed Roman; new combinations are formed, and it is not unexpected that measurements would vary in such mixtures. All in all, it seems probable that physical traits are generally stable; environment may cause individual variations but can exert little direct influence on a racial type.

Complexion.—Skin color is the most conspicuous of racial traits and in earlier days was used almost exclusively as a basis for major racial distinctions. The pigment in all human skins is the same. It consists of dark brown granules of a substance called melanin, imbedded in cells of the skin. If abundantly present the pigment gives a black effect. Smaller amounts give brown or yellowish effects. If little pigment is present a "white" appearance results, while a plentiful blood supply to the skin gives a ruddy tinge. The Negroes are, of course, notable for high amounts of pigment; the Australians also tend toward a black; brownish to olive or yellowish complexions characterize the greater part of the remaining races, while blond tendencies are most noticeable in northern Europe.

To be distinguished from ordinary blonds are instances of albinism, in which individuals have a congenital lack of pigment. An unusual example of this is the case of a group of "white Indians" in Panama among whom there is a strong albinistic strain, although they are closely related to the copper-skinned Indians of the same region. It is frequently assumed that skin color is affected by the environment—people living in the tropics tending to take on a heavier pigmentation as the result of natural selection. This may be to some extent true; but the albino Indians just cited are an extreme example of the fact that color and environment are not necessarily related.

Primates generally have a considerable amount of skin pigment, and hence it is reasonable to assume that primitive man was "dark-complected." But whether the pigmentation was primitively as heavy as in the modern Negro seems very doubtful; and there is, of course, no evidence on this point regarding older races.

Eyes present important characteristics. The varying eye colors depend upon the pigmentation of the delicate iris, the diaphragm covering the margins of the lens. Some pigment always present in the deeper layers of the iris reflects a purplish tint. In blue-eyed persons there is no other pigment in the iris. If, however, pigment is present more superficially, the blue is masked, and other colors appear. A slight amount of dark pigment will produce a gray or hazel color if the blue is not completely masked; heavier concentrations of pigment tend to give yellowish, brown, or black shades. In most races brown to black eyes dominate; blue eyes are in general characteristic only of northern Europeans or those who have interbred with them.

The upper eyelid frequently bears a fold which may project downward and, when the eye is widely open, cover the free, lash-bearing margin of the lid to a greater or lesser degree. Such a fold may occur at the inner corner of the eye (apparently an infantile character) or at the middle or the outer margin (regions in which folds tend to be more pronounced in old age). In the more characteristic Mongoloids of northeastern Asia this fold may extend clear across the eye.

The face.—Noses vary greatly in length, in breadth, and in profile. Breadth of nose was, as we have seen, characteristic of early human types, and breadth and shortness is today characteristic of Negroes and Australians. Mongoloids tend to be rather intermediate in nose shape, with a concave profile, while western Asia and Europe are regions where high, narrow, and long noses predominate. In narrow-nosed types the contour is frequently convex, with the peak either high up the nose, as in the "eagle beak," frequently seen in Europe, particularly northern Europe, or with a general pronounced convexity, as in the Armenoid type common in Asia Minor and in somewhat modified form in the Balkans. Curiously, a somewhat similar nose form is found among the Oceanic Negroes, and many North American Indians are hawk-nosed. Lips, in general, give little basis for racial characterization. In Negroes, however, we find an exceptional condition in rather thick and out-turned lips.

The proportions of the face are frequently characteristic of racial groups. In a child the face is always relatively short and broad; in adults these proportions are sometimes retained, but in many cases the depth of the facial region of the skull and of the jaw increases to give long "horse-faced" types. The short face seen, for example, in many Negroes and Australians is perhaps primitive; long faces tend to be common, on the other hand, in northwestern Europe. In some broad-faced races the cheek bones jut out prominently; this is a characteristic Mongoloid feature. Also useful in classification are features of the face as seen in side view. Prognathism is a term used to characterize a forward-jutting of the jaws such as is seen, in our own species, most commonly in Negroes and Australians, in contrast with the more vertical plane of the face in, for example, Europeans. We have noted variations in chin projection in the older evolutionary story. Similar but less pronounced variations are present among existing races.

Hair.—Perhaps the most useful bases for racial classification lie in the characters of the hair. As regards form, one extreme type is that found in Mongoloid races, where the hair is very coarse and straight. In Alpine "whites" and Negroids the hair is coarse to medium in texture, while fine hair is characteristic of Mediterranean peoples. In marked contrast to the straight hair of Mongoloids is that found in typical Negroids, markedly coiled in tiny spirals, giving a woolly appearance. In the Negroid type the root portions of the hairs beneath the skin are curved to begin with, and the shafts are flattened, comparable to a ribbon of tooth paste, allowing curling to take place readily. In the Mongoloid type the roots are straight and the hairs have a round cross-section, making twisting difficult. Between these two types there are many intermediates. In many Negroids the hair is longer and less tightly coiled than in extreme forms; "kinky" or "frizzly" are here more appropriate terms. Many peoples, such as Europeans and Polynesians, for example, have wavy hair. Here the hair tends to be long, moderately fine, and with an oval cross-section which allows some degree of twisting. Although hair is not preserved in prehistoric remains, the present distribution of hair types suggests that considerable curling may have been characteristic of early man.

There is great diversity in the amount of hair present in various races. In both Negroids and Mongoloids there is little beard or body hair. On the other hand, wavy-haired races tend to have relatively heavy beards and much body hair. The "aboriginal" Ainus of Japan have the reputation of being the hairiest of living races. Among Western peoples the short-headed types, "Alpines," in a broad sense, tend to have heavier beards and more abundant body hair than long-headed races.

Hair color is, for the most part, dependent upon the amount of melanin (similar to that of the skin) present in the hairs; hair and skin colors are typically correlated. The pigment is very abundant in Negroes and Mongoloids, giving a black color; little pigment is present in blond northern Europeans; an intermediate condition yields brown hair. In addition to the melanin a second, red-gold, pigment is sometimes present; if the melanin is reduced in amount red hair may result. In many light-haired races the amount of black pigment in the hair tends to increase with age, so that blond or red-haired children tend to have more brownish and darker hair as adults.

Stature unquestionably differs greatly from one racial group to another. However, it must be used with caution for racial discrimination, for it is obvious that there are great individual variations within groups, and we have noted that nurture may be responsible for marked individual variations in height. We must, in addition, discount the sporadic appearance of gigantism, cretinism, and other abnormalities due to improper functioning of the secretions of the pituitary and thyroid glands.

The adult male may be taken as a standard for measurement. Average human stature on this basis would be in the neighborhood of five feet five inches. The shortest groups of mankind are the Negroid Pygmies, in which a five-foot stature is about the maximum, and many individuals may be half a foot or more short of this. Short stature, although not so extreme, is common in the general Malay region; the Bushmen and Hottentots of South Africa, the Ainus of Japan, the Lapps of northern Europe, the Eskimos, and a few Indian tribes are also markedly short. On the other hand, height averages of about five feet seven inches or higher are found in a number of regions, notably in various types in northwestern Europe and the Balkan mountains, a number of tribes of central and southeastern Africa, some Indian groups of eastern North America, and a few other scattered regions.

Skull proportions.—The cephalic index, the ratio of breadth to length of the braincase, is a universally used measurement discussed in the last chapter. Extinct human types, we have noted, were almost all dolichocephalic; very probably brachycephaly is a comparatively late human development. In Europe broadheads were rare until quite late prehistoric times; in America there is much evidence that long-headed folk preceded the more typical brachycephalic Indians. Among existing Old World races we find that dolichocephaly is prevalent in Africa and Australia and the "fringes" of Europe and Asia; brachycephalic skulls are dominant in all the more central portions of Eurasia, suggesting a newer development.

But while this index measurement is of interest, there are many other features of the braincase to be kept in mind. Absolute size of the head, slope of the forehead, height of the cranial vault, are all items of interest. Further, broad skulls are not always well rounded; the brachycephalic people of southeastern Europe and Asia Minor tend to have flattened occiputs and angular skull contours. The capacity of the braincase interior is a measurement to which we have referred earlier. As noted, this figure is of interest and appears to show racial variants but cannot be used as an index to mentality. European male capacities average about 1,500 cubic centimeters; some small-bodied races average 300 cubic centimeters less without obvious harm to the intellect.

Blood groups.—When, long ago, attempts were first made to give blood transfusions, it was discovered that the blood was not the same in all men and that frequently after the introduction of blood from another person the new red blood cells tended to stick together in clumps (agglutinate) in the host, with disastrous results. Research has given us the reasons for this effect. In many cases the surface of the red blood cells of an individual contain one or both of a pair of complex chemical structures (proteins) which are arbitrarily called A and B. There are thus in any human group four possible types of individuals, which (if absence be designated by O) can be called O, A, B, and AB. It is a curious but unexplained fact that a person whose cells have the A material contains a second type of protein in his blood serum which, if blood of a B-type individual is put into his circulation, tends to unite with the B particles of the transfused blood cells and causes clotting. Similarly, B individuals in their serum carry a substance which will clot introduced cells from an A individual. People whose cells have both A and B substances obviously lack a clotting material of either type and hence can take in blood from anyone without danger—they are *universal receivers*. On the other hand, O people have blood cells that lack both A and B; such blood is not susceptible to attack in any case and can be transfused into anyone. People with O-type blood are *universal donors*.

The recognition of substances A and B is of interest not only to the physician but also to students of heredity and evolution. These substances are inherited by normal genetic processes. And, further, it is found that the percentages of A and B present in individuals vary greatly in different areas and races. Tests of the blood have been made in many thousands of persons from all regions of the world.

The A material is, on the average, present in 20–25 per cent of individuals in the Old World. In parts of Australia, however, it may be present in as much as 40 per cent of the natives. In the Americas substance A is apparently very rare in natives of South America and much of the southern and eastern regions of North America. In the Indians of Canada and the Rockies, however, much higher percentages are present, and the Blood and Blackfeet Indians have this factor present in more than half the individuals. Substance B is, in the Old World, present in many areas in 15–20 per cent of individuals, with the highest percentages mainly in central Asia. In certain peripheral groups, however, including the Australians, the Bushmen of South Africa, and people on the western fringes of Europe, B is quite rare. And in American Indians, except for one or two dubious records, B is almost unreported. In full-blooded American Indians, as in Australians, Filipinos, and isolated Negro groups, O-type runs very high.

Such items of distribution of blood groups as those noted above are interesting and suggestive. Blood groups appear to give one more clue to aid in unraveling the snarls of racial relationships and histories.

Classification of races.—No general agreement exists among students of race as to the criterion which should be used as the major feature in racial distinctions; skull shape, hair texture and color all have their advocates. The schoolbooks of our grandparents used skin color as the major factor and divided mankind into the whites of Europe, the blacks of the Old World tropics, and the yellow peoples of Asia, to the last of which were usually added the "red" Indians of America and the brown peoples of southeastern Asia and the Pacific islands. This classification gives a fairly neat geographical arrangement, but it tends to obscure the facts of migration and presents an unduly simplified picture. We shall here use a

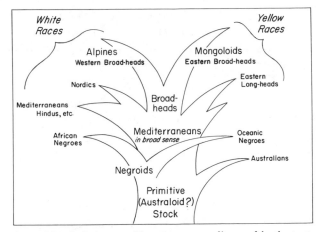

A provisional phylogeny of human races as discussed in the text

slightly different arrangement, as in the accompanying diagram, and attempt to take into account factors of racial migration and evolution. In this, skull proportions, hair characteristics, skin color, and nose proportions are used as readily identifiable key characters:

I. *Australoids:* Long-headed, wavy hair, dark complexions, flat noses; the inhabitants of Australia and certain primitive tribes of southern Asia.

II. *Negroids:* Long-headed, woolly hair, dark complexions, flat noses; the Negroes of Africa and New Guinea and the Pygmies are the major members of the group.

III. *Caucasoids:* The "white" races, with variable head shapes, brown to light complexions, wavy hair, narrow noses. They include:

A. *Mediterraneans* (in a very broad use of that term): Long-headed, brown complexions; the long-headed peoples of southern Europe and the

Mediterranean region, southern Asia to India, and traces of similar populations farther to the east.

B. *Nordics:* A northern European variant of the last, with blond hair and fair skin.

C. *Alpines:* Broad-headed, brown to light complexions; the broadheads of Europe and western Asia.

IV. *Mongoloids:* Broad-headed, straight hair, "yellow" or "red" complexions, noses of intermediate type; most of the peoples of eastern Asia and America.

Our present knowledge of human history suggests that long heads, curly hair, dark complexions, and flat noses were primitive human characteristics and that short skulls, straight hair, light complexions, and thin noses are more specialized (although by no means necessarily "higher") features.

AUSTRALOIDS

Logically one should, in a discussion of human races, treat them in order from the most primitive types—those who have departed least from the presumed ancestral stock—to the most advanced or specialized forms. But which races are truly primitive? As our survey in the last chapter shows, we do not have enough fossil data to work out the pedigree of our modern races at all satisfactorily. Further, it is probable that, did we know this pedigree well, we would find that as far back as we could trace our own species there was always some racial variation and that there never was a single, unique, primitive stock. And even had such a type existed, we would hardly expect that any race living today would adhere strictly to the ancestral pattern.

But despite these difficulties, we can to some extent sort out modern racial groups into relatively primitive and relatively more advanced and specialized types. It will be generally agreed that the "white" and Mongoloid groups which dominate the Northern Hemisphere in the Old World, and the New World as well, are relatively specialized or "advanced" types and that we must look to the southern portions of the Old World—Africa south of the Sahara, the East Indies, Australia—for types closer to our concept of what the ancestral *Homo sapiens* may have been like.

In these peripheral southern areas we find as native populations two major groups, each of which has some claim to primitiveness—the natives of Australia and related races, which may be termed "Australoids," and the Negroids. The two have some common features which may be primitive. Both have well-pigmented skins, dark eyes, and dark hair, which is wavy to tightly curled (light colors and absence of hair wave are probably advanced or specialized conditions). Both tend to have dolichocephalic skulls and broad flat noses. In both the face is more projecting—prognathous—than in the northern races, and this again can be reasonably interpreted as primitive. Both have broad lips, which possibly is a primitive trait although not certainly so (this character is less developed in Australians).

Beyond these common features, however, the Australoids show a number of features which seem to make them the better candidates for a primitive position. The Australoids have low foreheads, prominent brow ridges, and considerable body hair—all probably primitive features. The Negroes, in contrast, are as advanced as any northern race in the development of a high and smooth forehead and reduction of body hair. The dark coloration of the typical Negro and the tight curling of his hair seem surely specializations. On the whole, then, we may reasonably regard the Australoids as the modern group which has departed least from human ancestors of Paleolithic days.

Best known of Australoids are the natives of Australia, found in full possession of that continent when discovered by Europeans. Today there remain about 60,000 "blackfellows," mainly concentrated along the northern coasts. The Australians have little "political" organization, there being seldom more than a banding together of a few families, but there is a very complex social system as to relationships and marriages and a high development of totems. As far as material culture goes, however, they are at a very primitive, essentially Paleolithic stage. They are hunters and food-gatherers, ignorant of metals, with spears, throwing-sticks, and clubs as weapons, without permanent shelters or habitations, and with little in the way of clothing or other belongings. There is no agriculture, and domestic animals are unknown, except for the dingo which, although now partly reverted to the wild, was probably introduced by them.

There is considerable variation in physical characters among living Australian natives, particularly in the northeast, where the differences may be due to an admixture of Negroid blood from adjacent New Guinea. We have noted above such features as the long head, the broad, flat nose, and the prognathous face, which they possess in common with Negroids. But in contrast with that group, as we have seen, the forehead is low, the brow ridges marked; the brain is frequently small, the chin poorly developed. The skin is well pigmented, but the coloration tends to give a brown effect rather than the black of the typical Negro. And the dark hair, abundant on body as well as head, is wavy or curly, rather than kinked in a Negroid fashion.

The extinct Tasmanians are perhaps to be regarded as a subgroup of the Australian type, less subjected to invasions by other races from the north and hence purer in nature than the Australians proper. This race, when first found by white men a century and a half ago, numbered only a few thousand. Proving troublesome to the settlers, most of the natives were hunted down and exterminated, a few pinned to a small reservation. The only survivors are a few individuals of mixed white and Tasmanian blood. Their extinction is an anthropological calamity. Like the Australians, they were in a Paleolithic stage of culture, and even their stone tools were of a very crude sort. In general characteristics they appear to have resembled their Australian relatives, but the nose was even broader and flatter, and there was a somewhat closer approach to the Negroids in a rather darker pigmentation and more tightly curled hair.

If we tentatively accept the thesis that the Australoids most closely represent the type ancestral to modern man, we would assume that similar types were once present in Asia, that a migratory wave of these peoples spread southward to Australia and survived there in relative isolation, and that their representatives in Asia became submerged by the development from them of more progressive human types.

Such an explanation would be closely in accord with the known facts of evolution in other animal groups; we have noted that Australia is a refuge for many archaic mammals whose relatives are almost entirely extinct in other regions.

If this explanation be true, it is not unreasonable to expect that there would be some trace at least, of the former presence of Australoids in Asia. This appears to be the case. Some survivors who may be termed "Indo-Australians" are found in a number of regions in southern Asia, particularly in central and southern India. There, among the more isolated hill dwellers, such as the Bhils, and the Veddahs of Ceylon, we find tribes—some still primitive hunters—which exhibit many of the features of the Australian native. These "Pre-Dravidians" may well represent isolated mainland survivors of this ancient human stock. There appears also to be a strong Australoid strain in the natives of New Guinea and Melanesia, and there are traces of such peoples in south Arabia. And it is even possible that some Australoid blood was present in the earliest human invaders of the Americas.

NEGROIDS

The characteristics of the typical Negroid are familiar: the hair is black, short, coarse, and of a kinky or woolly texture, and there is little hair on the face or body; the skin is black, the eyes dark. The face protrudes markedly (prognathism), and the lips are thick and out-turned. In typical members of the group the skull is long, the nose short and broad, the stature medium to high. In contrast to Australoids the forehead is typically high and smooth, with little development of brow ridges.

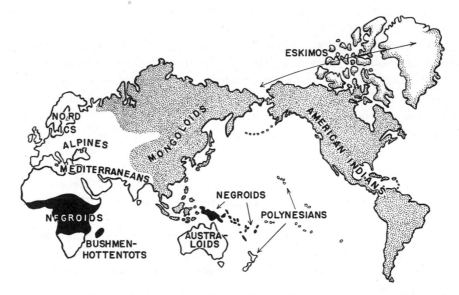

A map of the world to show some features of the distribution of races before European colonization. Australoids, "whites," and Bushmen-Hottentots, *unshaded;* Mongoloids, including American Indians and Eskimos, *stippled;* Negroids, *black.*

African Negroes.—The major homeland of the Negroids today is Africa south of the Sahara, where, except for the Hottentots and Bushmen mentioned later, they are the only native races. The Negroes are a flourishing group and are not only successful in the African environment but, as has been proved, can withstand transplantation to new regions. Culturally the Negroes, before being directly influenced by white civilizations, had progressed far beyond the Paleolithic level and had for the most part reached a full Neolithic level, with well-developed agriculture, domestic animals, weaving, pottery, and even some knowledge of metals. In a number of cases social development had progressed to the organization of powerful tribal structures, but there were few stable or highly organized communities except in isolated cases such as the city-building Fulas of the Niger region and perhaps the unknown builders of Zimbabwe in Rhodesia. Among the typical Negroes of Africa (and for the moment neglecting the Pygmies), a number of subgroups may be distinguished. The belt of grasslands running from the Nile region westward to the Atlantic along the western "bulge" of Africa is the home of the Sudan Negro, characterized by great height and powerful build. Many of the tribes in this northern region show distinct traces of the infusion of blood from "Mediterranean" whites. They are agriculturalists or cattle-herders. South of these grasslands is the region of the tropical rain forests, including much of the Congo and a strip to the northwest along the coast. Cattle and grain do not do well in this environment, but there are pigs and goats, yams and taros and plantains (much as in southeastern Asia, curiously). The forest Negro is a comparatively broad-headed, short-legged, stocky type, which contrasts rather widely with his northern neighbors. From this and the preceding group came most of the slaves imported into America. Somewhat intermediate in character are the Bantu-speaking inhabitants of eastern and southern Africa, from the great lakes south to Cape Colony; cattle are the dominant element in their economy. These sturdy and frequently warlike peoples include such well-known tribes as the Zulus and the various Kaffirs. They appear to have been spreading southward from a northern center when first encountered by Europeans, and had forced the Hottentots and Bushmen out of most of the considerably larger territory which they once occupied. Centering in southern Abyssinia is a series of tribes in which there appears to be a considerable admixture of Mediterranean blood from their neighbors to the north. In consequence, we find less kinky hair, lighter skin, and higher and narrower noses.

Oceanic Negroes.— A second area occupied by Negroes lies far from Africa, in Melanesia, the "black islands," an island group which begins at the west with New Guinea (or Papua) and extends on eastward via the Solomons, New Hebrides, and New Caledonia to the Fiji Islands. In this region the major elements of the population are the Oceanic Negroes, typically agricultural village dwellers. In almost all major features these Negroids are exceedingly similar to their presumed African relatives; however, the jaws tend to be rather less projecting, the lips not as much thickened, the hair less kinky, and (curiously) there is frequently a strongly hooked nose like the Armenoid type rather than that characteristic of other Negroes. These Oceanic Negroes are none too well known, for they are, in general, decidedly unfriendly and have a reputation for treachery and bloodthirstiness.

The racial setup in Melanesia is far from simple, for while some of the tribes appear to be relatively "pure" Negro types, there are numerous variants. In interior New Guinea there are Negritos, but they are not as small in stature as typical Pygmies and somewhat resemble their Negro neighbors. In New Guinea some tribes appear to have an infusion of Malay blood. In various parts of Melanesia there are groups which resemble Australians as much as Negroes and may represent a fusion of Negroes with an older Australian element in the population. And in some of the outer parts of the Melanesian island chain there are distinct traces of the Polynesians, whose major home lies on the Pacific islands to the east.

Pygmies.—Still more discontinuous in geographical distribution are the pygmy Negroids, for whom the term "Negritos" is frequently used. These little people are of extremely short stature, with unusually flat and broad noses, and with a tendency toward brachycephalic skulls. Otherwise, however, they exhibit typical Negro features. They are found only in a limited number of much-restricted areas, including parts of the deep Congo forests, the Andaman Islands in the Bay of Bengal, and spots in the interior of the Malay Peninsula, New Guinea, and the Philippines. The African Pygmies dwell in the forests around the villages of the forest Negroes, with whom they are on friendly terms. This is not the case in the other regions where Pygmies are found; the areas in which they dwell appear to be essentially isolated hiding places where they keep out of danger from aggressive and more powerful neighbors. Like the Australians, they are essentially in a Paleolithic cultural stage, simple hunters and food-gatherers. Their extremely scattered distribution has suggested to many authors that they represent survivors of a human stage still farther down the evolutionary line than that of the typical Negroes. There is at present, however, no fossil evidence that man ever went through a pygmy human stage in his evolution.

Bushmen and Hottentots.—For want of a better place, we may mention these quite isolated little groups here. The Bushmen of South Africa are a vanishing race of which only a few thousand survive in the Kalahari Desert region of South Africa. Culturally they are quite primitive, being still in a hunting, food-collecting stage analogous to that of the Upper Paleolithic or Mesolithic peoples of prehistoric Europe. Their hunting equipment shows one major advance over that of the Australians, for they possess

Archaic living types. *Left*, a native Tasmanian lady (by the name of Betsy). (Courtesy Beattie Studios, Hobart, Tasmania.) *Right*, a Bushman of South Africa. (Photograph by George Leith, courtesy Peabody Museum, Harvard University.)

Australoids. *Left*, a Veddah of Ceylon, member of an archaic group probably related to the Australians. (After the Sarasin brothers.) *Right*, an Australian from Cooper Creek, South Australia. (Photograph courtesy Peabody Museum, Harvard University.)

Eastern races with Mediterranean blood. *Left*, a Polynesian. A "dark white" stock with Mongoloid and Negroid strains. (Courtesy American Museum of Natural History.) *Right*, an Ainu. Essentially archaic "white" with some Mongoloid admixture, Ainus are usually excessively hairy, but this young man would pass unremarked on a European street. (From Kubo.)

Negroids. *Above*, two types of African Negro. *Left*, a Zulu and his wife. The once warlike Zulus are typical members of the Bantu group of central and southern Africa, Negroes with an admixture of "dark white" blood from the north. *Right*, a Kru from Liberia, a typical Negro similar in stock to victims of the slave trade. Two oceanic Negroes are shown. *Upper right*, a Papuan from New Guinea, of a tribe more than suspected of cannibalism. *Opposite*, a Melanesian from the Admiralty Islands. Also cannibalistic and "reported to be unfriendly to strangers." *Below*, Pygmies. *Left*, African Pygmies from the forests of the Belgian Congo with a European of normal stature. *Right*, an Aeta girl from the Philippines. (African Negro photographs courtesy Peabody Museum of Anthropology, Harvard; Papuan courtesy J. D. Ripley; Melanesian and Aeta girl courtesy American Museum of Natural History; African Pygmies courtesy H. C. Raven.)

the bow and hunt large game with poisoned arrows. They are very short in stature, with an average height well under five feet. In many regards they resemble the Negroes, for the hair is black and exceptionally kinky, and the nose broad and flat. In other respects, however, they depart considerably from the Negroid type. The skin has relatively little pigment and is rather yellowish; the eye form is rather suggestive of the Mongolian type; the cheek bones are prominent, the face flat; a racial peculiarity, particularly of the females, is the pronounced steatopygia, an excessive development of fat on the buttocks.

The Hottentots, also of South Africa, are much more numerous and, as cattle-herders, culturally more progressive. They show Bushman features but to a less pronounced degree. While there is no accord on the matter, the majority of workers believe that the Hottentots represent a Bushman stock with a considerable admixture of Negro blood.

Archeological finds indicate that the Bushmen are surely old inhabitants of southern Africa and that they once occupied a much larger area than that in which they are found today. Their antecedents are unknown and are matters for speculation; we have noted earlier some fossil skulls showing Bushman-like features. Paintings and engravings on rock made by the Bushmen are suggestive of those found in North Africa and Spain in late Paleolithic or Mesolithic times, but there is no evidence of bodily similarity between Bushmen and ancient men of those regions. The low stature and Negroid hair and nose bring to mind the Pygmies, but there are no other marked similarities. The complexion and eye shape suggest Mongolian relationships, but geographical considerations render this quite unlikely.

We have considered under the general category of Negroids a variety of primitive peoples—African Negroes, Oceanic Negroes, Negritos, Bushmen, and Hottentots. Taken together, we may not be far from the truth in considering this assemblage of races as representative of a series of early branches from the main line of evolution of *Homo sapiens.*

Negroid distribution.—The present distribution of the Negroids offers an interesting problem. They are found in two separate areas—Africa, on the one hand, and southeastern Asia and Australasia, on the other. How has this discontinuous distribution come about? One suggestion is that one of the two areas—Africa, let us say—was the homeland, whence a migration took place by sea to the other. But pure Negroids are almost never ambitious seafarers. The most reasonable explanation at present is one which draws suggestions from the history of lower mammalian groups. Among mammals, there are in Africa or the East Indies today many forms once present in Eurasia but now extinct there. Perhaps the Negroids were, at an early stage of human history, present in continental Asia, migrated thence both east and west to their present territories, but became supplanted by other races in their original home. If this were true one might reasonably expect to find some trace, at least, of Negroid blood in southern Asia. Apparently this is the case, for anthropologists have noted Negroid characters among the more obscure peoples of India, Burma, Assam, Persia, and Arabia, perhaps the last vestiges of an ancient Negroid population.

MEDITERRANEANS AND RELATED STOCKS

We have now, through a consideration of the Negroid and Australoid groups, accounted for the peopling of the more southerly regions of the Old World, including most of Africa and Australasia. These racial groups, however, have left little imprint on Asia, Europe, or the Americas. These areas are occupied by groups generally lumped as "white" and "yellow," —Caucasoid and Mongoloid. While diverse in character, these northern groups are certainly more specialized than members of either of the major evolutionary stages so far discussed. If we attempt to find their "lowest common denominator," a type of man from which the remainder of mankind could have evolved, we find that in all probability he should possess essentially some such set of characters as follows: As in Australians, the head might be persistently long, the hair wavy and brown to black in color, the face and body at least moderately hairy, the eyes brown. On the other hand, we would expect a more advanced condition in a lighter brown skin and a narrower and higher nose than that of typical Negroids and Australoids.

We have no difficulty in finding a series of races which fill this bill of particulars; they are, and long have been, among the most numerous and important peoples of the world. Extending from Spain and Morocco eastward along both sides of the Mediterranean, through Arabia and Persia to India, we find race after race of brown, long-headed peoples who closely fit the diagnosis given above. The western members of the group are generally termed Mediterraneans, and it is not inappropriate to apply this term to the series as a whole. Few major changes are needed, as has been noted, to derive them from ancestral Australoids, and it is reasonable to assume that they represent a progressive step in the advance of *Homo sapiens* in an Asiatic homeland. Here, however, we have reached a group late enough in appearance to be still represented close to its area of origin.

Mediterraneans of India.—Northern India is a region which has been subject to repeated racial invasions and in which many racial types and mixtures are to be found; peninsular India has been much less subjected to external influences. In this region there are, as we have noted, surviving vestiges of Negroid and Australoid groups. The dominant type, however, the typical Hindu, is a more advanced one, with the essential Mediterranean features. Presumably they represent the Mediterranean close to his native home.

A curious offshoot of the native Indians is that nomadic group known as the gypsies, or Romanies. Despite the implications in their usual name, these swarthy people appear to be descended from pariah tribes of India which have wandered westward to appear in Europe during the Middle Ages and attach themselves to Western peoples. Their racial type appears to be fundamentally that of the Indian Mediterraneans, although an admixture of Western blood appears to be present; their speech is basically a Sanskrit modification, although there are many loans from Western languages.

Far East Mediterraneans.—The westward expansion of the Mediterraneans is readily followed and will be discussed shortly. What of their history to the east? Most of central and eastern Asia is today occupied by the Mongoloids, peoples which differ markedly from the Mediterraneans in such features as their round skulls and straight hair. But, despite their present dominance, there are many indications that they are late-comers and were preceded in many eastern regions by an earlier wave of brown, long-headed, wavy-haired Mediterraneans.

India is surrounded on the north and east by Mongoloid groups, but traces of the earlier presence of brunet Mediterraneans in these areas seem to be shown by mixed types found in Burma and east into Indochina. The inhabitants of these regions, although classed as yellows, show Mongoloid features in a much less marked manner than is the case farther north; probably they represent a blend with earlier Mediterraneans. Farther south, in the Malay Peninsula and the East Indies, the characteristic Malay populations, while dominantly Mongoloid, show some indications of mixture with a preceding Mediterranean "wave." In various isolated groups in the East Indies (such as the Murut people of Borneo) these Mediterraneans have survived in relatively pure form. They tend to retain their primitive long heads and wavy hair, although showing some Mongoloid features and perhaps some Negroid admixture. To these peoples the term Indonesian is applied. As would be expected, these survivors are inland dwellers, with the more Mongoloid Malays occupying the coasts.

Polynesians.—We have, then, many indications of an early eastward "push" of the Mediterraneans into southeastern Asia and beyond this into the East Indies. But this was by no means the end of their migrations in this direction.

The numerous island groups scattered over the open Pacific from Hawaii to New Zealand are known collectively as Polynesia, and their inhabitants form the Polynesian race. These tall, fine-looking people are expert seamen, as they must have been to have reached their isolated homes, and, although many cultural elements which we find in "civilized" groups are absent, these lacunae are mainly explicable in terms of the peculiar environment under which they live. It is certain that they reached these islands from the west, probably in a series of waves, of which some are of quite recent date and are well recorded in Polynesian tradition. Presumably their starting point lay somewhere in southeastern Asia or Indonesia, but there is no trace of them today on the mainland and almost none along the island chains of Indonesia and Melanesia. Very probably their migration route lay north of the major islands; the inhabitants of the little islands of that region—the Marianas, Marshalls, etc., known collectively as Micronesia—appear to be somewhat allied to the Polynesians.

Some racial admixture appears to be present in the Polynesians. In their southern members, particularly the Maoris of New Zealand, there are suggestions that a Negroid strain was picked up during their travels. There are more distinct indications of Mongoloid blood, for the cheek bones tend to be prominent, the head shape, although variable, is often brachycephalic, and body hair is scant. But in most features, such as the brown and wavy hair, brown skin, and brown eyes, it seems apparent that the Polynesians are essentially a far-flung eastern branch of the Mediterranean stock.

The Ainus.—The Mediterraneans, in the broad sense of a primitive, long-headed, wavy-haired, brunet white stock, spread in early days to the northeast as well. Most of the northern half of eastern Asia is filled with typical Mongoloids. But there are long-headed strains among various Siberian tribes to the far northeast, which suggests the early presence there of Mediterraneans. More positive evidence is found in Japan, where the Ainus are an archaic race of strong affinities with this type. Now few in number and restricted to parts of the bleak northern island of Yezo (Hokkaido), where they are "preserved," and farther north, in Sakhalin, the Ainus appear to have once occupied all of Japan. The skulls tend to be dolichocephalic, the hair wavy and abundant (unusually so), the eyes brown, the skin a light brown. These are features which suggest a relationship to the brunet Mediterraneans or perhaps to some even older "white" strain (still another suggestion, founded on their hairiness, is that they are direct Australoid derivatives). Most of the Ainu traits are in marked contrast with those of their Asiatic neighbors. There is, however, some suggestion of a bit of Mongolian blood in the prevalence of black hair and rather thickset bodies and prominent cheek bones.

American longheads.—We have every reason to believe that long-headed brunets, essentially Mediterranean, passed still farther to the northeast to become early, perhaps the earliest, inhabitants of the Americas. The Indians are usually classified as a subdivision of the Mongoloids, and rightly so, it seems, as regards the majority of the population. But there are Indians and Indians. In many regions there are groups which show few if any of the characters of typical Mongoloids and tend to have long heads, cheek bones less prominent than in typical Mongols, and other features suggesting the basic Mediterranean type. The earliest known Indians of the Southwest, the basket-makers who preceded the present tribes of that region, are known to have been markedly dolichocephalic; long-heads are abundant among the Indians of the Amazon basin, and the Yahgans of far Tierra del Fuego are apparently of this type. Even in North America the eastern Indians, now nearly extinct, lacked a full Mongoloid development.

Presumably the peopling of the Americas did not consist of a single migration but of a series of successive waves of invasion. The last immigrants were definitely Mongoloid; the earlier ones, however, appear to have been essentially Mediterraneans (and there may even have been some still earlier comers representing Australoid types). As would be expected, traces of these earlier settlers are to be found today mainly in areas far off the beaten track of later migrations.

European Upper Paleolithic survivors.—In the last chapter we noted the presence in the Upper Paleolithic of Europe of men of our own species —mainly long-headed individuals—who show little of either Negroid or Mongoloid features and hence might well be considered white. Were they early forerunners of a westward migration of Mediterraneans? Probably not in any direct sense. A few of them were rather similar to the Mediterraneans, but most, including the typical Cro-Magnards, were quite tall, rather rugged specimens, contrasting sharply in many respects with the typical Mediterraneans, who are relatively small and less robust. One possible explanation of their nature calls upon the evidence from Palestine which suggests an interbreeding between Neanderthal man and *Homo sapiens*. Possibly the Upper Paleolithic men of western Europe do represent a blend between early westward-traveling Mediterraneans and Neanderthals, with the *sapiens* strain dominant, the Neanderthal mixture giving a touch of crudity to the skeleton.

Concerning the later history of Cro-Magnon man and his contemporaries, there is much speculation and little knowledge. A touch of this blood may persist in northern Africa, where Upper Paleolithic populations were numerous, and in the Canary Islands off the western shore of that continent the people show many features suggesting partial derivation of this ancient stock. It seems reasonable to believe, however, that most of this population would have tended to drift northward as the Ice Age came to a close. In the following Mesolithic cultural period skeletal remains which are suggestive of Upper Paleolithic types are found farther north along the European coast, as far as southern Scandinavia, and it is highly probable that a strain of Paleolithic blood may persist today in some northern regions. This is especially the case in the British Isles (particularly Ireland), northern Germany, and the Atlantic coast of Scandinavia, regions where otherwise the Nordic type is dominant. Some of these "suspects," particularly in Norway, differ little from their Nordic neighbors except in a more massive build. Others, especially in north Germany, tend toward darker hair and rounder heads, the latter perhaps inherited from the (relatively rare) brachycephalic members of the Upper Paleolithic races.

Western migration of the Mediterraneans.—Whatever the nature of the Upper Paleolithic peoples, the true Mediterraneans spread westward at an early date, and widely. As early as Mesolithic times these southern whites were present over a vast territory. They were present then in Palestine. In Africa Mediterraneans which appear to be of this age are found not only in Egypt but far to the south, near the headwaters of the Nile. To the west Mediterraneans had reached the shores of the Atlantic in Portugal. By Neolithic days they were in possession of the entire Mediterranean region and had spread far to the north. People who appear from their skeletons to have been Mediterraneans inhabited the Danube

"Dark whites" of Southern Asia.
Although India is a land of varied racial components, the typical Hindu (*left*) is in essence a dark, long-headed, wavy-haired type related to the Mediterraneans. East of India strong admixtures of this Mediterranean strain are to be found in the races of southeastern Asia and the East Indian islands. The Muruts of Borneo (*right*) appear, for example, to be a basically Mediterranean people with an infusion of Mongoloid blood. (Hindu courtesy Griffith Taylor; Muruts courtesy Peabody Museum, Harvard University.)

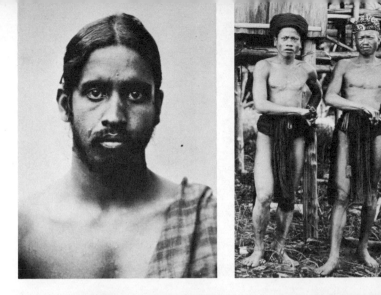

A Mediterranean from North Africa.
Hammu ben Haddu, from the Rif country of Morocco, seen in profile and full face. Somewhat lighter and with thinner lips but in general much the same type as the Indian native shown above or the Portuguese below. (Photograph courtesy Carleton S. Coon.)

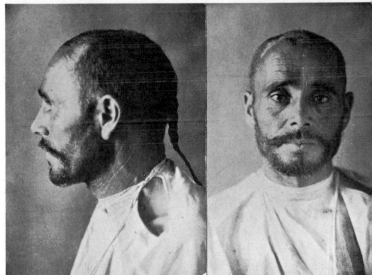

European longheads. *Left*, a Mediterranean from Portugal, not distant in physical type from the North African native shown above. *Right*, a Nordic, a native of Sweden, with light complexion, blue eyes, and (despite the appearance in the photograph) blond hair. (Photographs courtesy Harry L. Shapiro and American Museum of Natural History, New York.)

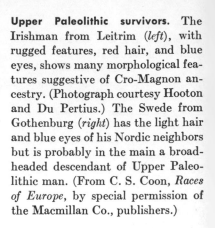

Upper Paleolithic survivors. The Irishman from Leitrim (*left*), with rugged features, red hair, and blue eyes, shows many morphological features suggestive of Cro-Magnon ancestry. (Photograph courtesy Hooton and Du Pertius.) The Swede from Gothenburg (*right*) has the light hair and blue eyes of his Nordic neighbors but is probably in the main a broadheaded descendant of Upper Paleolithic man. (From C. S. Coon, *Races of Europe*, by special permission of the Macmillan Co., publishers.)

European broadheads. *Left*, a typical Alpine (from Bavaria), with brown eyes and hair, nose not particularly prominent, and a stocky bodily build. *Right*, a Dinaric from Albania, with taller stature and prominent nose. (Photographs courtesy Carleton S. Coon.)

European broadheads (*continued*). *Left*, an Armenoid from the Lake Van region of Armenia, with the very prominent nose characteristic of that type. *Right*, an east-European broadheaded type (from the Ukraine), with brown hair and blue eyes. (From C. S. Coon, *Races of Europe*, by special permission of the Macmillan Co., publishers.)

basin and, in a somewhat taller phase, the plains farther north from the Rhine to southern Russia. Farther east, as well, the Mediterraneans appear to have pushed north into the Eurasian steppes, for Neolithic skeletons from Turkestan are of Mediterranean types. The Mediterraneans thus constitute the oldest known population of the Near East and were apparently the main factor in the peopling of Europe and northern Africa.

The contributions of the Mediterranean peoples to the evolution of culture cannot be overestimated. We have noted that Neolithic culture appears to have originated in the Near East; this was an area inhabited by typical Mediterraneans. They appear, further, to have been the peoples who carried this culture to Europe and westward along the Mediterranean and northward by various routes to the Danube and the northern plains— for almost everywhere the villages of the Neolithic settlers reveal skeletons of this type.

Linguistically, however, they are of lesser historical importance. Many of them now speak Indo-European languages akin to those of other European peoples. But this group of languages is not native to the Mediterranean world. Many Mediterraneans today speak Semitic languages, such as Arabian and Hebrew, or related "Hamitic" African tongues, such as ancient Egyptian (and its modern Coptic derivative), and Berber. In addition, however, it is strongly suspected that the older Mediterraneans possessed still other languages, most traces of which have been lost; Basque is a relic.

Present distribution of Mediterraneans.—Today this great racial group of darker, long-headed, southern whites occupies a long strip of territory extending west from India to the Atlantic. An eastern group, Irano-Afghans, rather tall and hook-nosed, occupies the plateaus and mountains of Persia and Afghanistan and extends northward into part of western Turkestan. In Arabia are found typical Mediterraneans rather shorter in stature; the basic Jewish stock was of this type. While many Arabs are stationary, the Bedouin of song and story is a desert nomad with a culture based on possession of camels and horses. Quite similar to the Arabs, but with broader faces and less prominent noses are the native Egyptians; they are a people who, as shown by archeological work, have changed hardly at all for some 5,000 years. To the south, the general region of Ethiopia is characterized by races such as the Gallas and Somalis which are essentially Mediterraneans but with a Negro admixture in many cases. In northern Africa west of Egypt the Mediterranean type also prevails but in a rather taller phase. There is an overlay of conquering Arabs, but the older native population is that of the Berbers; quite in contrast with most Mediterraneans, many of the latter are still in an essentially Neolithic phase of culture, while others, the Tuaregs, have become camel-owning Saharan nomads. Shorter men, however, reappear in Spain, which is a nearly pure Mediterranean country, with bodily features closely comparable to those seen in Arabia.

North of the Mediterranean Sea the area once occupied by this race has been much restricted because of racial changes and the great increase, in relatively late times, of round-headed types. In Asia Minor the original population appears to have consisted of Mediterraneans with rather prominent noses, but they have been swamped in great measure by brachycephalics. In Greece, however, there is still, as in classical times, a very large proportion of Mediterranean blood, and a pronounced Mediterranean strain is to be found from here northward through the eastern Balkans to the Russian coasts of the Black Sea.

Farther west the original Mediterranean population of Italy is still strongly represented in Sicily and the south of the peninsula; as we progress northward, however, brachycephalic types increase in proportion, and there is little trace of Mediterraneans in northern Italy. This race has left a strong impression on the populations of central and southern France, but the French are mainly brachycephalic.

In central and northern Europe this stock has almost completely disappeared, in its original form at least. In Great Britain the Neolithic population was mainly one of Mediterranean longheads, and there are still marked traces of them in the population of Wales and the west coast of Scotland. Otherwise Mediterraneans in northern Europe are non-existent.

Early Nordics.—The Mediterranean blood does survive in the north, but in a modified form, in the Nordics. Typical members of this race, found in its purest form in Scandinavia and the east Baltic shores, are different in many respects from the typical Mediterraneans. They are tall, with light skin, hair which tends to be blond, and eyes which are blue or gray, in contrast to the brown of the Mediterraneans. But like that group they are essentially dolichocephalic, and there are no reasons why they cannot have evolved by a "bleaching-out" of early Mediterranean migrants to the North.

The prehistoric archeology of Europe furnishes much evidence which is in agreement with this theory, although of course the evidence from skeletons cannot tell us anything of pigmentation. We have noted that the Neolithic Mediterraneans sent strong colonies north into central Europe to the Danube basin and into Germany and the Russian plain. This was perhaps 5,000 years ago. A thousand years or so later bronze begins to be found in these areas. During the next millennium, the Bronze Age in this region, there were some cultural shifts and evidences of migrations, but essentially the same people seemed to have remained there. When the Iron Age arrived, about 1000 B.C., historical evidence abundantly shows that the northern Europeans were Nordics, with the blond features, blond hair, and blond eyes of the modern Nordic. They were not, however, immigrants, as far as we can tell, but the descendants of the browner Neolithic Mediterraneans, who had arrived there several thousand years before. Here, in central and northern Europe, it appears, occurred the loss of pigment which is characteristic of the Nordic group.

Best-known centers of this early Iron Age Nordic race were southern Germany and the Danube basin. But even in early days the Nordics occupied a much larger territory. To the west the Rhine Valley appears to have been their approximate boundary. They spread far to the north, into Scandinavia, where southern Sweden became an important Nordic focus. Eastward, the steppes of south Russia appear to have been Nordic country in Iron Age days. There is also good evidence of the early presence of northern blonds in southern Siberia and western Turkestan—country now inhabited by Asiatics—and there is even evidence of Nordic penetration far to the east into the Tarim basin of central Asia.

These early Nordics were not particularly progressive culturally. They adopted the use of iron with avidity and used it energetically (if perhaps somewhat bloodily). They were capable of agricultural pursuits but were never city-builders. They seem, on the contrary, to have been a rather restless race, and many, the more eastern groups particularly, were essentially nomadic, with the horse an important element in their activities.

In another respect, however, the Nordics played an important role in later European culture, namely, as language-bearers. Today a great majority of Europeans, as well as their overseas colonists and the people of Iran and large parts of India, speak languages of the Indo-European group. This great variety of tongues—from Gaelic and English on the west to Sanskrit and its modern derivatives in India—shows many fundamental similarities which indicate that they had a common origin. There was at one time a tendency to assume that all speakers of these languages were racially similar, members of an "Aryan" race (the name is derived from that of the Sanskrit-speaking invaders of India), but this thesis can no longer be maintained.

We can, however, reach with considerable certainty some idea of the region in which this language group originated, using historical and linguistic evidence. Quite surely this was in the Great Plains region of Europe from eastern Germany and Poland eastward through southern Russia. This region, as we have just seen, is the center of the prehistoric Nordics. We are thus justified in believing that these long-headed northern blonds were the originators as well as the disseminators of Indo-European tongues.

Although the early Nordics were too barbarous to have left us written records, archeology and historic accounts by more southern peoples can give us some idea of the distribution of the various Nordic groups a thousand years or so before Christ. At that time much of the Danube region was occupied by Nordics, generally termed Illyrians and Thracians, rather poorly known but important as the peoples who were probably responsible, through invasion, for the introduction of both Indo-European languages and a blond ruling caste into Italy and Greece. The western Nordic territory, roughly western Germany today, was occupied by the Celts. To the north, in southern Scandinavia, was the early Germanic center. Southeast of this, in Poland and toward the Ukraine, were Nordics who spoke ancestral Slavic and tongues from which have descended Baltic languages, such as Lithuanian.

Farther east, in the steppes of south Russia and western Asia, there is relatively little data, either archeological or historical, as to the early nomadic Nordics of these regions, but some general facts can be made out. In early Greek days the southern Russian plain was occupied by the Scythians—trouser-wearing, horse-riding nomads who were early users of the covered wagon. They were apparently blond Nordics who later fused with other peoples and made little historical impression. To the east of them, into Asia, stretched other horse-nomads, still more poorly known, who were apparently speakers of tongues now found in Iran and India.

Nordic migrations.—These Nordics appear to have been rather restless and aggressive peoples. About them, mainly to the south, were wealthier and seemingly less warlike nations. Given these circumstances, it is not unnatural that much of Nordic history has been one of movement, war, and invasion, from early times down to the Viking raids of the early Middle Ages.

Some of the movements are essentially prehistoric, although often recorded by tradition. Some took place during the Bronze Age, more at a later date, when these barbarians had iron weapons and horses. The Aryans moved south into India, as the Sanskrit records indicate, at about 1400 B.C.; the Persians' southern movement into Iran appears to have been of about the same date, and the fact that an Indo-European language appears in the Hittite empire indicates an early wave of Nordics into Asia Minor. Farther west, too, there were early southern movements. As noted above, Nordics made their appearance in Greece and Italy in prehistoric times, and the Hyksos and "sea-peoples" who annoyed the Egyptians equally early were not improbably far-flung Nordics.

Centuries later the Celts "erupted." This group, then centered in southwestern Germany, was apparently not purely Nordic—a shade browner, a bit more round-headed than the true type. By about 500 B.C. they appear to have reached a flourishing condition and began to expand. This expansion went in various directions—east as far as Asia Minor, where Galatia received its name from Gaulish invaders, south into Italy to conquer the Po Valley and sack Rome. Mainly, however, they moved westward, where the Gauls dominated France in Caesar's day, and on into Spain and the British Isles, which had become essentially a Celtic land when invaded by the Romans.

Then followed the tribal wanderings of the German peoples, most of which took place in the full light of history. As has been said, the early Germans centered in southern Scandinavia. From that area came several prominent groups which broke into the Roman Empire—the Burgundians and the Goths, who first wandered southeast to the Russian plains and then, in two divisions, west to Italy and Spain; and the Vandals, who even reached Africa, leaving on their route such destruction as to give rise to the term "vandalism." Other German tribes followed the westward movements of the Celts to make Germany German for the first time. To the west the Franks extended into Gaul to give their name to modern France. Anglo-Saxons, Germans from the North Sea Coast, crossed to Great Britain. And for many centuries afterward the Vikings from the Scandinavian fiords plied the seas, making their greatest influence on the British Isles, but wandering south to Africa and the Mediterranean and on the west venturing to Iceland, Greenland, and even "Vinland."

Last of all Nordic dispersals was that of the Slavs. This, however, had a curious history. The Slavs were originally Nordics who appear to have had their homes in the forests and marshes lying northeast of the Carpathians in southeastern Poland and the adjacent part of the Ukraine. There they remained—obscure and relatively undisturbed—until most of the great racial movements were over. Then, beginning about 500 A.D., began an expansion which carried Slavic languages far and wide. To the south such languages are spoken in much of the Balkans. To the west the Slavs reached Bohemia and at one time penetrated Germany as far west as Berlin and Saxony. To the northeast they spread over much of modern Russia, driving Finnish tribes before them, and today Slav-speaking immigrants are playing an increasingly important part in northern Asia.

But this has been a conquest of language, not of race. As we have noted, the Slavs were originally long-headed Nordics. Today relatively few speakers of Slavic tongues are longheads. The primitive Slav blood has been submerged by brachycephalic races.

The present distribution of Nordics is far more limited than our recital of their conquests would suggest. Today their main center is close to the original Germanic home. In pure form they are mainly concentrated in eastern Norway, central and southern Sweden, Finland, and the little Baltic States, while the Low Countries and Great Britain also show a large percentage of Nordic types. Outside of these districts Nordic blood tends to be rare. In coastal Norway, Denmark, and northern Germany the Nordic strain is clearly seen in the predominantly blond types encountered, but we are dealing with a blended people with more brachycephalic skulls—the squareheads of popular terminology. There are still moderate amounts of Nordic blood present in central Russia, Saxony, Austria, and parts of Switzerland and northern France. Elsewhere, however, in all the Mediterranean region and in Asia, the Nordic is almost non-existent.

The Nordic has been of interest linguistically and as a fighter but is not particularly remarkable in other fields of activity. Despite this, there tended to grow up a great cult of worship of the Nordic, the essence of which is seen in Hilaire Belloc's verses:

> Behold, my child, the Nordic man,
> And be as like him as you can:
> His legs are long—his mind is slow
> His hair is lank and made of tow.
>
> And here we have the Alpine race.
> Oh! What a broad and brutal face.
> His skin is of a dirty yellow;
> He is a most unpleasant fellow.
>
> The most degrading of them all
> Mediterranean we call.
> His hair is crisp and even curls
> And he is saucy with the girls.

It is amusing that the center of this Nordic worship has been Germany, in which, as we have seen, true Nordics are relatively rare. It may be noted, however, that emphasis in the Nazi period shifted from Nordic to Aryan. Since this shift was from a racial term to one associated instead with speech or culture, we need not, as biologists, be further concerned with it here.

ALPINES

The races so far considered have been, almost without exception, long-headed. A final major event in the evolution of man would seem to have been the appearance of brachycephalic skulls. That this event was a relatively late one is suggested on grounds both of distribution and of archeology. A map of the distribution of head shapes in the Old World today shows that broadheads are overwhelmingly preponderant in central and eastern Asia and in eastern and central Europe; the longheads dominate in the "pendant" southern continents of Africa and Australia but in Eurasia are found only about the peripheries of that continental mass. This strongly suggests that brachycephaly is a relatively new human feature, the bearers of which have spread until the longheads have been forced out into marginal regions. Archeologically our evidence is meager, apart from Europe, but the record there seems to agree with this conclusion. Roundheads, as we have seen, are very rare there in the Upper Paleolithic, not common until the Bronze Age, and have but recently attained their present dominance.

The roundheads are generally regarded as sharply divided into two groups. The western European members are usually regarded as whites, closely related to Mediterraneans and Nordics; the eastern brachycephalics, on the other hand, are the typical Mongoloids. We find, however, that along the borders of Asia and Europe there are types of roundheads which are essentially "neutral," which do not appear to be mixtures, and yet are hard to classify as either Europeans or as true Mongoloids. Nevertheless, the histories of these two types have been in the main discrete, and we may for convenience treat of them separately, beginning with the western roundheads, to whom the term "Alpine," in a broad sense, is generally applied. The name is due to an interesting fact of distribution. Excluding for the time being the Russian area, most of the European broadheads tend to cluster along a line running from France east along the Alps, down the chains of the Balkans, and on east into the mountains of Asia Minor. These roundheads have thus been situated during historic times in an east-to-west zone between the Mediterraneans and the Nordics on the other side of the Alps. Even if originally a "pure" race, it is obvious that, in close apposition with other types to the north and south, they might readily have mixed with them and taken on characters of their neighbors. This apparently has occurred, and there is little we can say of features common to all Alpines, except for the brachycephalic condition and the probability that they have been, in general, types of moderately brown pigmentation.

Alpines of western Europe.—A large proportion of the population of France is essentially Alpine. The typical Frenchman of this type is a short sturdy individual, with a broad face and short nose, a round head, and a moderately brown complexion. Such types are found farther east along the Alpine mountain ranges, but relatively rarely. Most of the rest of this region is occupied by roundheads whose physical makeup appears to have been modified in various regards through the influence of neighboring populations. One such group is found north of the Alps in Bavaria, Bohemia, and the Saxon region of Germany, where a prominent type among the inhabitants is a round-headed blond. Presumably these people represent a cross between the Alpines and the Nordics which once peopled this region. This race has been termed "Noric," after the name of an old Roman province in this region, Noricum.

Dinarics.—More important is a section of the Alpine peoples whose center of distribution lies in the eastern Alps—from east Switzerland through the Tyrol and south through Yugoslavia to Albania. These mountaineers are the Dinarics—tall and rather brownish men with prominent faces and noses. They appear to carry a considerable amount of Mediterranean blood. Beyond the mountains they extend to the northeast, where they are prominent in the Hungarian region, while to the south they are the major stock in the Po Valley and have introduced a considerable brachycephalic strain into Italy as a whole.

Armenoids.—Eastward beyond the Bosphorus there are strong Dinaric influences in the dominantly brown and round-headed populations of Asia Minor and Syria. Here, however, we find an admixture of a further related brachycephalic racial type, the Armenoid, so called because it is found in its purest state in the territory of that harried race, the Armenians. The Armenoids are similar to their Dinaric relatives but tend to be a bit darker and with a still more prominent face and the large convex nose common in the Levant. Beyond Armenia the Alpine occupation of the highland chain ceases; but there appear to be traces of typical Alpines still farther east, in Turkestan.

Alpine origins.—The origin of these Alpines is a question to which we have as yet no thoroughly satisfactory solution. Archeology, as far as the facts go, indicates, as said above, that roundheads were rare in these regions in early prehistoric times, with a continued and accelerated increase in their numbers in later periods of history, particularly during the Bronze Age. Can they simply have evolved on the spot by a gradual change of head shape in the old Mediterranean populations? Possibly, but it is difficult to see why such a shift should have occurred only in this one region. The striplike distribution, combined with the enormous increases in Alpine numbers in later times, suggests invasions of brachycephalics from the east along the mountain belt. But the evidence shows no indication of major movements of this sort. Possibly the truth may lie between these extremes. There may have been a certain amount of east-to-west migration from time to time, bringing in a relatively modest amount of roundheads who frequently interbred with their neighbors, particularly the Mediterraneans. It is possible—although incapable of proof—that the great increase in numbers of brachycephalics has been due not to great original numbers but to greater fecundity and to a dominance of roundheadedness in racial crosses.

East European roundheads.—But the Alpines of the Alpine strip just discussed are not the only round-headed peoples of Europe. In northwestern Europe, in the general Nordic area, we have already noted the presence of brachycephalic types which, it has been suggested, may represent the survival of very early roundheads. More important is the situation in eastern Europe. There, particularly in Poland and most of Russia, we find the peasants to be a nearly uniform type of individual. They are round-headed, short and stocky, as are many Alpines. But they tend to have broad faces, low short noses, hair which tends to be straight and coarse—rather suggestive of more eastern peoples. And finally, as in Nordics, skin, eyes, and hair tend to be light in color. No generally accepted term for this race exists. One usually thinks of them as the characteristic Slavs; but this is a linguistic rather than a racial term, and we have already noted that the original Slav-speakers were Nordics. East European is a more non-committal and more useful term.

The characters of this race suggest that it has been formed by a fusion of two types. One presumably was that of the Nordics, who formerly inhabited much of this region and supplied both blond complexion and language. The second was presumably a brachycephalic race with characters not as markedly European as in Alpines and hence very probably from some more easterly homeland.

Turanians.—Along the boundaries of Asia and Europe, particularly in west Turkestan between the Caspian and the mountains of central Asia, there are found today various racial groups to which the terms Turanians or Turco-Tartars are often applied. These are round-headed people, typically with short broad bodies, broad faces, reduced noses, and straight hair as in Mongoloids. But the more marked Mongoloid features are absent, and the general brown pigmentation suggests comparison rather with western brachycephalics. It may be that in some cases this type has resulted from a cross between Mongoloids and Alpines, but certain of these people do not suggest a mixed heritage. Not improbably we are dealing here with a central, halfway group of roundheads in the center of the area of origin of brachycephalic men who have never possessed the more specialized features of either eastern or western varieties of the roundheads.

One may reasonably believe that the people who have contributed the brachycephalic element to the eastern European race were members of this general group who spread westward. In this general category, too, may be placed the little Lapps of northern Scandinavia, who are known to have reached their Arctic home as the result of a long migration from the western Asiatic steppes.

MONGOLOIDS

The vast majority of the inhabitants of eastern Asia and of the Americas (before European settlement) are generally grouped as Mongoloids, including the yellow and red races of popular classification. In its most characteristic development the Mongoloid physical type is readily recognizable. The skull is brachycephalic, the face broad but flat, with prominent cheek bones, the nose small, the body stockily built. The skin has a yellowish tinge, which may verge on brown; the eyes are dark in color. Hair is long and abundant on the head, but beard and body hair is poorly developed, and the hair tends to be notably coarse and straight. A specialization frequently seen is that of the eye-opening, which is a slanting slit with a characteristic fold covering the upper eyelid.

As has been noted, the geographic distribution suggests that the Mongoloids are a relatively recent group; but we have little archeological evidence as yet concerning their racial history, and their classification into subgroups is still imperfectly worked out.

Mongolia and eastern Siberia are regions in which Mongoloid characters are seen most highly developed. It is suggested that some of the more striking Mongoloid features have emerged as adaptations to the intensely cold climate of eastern Siberia, but these peoples do well in any area. The Mongoloids of northeastern Siberia are hunting and fishing tribes. To the west of them are nomads—Yakuts and Samoyeds—with reindeer herds. We have noted that in earlier days the western Asiatic steppes appear to have been largely populated by blond Nordics. Today much of this territory has been taken over by Mongoloids. Like the more westerly white nomads, the steppe Mongoloids took to horseback and made forays not only south into China but westward; the Huns and the Mongols are names familiar as western invaders. In west-central Asia Mongoloids have replaced European types to a considerable degree, but, except for a minor Mongoloid element in Russia, these orientals have had little influence on the population of the Western world.

Most of the peoples of eastern Asia are to be classified as Mongoloids, but the more characteristic features of this group become less prominent as one goes southward through China. There are, as was noted earlier, many peoples in southeastern Asia, brown skinned and usually of small stature, such as the Cambodians, Siamese, and Burmese, who are difficult to classify and seem to partake of both Mediterranean and Mongoloid characteristics. Possibly they may represent mixtures of the two; but equally possibly it may be that they have come from ancestors in which the full complement of Mongoloid features had never developed. In this region the Mongoloid type is best developed in the Malays, who differ from their northern relatives in their still shorter stature and browner skin. They are the typical inhabitants of the Malay Peninsula and Indonesia— that is, the East Indies short of New Guinea. If the ideas of racial history discussed in earlier pages are correct, Australoid and Negroid peoples must once have been present in Indonesia; but except for a few Pygmies, all traces of earlier populations have vanished with the spread of the Malays here. Variants of the Malay type extend along the islands fringing the eastern coast of Asia, to the Philippines and Formosa; the Japanese are fairly recent arrivals in their islands and appear to have a large amount of Malay blood.

In the sphere of cultural achievements the Asiatic Mongoloids have been exceeded only by the Mediterraneans; for while this group includes many a tribe of hunters, fishers, and pastoral nomads, the Chinese and other southern groups advanced far in civilization, and recent decades have shown eastern Asiatics to be readily capable of adapting modern technology to their own needs. And although it appears to have been the Mediterraneans who were responsible for the development of Neolithic culture in the Western World, the Mongoloids of southeastern Asia appear to have developed, in parallel, a second early type of agricultural life suitable to more tropical regions, with yams, taro, and later, rice as the staples rather than the wheat and barley of the west.

American Indians.—It was suggested earlier that Mediterraneans may have been early invaders of the Americas. However that may be, the great majority of the native American population is of Mongoloid descent. In most American Indians many of the characters are those of the true Mongoloid, such as the brachycephalic head, prominent cheek bones, the color, form, and distribution of the hair. There are, however, some differences from the Mongoloid pattern; the skin has more of a brownish or reddish tint; the nose, in sharp contrast to the typical Mongoloid, is frequently high bridged and convex, and there is seldom any marked trace of the Mongolian fold of the eyelid.

It is probable that the Mongoloid invasion included a number of successive waves traveling down from Alaska, spreading out over the two Americas, and blending with earlier migrants. In agreement with this, we have seen that traces of older stocks are most frequently off the main migrational tracks, while, on the other hand, certain Indian tribes of western North America show fairly close physical and even linguistic similarities to some Siberian tribes. But on the whole there are few traces among American Indians of the more specialized Mongoloid features (such as the eye fold) which are prominent today along the migration route in eastern Siberia, so that it is probable that much of the immigration took place many thousands of years ago.

Our usual conceptions of Indian build are based on the type seen in the western tribes of pioneer days, the "noble redskin" as portrayed on the nickel—tall, powerful, copper colored, with craggy features and prominent nose. But from this there are variants so numerous that we shall mention only a few. In Central America, for example, the descendants of the Indians who were responsible for Mayan culture are much shorter in stature, the features much less rugged. The forest Indians of South America tend to differ still more; the skin tends to be darker, the nose short; the head tends to retain a dolichocephalic contour; the hair sometimes is persistently wavy rather than straight.

Although many groups of Indians, such as Mayas, Aztecs, and Incas,

Mongoloids. *Right*, a girl from Inner Mongolia. Note the typical Mongoloid eye fold. *Next below*, a Chuckchi from the Kolyma district of northeastern Siberia. This race shows resemblances to certain North American Indian groups. *Lower right*, a Malay. The Malays show only partial Mongoloid characteristics and are to some extent similar to the "dark white" Mediterraneans. *Lower left*, a Visayan girl from the Philippines. The Visayans are typical Filipinos, related to the Malays and, on the other hand, showing resemblances to the Japanese, in whom a strong Malayan strain exists. (Mongolian and Visayan courtesy American Museum of Natural History, New York; Malay and Chuckchi courtesy Peabody Museum, Harvard University.)

American Mongoloids. *Left*, Black Heart, a Blackfoot Indian. *Lower left*, a group of Ona Indians from Tierra del Fuego, at the southernmost extremity of South America. This is a tribe in a low stage of culture and now nearly extinct. These men are characterized by unusual height. A windbreak of hides was a usual shelter. They were expert archers; their main food supply was the guanaco (a wild relative of the llama). *Below*, a Mamayauk Eskimo girl from Langton Bay (on the Arctic Ocean east of the Mackenzie River). (Photographs courtesy American Museum of Natural History, New York.)

had reached a moderately high degree of civilization before being harassed by Europeans, it seems fairly sure that relatively little cultural equipment was brought into this hemisphere by the various Indians in migrations. Some—as the tribes of California and the Great Basin, of part of the Amazon basin, and of far Patagonia—were still purely hunters or food-gatherers when first encountered. Most groups had become more or less agriculturalists, but except for the dog, such few animals as were domesticated were American forms, and so also (with a few debatable exceptions) were the domestic plants, many of them promptly adopted into European culture. This suggests that the settlers had been derived from Mesolithic or even Upper Paleolithic ancestors; however, it may be pointed out that little in the way of advanced culture could have survived long tribal journeys through the subarctic regions of Siberia and Alaska.

Eskimos.—Sharply marked-off from the normal Indian types are the Eskimos, who are widespread along the northern lands from the northeastern tip of Asia through northern Alaska and the Arctic shores and islands of Canada to Labrador and Greenland. Physically they are closer to the Old World Mongoloids than the typical Indians but have one notable peculiarity—a remarkably narrow head. Their culture is a highly specialized one, admirably fitting them to live in cold northern climates, where they make a living primarily by catching seals and other forms of marine life. Their history is not well known, but there appear to be some close relatives among northern Siberian tribes. Probably they migrated to the Americas later than most, if not all, of the true Indians and spread eastward along the Arctic coasts within the last ten thousand years.

A Synoptic Classification of Chordates

In the brief classification below of the vertebrates (and their lower chordate relatives) a number of minor groups have been omitted and subdivisions of major groups have been made only where necessary for the purposes of the present volume.

SUBPHYLUM HEMICHORDATA

(little or no development of notochord or dorsal nerve cord)

CLASS PTEROBRANCHIA

(simple, sessile, colonial plantlike animals, gathering food by ciliated "arms")

CLASS ENTEROPNEUSTA

(worm-shaped burrowers, with well-developed gill apparatus—the acorn worms)

SUBPHYLUM UROCHORDATA

(tunicates, with notochord and nerve cord well developed in larvae in many cases but sessile or floating when adult, consisting mainly of an elaborate gill apparatus)

SUBPHYLUM CEPHALOCHORDATA

(*Amphioxus*, with notochord, nerve cord, and gill apparatus all well developed in the adult stage)

SUBPHYLUM VERTEBRATA

(generally developed backbones and other advanced structural features characteristic of true vertebrates)

CLASS AGNATHA (jawless vertebrates)

ORDERS OSTEOSTRACI, ANASPIDA, HETEROSTRACI (ostracoderms of the Silurian and Devonian periods)
ORDER CYCLOSTOMATA (living cyclostomes)
SUBORDER MYXINOIDEA (hagfishes)
SUBORDER PETROMYZONTIA (lampreys)

CLASS PLACODERMI (Paleozoic jawed vertebrates)

ORDER ACANTHODII (spiny sharks)

ORDER ARTHRODIRA (arthrodires; jointed-necked fishes)

ORDER ANTIARCHI (related to arthrodires, but with peculiar bony "arms")

ORDER MACROPETALICHTHYIDA

ORDER STEGOSELACHII (The last two orders are modified arthrodire relatives which are suspected of being transitional, with reduction of bony armor, to the sharklike fishes.)

CLASS CHONDRICHTHYES (cartilaginous fishes)

SUBCLASS ELASMOBRANCHII (sharks and related forms)

ORDER CLADOSELACHII (primitive Paleozoic sharks)

ORDER SELACHII (typical sharks, Paleozoic to Recent, with claspers, narrow-based fins)

ORDER BATOIDEA (skates and rays)

SUBCLASS HOLOCEPHALI (differing from sharks in having a gill covering, upper jaws fused to skull)

ORDER BRADYODONTI (Paleozoic forms, poorly known, mainly represented by tooth plates)

ORDER CHIMERAE (chimeras)

CLASS OSTEICHTHYES (higher bony fishes)

SUBCLASS ACTINOPTERYGII (ray-finned fishes)

SUPERORDER CHONDROSTEI (primitive ray-finned fishes, with heterocercal tails; represented by fossil paleoniscoids, mainly Paleozoic, and three modern types: *Polypterus*, the sturgeons, and the paddlefish)

SUPERORDER HOLOSTEI (dominant ray-finned forms of the Mesozoic, with abbreviate heterocercal tails; living forms include the gar pike and *Amia*, the "bow fin")

SUPERORDER TELEOSTEI (dominant fishes of Cenozoic and Recent times, with a homocercal tail; many thousands of forms, classed in a number of orders)

SUBCLASS SARCOPTERYGII (CHOANICHTHYES) (with fleshy fins and, usually, internal nostrils)

ORDER CROSSOPTERYGII (broadly ancestral to land vertebrates; mainly Paleozoic fossils; one aberrant living form)

ORDER DIPNOI (lungfishes, including three living genera; many similarities to ancestral types, but aberrant in teeth, skulls)

CLASS AMPHIBIA (tetrapods, but without amniote type of egg)

SUBCLASS APSIDOSPONDYLI (vertebral centra primitively formed by distinct blocks of bone or cartilage)

SUPERORDER LABYRINTHODONTIA (stem amphibians, extinct, but dominant in late Paleozoic and Triassic times; many dozens of fossil genera, arranged in several orders)

SUPERORDER SALIENTIA

ORDER ANURA (frogs, toads; living forms have highly specialized limbs and shortened trunks; vertebrae much modified)

SUBCLASS LEPOSPONDYLI (vertebral centra formed as single structures, often spool-shaped)

    ORDERS AISTOPODA, NECTRIDIA, MICROSAURIA (late Paleozoic fossils; the last perhaps ancestral to two living orders)

    ORDER URODELA (salamanders and newts; typical body form but many degenerate characters)

    ORDER APODA (or GYMNOPHIONA) (wormlike, burrowing types)

CLASS REPTILIA
(amniotes, but without advanced avian or mammalian characters)

SUBCLASS ANAPSIDA (without temporal opening)

    ORDER COTYLOSAURIA (archaic "stem reptiles" of the late Paleozoic and Triassic)

    ORDER CHELONIA, TESTUDINATA (turtles)

SUBCLASS EURYAPSIDA (extinct groups; single temporal opening)

    ORDER PROTOROSAURIA (various obscure Permian and Mesozoic reptiles)

    ORDER SAUROPTERYGIA (plesiosaurs and their relatives; marine Mesozoic reptiles, with limbs transformed into powerful paddles)

SUBCLASS ICHTHYOPTERYGIA

    ORDER ICHTHYOSAURIA (ichthyosaurs; highly specialized for marine life)

SUBCLASS LEPIDOSAURIA ("diapsid"—two-arched temporal region; without archosaur specializations)

    ORDER EOSUCHIA (Permian and Triassic ancestral diapsids)

    ORDER RHYNCHOCEPHALIA (living *Sphenodon* of New Zealand; fossil relatives)

    ORDER SQUAMATA (lizards and snakes; temporal arches reduced)

SUBCLASS ARCHOSAURIA ("ruling reptiles"; diapsid temporal region, with specializations tending toward bipedal life)

    ORDER THECODONTIA (Triassic ancestors of dinosaurs, birds, and others)

    ORDER CROCODILIA (crocodiles and alligators; degenerate amphibious archosaur survivors)

    ORDER PTEROSAURIA (extinct flying reptiles; membrane wing)

    ORDER SAURISCHIA ("reptile-like" dinosaurs, with triradiate pelvis; carnivores and large amphibious forms)

    ORDER ORNITHISCHIA ("birdlike" dinosaurs, with tetraradiate pelvis; herbivorous; bipeds, including duckbills and armored and horned quadrupeds)

SUBCLASS SYNAPSIDA (lateral temporal opening; forms leading to mammals; extinct)

    ORDER PELYCOSAURIA (primitive Permian mammal-like reptiles; close to stem reptiles)

    ORDER THERAPSIDA (advanced mammal-like forms of late Permian and Triassic)

CLASS AVES
(birds—archosaur descendants; with feathers, temperature control)

SUBCLASS ARCHAEORNITHES (Jurassic fossil birds; many reptilian characters)

SUBCLASS NEORNITHES ("modernized" birds)

    SUPERORDER ODONTOGNATHAE (toothed birds of the Cretaceous)

    SUPERORDER PALAEOGNATHAE (mainly ostrich-like birds, or ratites; relatively primitive)

    SUPERORDER NEOGNATHAE (all remaining birds; many orders, similar anatomy)

CLASS MAMMALIA (hair, nursing habit, brain of advanced type)

SUBCLASS PROTOTHERIA (egg-laying mammals)
ORDER MONOTREMATA (platypus and spiny anteater—Australia)

SUBCLASS THERIA (live-bearing)
INFRACLASS AND ORDER PANTOTHERIA (small Jurassic mammals, probably ancestral to remaining groups; there are two further Jurassic orders of obscure relationships)
INFRACLASS ALLOTHERIA, ORDER MULTITUBERCULATA (extinct group of primitive but aberrant mammals, perhaps comparable in habits to later rodents—Jurassic to Eocene)
INFRACLASS METATHERIA, ORDER MARSUPIALIA (pouched mammals, such as opossum and many Australian types; young born alive but at immature stage)
INFRACLASS EUTHERIA (higher mammals, with an efficient placenta)
ORDER INSECTIVORA (ancestral placentals and modern descendants, such as shrews, moles, hedgehog)
ORDER CHIROPTERA (bats)
ORDER PRIMATES (arboreal offshoot of primitive placental stock)
SUBORDER LEMUROIDEA (tree shrews, lemurs)
SUBORDER TARSIOIDEA (*Tarsius* and extinct relatives; transitional between lemurs and monkeys)
SUBORDER ANTHROPOIDEA (monkeys, apes, man)
INFRAORDER PLATYRRHINI (South American monkeys, with nostrils opening sideways)
Family Hapalidae (marmosets)
Family Cebidae (typical South American monkeys)
INFRAORDER CATARRHINI (Old World monkeys, apes, man; nostrils open downward)
Family Cercopithecidae (Old World monkeys)
Family Simiidae (manlike apes)
Family Hominidae (man)
ORDER CARNIVORA (carnivores)
SUBORDER CREODONTA (extinct archaic carnivores)
SUBORDER FISSIPEDIA (modern land carnivores)
INFRAORDER EUCREODI (extinct ancestors of modern types)
INFRAORDER ARCTOIDEA (dogs and relatives)
Family Mustelidae (primitive; weasels, skunks, badgers, otters, and so on)
Family Canidae (dogs, wolves, foxes)
Family Procyonidae (raccoons, pandas, kinkajous)
Family Ursidae (bears)
INFRAORDER AELUROIDEA (cats and relatives)
Family Viverridae (civets, mongoose, and the like; primitive Old World aeluroids)
Family Hyaenidae (hyenas)
Family Felidae (cats, lions, tigers)
SUBORDER PINNIPEDIA (marine carnivores—seals, sea lion, walrus)

ORDER CONDYLARTHRA (primitive extinct ungulates)

ORDERS AMBLYPODA, DINOCERATA, EMBRITHOPODA, ASTRAPOTHERIA, LITOPTERNA. NOTOUNGULATA (extinct orders of ungulates, mainly of archaic character, the last three characteristic of South America)

ORDER PERISSODACTYLA (odd-toed ungulates)

    Family Equidae (horses, asses, zebras)

    Family Titanotheriidae (titanotheres—large, ungainly, horned fossil forms)

    Family Chalicotheriidae (extinct forms related to the last two, but with claws, not hoofs)

    Family Tapiridae (tapirs)

    Family Rhinocerotidae (rhinoceroses)

ORDER ARTIODACTYLA (even-toed ungulates)

  SUBORDER PALAEODONTA (earliest, most primitive fossil artiodactyls)

  SUBORDER SUINA (relatively primitive types, with simple stomachs and simply built teeth, including the following living families as well as several extinct ones):

    Family Suidae (pigs of the Old World)

    Family Dicotylidae (peccaries of the New World)

    Family Hippopotamidae (hippopotamus)

  SUBORDER RUMINANTIA (cud-chewers, with complex stomach, crescent-shaped tooth cusps)

    INFRAORDER TYLOPODA (primitive cud-chewers, including, in addition to early extinct families):

    Family Camelidae (camels, llamas)

    Family Oreodontidae (oreodons, short-legged ruminants, abundant in North American fossil deposits)

    INFRAORDER PECORA (advanced ruminants, mostly with horns or antlers, including, besides extinct forms):

    Family Tragulidae (chevrotains—tiny, hornless, deerlike animals of the tropical Old World)

    Family Cervidae (the deer tribe)

    Family Giraffidae (giraffe and okapi of Africa)

    Family Antilocapridae (American pronghorn)

    Family Bovidae (cattle family—mainly Old World forms, including bison, sheep, goats, numerous types of antelopes)

ORDER HYRACOIDEA (conies of Africa and Syria; rabbit-like in habits but actually subungulate along with next two orders—related, and probably of African origin)

ORDER PROBOSCIDEA (elephants and fossil relatives, mammoths and mastodons)

ORDER SIRENIA (sea cows—manatee and dugong; an aquatic offshoot of an ungulate stock)

ORDER CETACEA (whales)

  SUBORDER ARCHAEOCETI (extinct ancestral whales)

  SUBORDER ODONTOCETI (toothed whales, porpoises, dolphins)

  SUBORDER MYSTICETI (whalebone whales)

ORDER EDENTATA ("toothless" mammals; South America)
  SUBORDER PILOSA (hairy edentates)
    INFRAORDER GRAVIGRADA (extinct ground sloths)
    INFRAORDER TARDIGRADA. Family Bradypodidae (tree sloths)
    INFRAORDER VERMILINGUA. Family Myrmecophagidae (South American ant-
      eaters)
  SUBORDER LORICATA (armored edentates)
    Family Dasypodidae (armadillos)
    Family Glyptodontidae (extinct giant glyptodonts)
ORDER TUBULIDENTATA (aardvark of Africa—an anteater, but not related to the
  preceding orders)
ORDER PHOLIDOTA (the Old World pangolin—an anteater, but not closely related to
  the last order)
ORDER RODENTIA (gnawing animals—except rabbit group)
  SUBORDER SCIUROMORPHA (squirrels, gophers, woodchuck, beaver, and others)
  SUBORDER CAVIAMORPHA (guinea pig and many other South American rodents,
    American porcupines)
  SUBORDER MYOMORPHA (rats, mice)
ORDER LAGOMORPHA (hares, rabbits—gnawing forms, but not closely related to the
  last order)

Bibliography

Below are listed a number of books and papers on vertebrates in general or on special groups. Those wishing to go more deeply into the study of these animals will find that most of these works include bibliographies leading further into the literature. For references to original scientific papers and monographs, the *Zoölogical Record*, published annually in London, gives complete lists, indexed by forms concerned and topics, of all books and papers published each year.

GENERAL ACCOUNTS OF VERTEBRATES

BLAIR, W. F., A. P. BLAIR, P. BRODKORB, F. R. CAGLE, and G. A. MOORE. 1957. *Vertebrates of the United States*. New York. 819 pp.

DEBEER, G. R. 1945. *Vertebrate Zoölogy*. London. 505 pp.

HORNADAY, W. T. 1909. *The American Natural History*. New York. 449 pp.

LYDEKKER, R. (ed.). 1895. *The Royal Natural History*. London and New York. 6 vols.

———. 1915. *Wild Life of the World*. London. 3 vols.

PARKER, T. J., and W. A. HASWELL. 1940. *A Text-Book of Zoölogy*. Vol. 2. Rev. by C. FORSTER COOPER. 6th ed. London. 758 pp.

YOUNG, J. Z. 1950. *The Life of Vertebrates*. London and New York. 767 pp.

SPECIAL TOPICS

COLBERT, E. H. 1955. *Evolution of the Vertebrates*. New York. 398 pp.

DARLINGTON, P. J., JR. 1957. *Zoögeography*. New York. 675 pp.

DEBEER, G. R. 1958. *Embryos and Ancestors*. 3d ed. Oxford. 197 pp.

DOBZHANSKY, T. 1955. *Evolution, Genetics and Man*. New York and London. 398 pp.

GOODRICH, E. S. 1930. *Studies on the Structure and Development of Vertebrates*. London. 837 pp.

GRAY, J. 1953. *How Animals Move*. Cambridge (Eng.). 114 pp.

GREGORY, W. K. 1951. *Evolution Emerging*. New York. 2 vols.

LORENZ, K. Z. 1952. *King Solomon's Ring*. New York. 202 pp.

MAYR, E., E. G. LINSLEY, and R. L. USINGER. 1953. *Methods and Principles of Systematic Zoölogy*. New York. 328 pp.

PROSSER, C. L. (ed.). 1950. *Comparative Animal Physiology*. Philadelphia. 888 pp.

REYNOLDS, S. H. 1913. *The Vertebrate Skeleton*. 2d ed. Cambridge (Eng.). 535 pp.

ROMER, A. S. 1945. *Vertebrate Paleontology*. Chicago. 687 pp.

———. 1955. *The Vertebrate Body*. 2d ed. Philadelphia. 644 pp.

SCHEER, B. T. 1948. *Comparative Physiology*. New York. 563 pp.

SIMPSON, G. G. 1949. *The Meaning of Evolution.* New Haven. 364 pp.

TINBERGEN, N. 1951. *The Study of Instinct.* Oxford. 240 pp.

———. 1953. *Social Behavior in Animals.* New York. 150 pp.

FISHES

BIGELOW, H. B., and W. C. SCHROEDER. 1953. Fishes of the Gulf of Maine. 2d ed. *U.S. Fish and Wildlife Service Fishery Bulletin,* No. 74. Vol. 53. 577 pp.

BROWN, M. E. (ed.). 1957. *The Physiology of Fishes.* New York. 2 vols.

CUNNINGHAM, J. T. 1912. Fishes. In *Reptiles, Amphibia, Fishes and Lower Chordata,* pp. 231–440.

GOODRICH, E. S. 1909. Cyclostomes and Fishes. In *A Treatise on Zoölogy,* Part IX: Vertebrata Craniata, Fasc. I (ed. E. R. LANKESTER). London. 534 pp.

GREGORY, W. K., and F. LaMONTE. 1947. The World of Fishes. *American Museum Science Guide,* No. 122. 96 pp.

HARMER, S. F., W. A. HERDMAN, T. W. BRIDGE, and G. A. BOULENGER. 1910. Hemichordata, Ascidians and Amphioxus, Fishes. *Cambridge Natural History.* Vol. 7.

JORDAN, D. S. 1905. *A Guide to the Study of Fishes.* New York. 2 vols.

——— and B. W. EVERMANN. 1923. *American Food and Game Fishes.* Garden City. 574 pp.

MOY-THOMAS, J. A. 1939. *Palaeozoic Fishes.* London and New York. 149 pp.

NORMAN, J. R. 1947. *A History of Fishes.* 3d ed. London and New York. 462 pp.

——— and F. C. FRASER. 1948. *Giant Fishes, Whales and Dolphins.* 2d ed. London. 360 pp.

ROMER, A. S. 1946. The Early Evolution of Fishes. *Quarterly Review of Biology,* Vol. 41, pp. 33–69.

THOMPSON, J. A. 1912. Cyclostomata, Lancelets, Tunicates. In *Reptiles, Amphibia, Fishes and Lower Chordata,* ed. J. T. CUNNINGHAM. London, pp. 441–494.

AMPHIBIANS AND REPTILES

BELLAIRS, A. D'A. 1957. *Reptiles.* London. 195 pp.

COLBERT, E. H. 1951. *The Dinosaur Book.* 2d ed. New York, 156 pp.

CUNNINGHAM, J. T., and G. A. BOULENGER. 1912. Amphibia. In *Reptiles, Amphibia, Fishes and Lower Chordata,* ed. J. T. CUNNINGHAM. London, pp. 157–230.

DITMARS, R. L. 1931. *Snakes of the World.* New York. 207 pp.

———. 1936. *Reptiles of the World* (rev. ed.). New York. 321 pp.

GADOW, HANS. 1909. Amphibia and Reptiles. *Cambridge Natural History,* Vol. 8, London. 668 pp.

HOLMES, S. J. 1927. *The Biology of the Frog.* 4th ed. New York. 386 pp.

LYDEKKER, R. 1912. Reptiles. In *Reptiles, Amphibia, Fishes and Lower Chordata,* ed. J. T. CUNNINGHAM. London, pp. 1–150.

NOBLE, G. K. 1931. *The Biology of the Amphibia.* New York. 577 pp.

OLIVER, J. A. 1955. *The Natural History of North American Amphibians and Reptiles.* New York and London. 359 pp.

POPE, C. H. 1955. *The Reptile World.* New York. 325 pp.

ROMER, A. S. 1957. *The Osteology of Reptiles.* Chicago. 772 pp.

SCHMIDT, K. P., and R. F. INGER, 1957. *Living Reptiles of the World*. Garden City. 287 pp.

SWINTON, W. E. 1934. *The Dinosaurs*. London. 233 pp.

WESTOLL, T. S. 1943. The Origin of the Tetrapods. *Biological Reviews*, Vol. 18, pp. 78–98.

WILLISTON, S. W. 1914. *Water Reptiles of the Past and Present*. Chicago. 251 pp.

BIRDS

EVANS, H. H. 1900. Birds. *Cambridge Natural History*. London. Vol. 9, 635 pp.

FISHER, H. 1957. *Birds as Animals: 1. A History of Birds; 2. Migration and Habits of Birds*. London. 2 vols.

HEILMANN, G. 1927. *The Origin of the Birds*. New York. 210 pp.

KNOWLTON, F. H. 1909. *Birds of the World*. New York. 873 pp.

PYCRAFT, W. P. 1910. *A History of Birds*. London. 489 pp.

TINBERGEN, N. 1954. *Bird Life*. Oxford. 64 pp.

WING, L. W. 1956. *Natural History of Birds*. New York. 539 pp.

WOLFSON, A. (ed.). 1955. *Recent Studies in Avian Biology*. Urbana. 479 pp.

MAMMALS

ANTHONY, H. E. 1928. *Field Book of North American Mammals*. New York. 625 pp.

BEDDARD, F. E. 1900. *A Book of Whales*. London. 320 pp.

————. 1909. Mammalia. *Cambridge Natural History*. London and New York. Vol. 10, 605 pp.

BOURLIÈRE, F. 1954. *The Natural History of Mammals*. New York. 363 pp.

————. 1955. *Mammals of the World*. New York. 233 pp.

FLOWER, W. H. 1885. *An Introduction to the Osteology of the Mammalia*. 2d ed. London. 383 pp.

FRASER, F. C. 1948. Cetaceans. In J. R. NORMAN and F. C. FRASER, *Giant Fishes, Whales and Dolphins*. London, pp. 203–349.

HAMILTON, W. J., JR. 1939. *American Mammals*. New York. 434 pp.

KELLOGG, R. 1928. The History of Whales. *Quarterly Review of Biology*, Vol. 3, pp. 29–76, 174–208.

SCOTT, W. B. 1937. *A History of Land Mammals in the Western Hemisphere*. 2d ed. New York. 786 pp.

SIMPSON, G. G. 1935. The First Mammals. *Quarterly Review of Biology*, Vol. 10, pp. 154–180.

————. 1937. The Beginning of the Age of Mammals. *Biological Reviews*, Vol. 12, pp. 1–47.

————. 1945. The Principles of Classification and a Classification of Mammals. *Bull. American Museum of Natural History*, Vol. 85, pp. 1–350.

————. 1951. *Horses*. Oxford and New York. 247 pp.

YOUNG, J. Z. 1957. *The Life of Mammals*. Oxford. 820 pp.

PRIMATES AND MAN

BOULENGER, E. G. 1936. *Apes and Monkeys*. London. 236 pp.

COON, C. S. 1954. *The Races of Europe*. New York. 739 pp.

————. 1954. *The Story of Man*. New York. 437 pp.

FLOWER, W. H., and R. LYDEKKER. 1891. *An Introduction to the Study of Mammals, Living and Extinct.* London. 763 pp.

HOOTON, E. A. 1947. *Up from the Ape.* 2d ed. New York. 788 pp.

HOWELLS, W. W. 1952. *Mankind So Far.* Garden City. 319 pp.

———. 1954. *Back of History.* Garden City. 384 pp.

LeGROS CLARK, W. E. 1934. *Early Forerunners of Man.* Baltimore. 296 pp.

———. 1955. *The Fossil Evidence for Human Evolution.* Chicago. 181 pp.

———. 1956. *History of the Primates.* 5th ed. London. 119 pp.

SMITH, G. ELLIOT. 1927. *The Evolution of Man.* Oxford. 195 pp.

———. 1931. *The Search for Man's Ancestors.* London. 56 pp.

SONNTAG, C. F. 1924. *The Morphology and Evolution of the Apes and Man.* London. 450 pp.

YERKES, R. M. 1925. *Almost Human.* London. 278 pp.

ZUCKERMAN, S. 1932. *The Social Life of Monkeys and Apes.* London. 356 pp.

Index

Numbers in italics refer to photographic inserts; the first number is that of the location of the insert, the second the number of the page in the insert. Generic and specific names of animals are printed in italics.

A

Aardvark, 307–8, 418, *306–1*
Aardwolf, 243, *242–1*
Abbevillean, 339, 343–44
Acanthodians, 47–48, 414
Acanthopterygii, 75–79
Acheulean, 339, 343–44
Acinonyx, 242–2
Acipenser, 68 1
Acorn worms, 28, 29, 413
Actinopterygii, 59, 67–86, 414
Addax, 277
Adrenal glands, 125
Aeluroidea, 416
Africa, 15
    early man in, 373
Agamid lizards, 162–63
Agnatha, 37, 38–42, 413
Agouti, 291
Ailuropoda, 248–2
Ailurus, 248–2
Ainus, 396–97, *392–1*
Air bladder of fishes, 58
Air-breathing, development of, 91–92
Aistopoda, 415
Albatross, 205, 206, *204–2*
Aldrovandia, 71, 72
Alligators, 182, 183, 415, *182–1*
Allotheria, 416
Alpaca, 270
Alpines, 387, 405–8, *398–2*
Amblypoda, 417
Amblystoma, 98–2
America, early man in, 374–76
Amia, 69, 414, *68–1*
Ammodytes, 78
Amniote egg, 132–36

Amphibia, 36, 87–101, 414–15
    ancestry of, 93
    contrast with fish ancestors, 90–92
    life-history of, 88–90
    locomotion of, 90
    modern groups of, 97–101
    older forms of, 90–97, *8–1, 98–1*
Amphibious dinosaurs, 100 01, *188 2*
Amphioxus, 25–26, 413, *38–1*
Amphisbaenids, 165–66, *166–1*
Amphiuma, 98–2
Anableps, 68–3
Anaconda, *166–2*
Anapsid reptiles, 142, 415
Anaspida, 413
Ancestry of vertebrates, 31–34
Angelfishes, 53
Angius, 162–2
Anglers, 77, 78, 80
Ankylosaurus, 196
Annelid worms, relation of, to vertebrate ancestry, 31–33
Anoa, 277
Anoles, 162, *160–1*
Anseriformes, 207–8
Antbirds, 215
Anteaters, 418
    Old World, 307–8, *302–2, 306–1*
    South American, 305, *306–1*
    spiny, 228, 416, *232–1*
Antelopes, 276–77, 417, *276–2, 278–1*
    American, 275, 417, *276–1*
Anthracotheres, 268
Anthropoidea, 317, 416
Antiarchs, 48, 49, 414
Antilocapridae, 417
Antlers, of deer, 272–73
Anura, 414
Apes, manlike, 322–27, 416, *322–2, 322–3, 322–4, 326–1, 326–2*

PRINTED IN U.S.A.